ELECTRICITY
FOR
ELECTRICIANS

ABRAHAM MARCUS

ELEC

Special printing for the
National Joint Apprenticeship and Training Committee
for the Electrical Industry

TRICITY

FOR

ELECTRICIANS

PRENTICE-HALL, INC.

Englewood Cliffs, New Jersey

ELECTRICITY FOR ELECTRICIANS

ABRAHAM MARCUS

Library of Congress Catalog Card Number 69-11871
Printed in the United States of America

Prentice-Hall, Inc., Englewood Cliffs, N.J.

Prentice-Hall International, Inc., *London*
Prentice-Hall of Australia, Pty. Ltd., *Sydney*
Prentice-Hall of Canada, Ltd., *Toronto*
Prentice-Hall of India Private Ltd., *New Delhi*
Prentice-Hall of Japan, Inc., *Tokyo*

Current printing (*last digit*): 16 15 14 13 12 11

72

PREFACE

For the second time in two centuries we are in the midst of a major industrial revolution. In the 18th century, sparked by such technological developments as the invention of the steam engine, the first industrial revolution swept away feudal methods of production. These were replaced by the more efficient and more highly productive factory system. Many people were displaced in the process and considerable unrest resulted. But the long-range benefits of the revolution were very great. Life for the average citizen became easier, more comfortable, and safer.

Perhaps the greatest impact was upon the minds of men. Not only were the old products produced cheaper, better, and in greater abundance, but men were stimulated to imagine and produce new products not dreamed of under the old feudal system.

The greatest advances, however, were achieved in the social structure. Stirred by the promise of a better life, men began to demand greater intellectual and political freedom. Democratic revolution swept side by side with the industrial revolution. In every way man became freer.

Now, two centuries later, a second industrial revolution is sweeping our world. The tremendous technological advances made within the past few decades are its stimulus. We have seen the development of the science of electronics, nuclear energy, the invention of the digital computer, and much more. Old methods of production are becoming obsolete and are rapidly being replaced by automation where machines perform the tasks formerly done by men.

As was true for the first industrial revolution, the changes being wrought by the second one are qualitative as well as quantitative in nature. Goods and services are cheaper and better under the new system of production. But above all, the winds of change are producing new products never envisaged under the older system. And, as in the case of the first revolution, social changes probably will flow from the second. It is too early, perhaps, to foresee the final results, but we may be sure that mankind will emerge the better for them.

This industrial revolution poses a serious threat and challenge to us. The new machines have taken over many of the jobs formerly performed by human workers. Not only are there millions of unemployed, but other millions, employed at present, are in danger of losing their jobs as the tide of automation reaches their industries.

Ironically, as the technological advance eliminates jobs, many other new jobs are created and remain unfilled due to the lack of suitably trained workers. There is a serious shortage of trained technicians — men and women who can operate, service, repair, and maintain the ever-growing number of new technical devices and machines. And the need for technicians will continue to grow as our technology expands.

We must be prepared to alleviate the hardships caused by the economic displacement of people caught in the maelstrom of change. One obvious step in this direction is the retraining of displaced workers to fill the new jobs. Such retraining is now being done in a number of on-the-job factory schools. In addition, there are governmental agencies which have been set up for this purpose. Undoubtedly, these retraining programs will be greatly expanded in the near future.

But the retraining of displaced workers alone will not be enough to meet the growing need for trained technicians. This need can be met only by the combined efforts of the various high schools, vocational schools, technical schools, and, in increasing numbers, the junior and community colleges of our country.

Before proceeding further, it is necessary to clear up a certain vagueness which has developed about just what is meant by a "technician." The dictionary describes a technician as one who is conversant with the technique of a particular subject. With so broad and ambiguous a definition, it is little wonder that the term "technician" has come to mean different things for different people. For some, a technician is merely a step above a handyman. For others, a technician is a step below an engineer. In fact, some regard a technician as an engineer who lacks only a college degree.

In this book, and in the other books of this series, a middle ground is taken. The technician is regarded as a person who is qualified to operate, maintain, and service the various devices and machines of our industrial technology. Unlike the handyman whose function it is to remedy obvious defects such as a loose screw, a blown fuse, or a dry bearing, the technician must understand the basic theories and principles behind the various machines and devices he encounters. Unlike the engineer, whose function it is to design these machines and devices, the technician need not have a mathematical, or quantitative, understanding of them. It is enough that he have a physical, or descriptive, understanding of them.

It is hoped, of course, that the technician will continue his studies to upgrade himself toward the engineering level. But first he must have the descriptive and physical understanding. It is this understanding that this series of books is intended to provide.

There are several questions to be answered if we are to train these much-needed technicians efficiently. First of all, what are we to teach them?

The machines of our factories are run by electricity as are our trains. We cook our food and heat our homes by electricity. Electricity is used for illumination and communication. In fact, there is hardly a facet of our lives in which electricity does not play a part. So widespread are the applications of electrical energy that most of the other types of energy — heat, chemical, and nuclear — are converted to electrical energy.

Hence electricity and its multitudinous applications constitute an all-important section of our technological civilization. And if the technician is to succeed in his task of servicing and maintaining the various machines and devices of our technology, he must have a good understanding of the theory of electricity and its various applications.

How are we to train these technicians? Of course, we may teach the student to operate, service, and maintain some specific device or machine. But such training suffers from certain serious defects. A technician trained for one type of machine may not be able to understand the operation of another type. Also, our technology is advancing so rapidly that the student may well find that the machine for which he had been training has become obsolete even before he has completed his training course.

A much better method is to teach him an understanding of the basic principles underlying whole classes of devices. Then, if a certain device becomes obsolete, the student will be able to understand the device which has taken its place since basic principles change very slowly. Even if the device is not obsolete, an understanding of its basic principles will enhance the effectiveness of the technician.

It is with all the above in mind that this book for technicians is offered. It deals with electricity and is divided into seven sections. Section I deals with early and modern theories attempting to explain what is meant by electricity and magnetism, as well as the structure of matter. Direct current is treated in section II; alternating current in section III. The various sources of electrical energy are dealt with in section IV; the transmission and control of electric power in section V; and practical applications of electricity in section VI. Section VII deals with practical power distribution systems. In addition, there are a number of appendices containing, among other things, glossaries of electrical terms, units, symbols, and formulas.

ABRAHAM MARCUS

CONTENTS

xi

IV

Sources of Electrical Energy   *227*

V

Transmission and Control of Electric Power   *309*

VI

Practical Applications of Electricity   *343*

VII

Practical Power Distribution Systems *465*

APPENDICES   *493*

ELECTRICITY
FOR
ELECTRICIANS

PROLOGUE

The ancient people of the world had a "magic." Digging in the earth or along seashores, they sometimes found yellow, glasslike pebbles. These pebbles consisted of **amber,** a fossilized form of resin which had oozed from a certain type of now-extinct pine tree. These amber pebbles were highly prized as ornaments.

Their magic was that, if they were rubbed with a cloth, they acquired the power to attract small pieces of straw, bits of paper, and such. As early as 600 B.C., a Greek philosopher, Thales of Miletus, wrote about experiments with amber, which the Greeks called "elektron." It is from this word that we get our modern terms "electricity," "electron," and "electronics."

The ancients had a second "magic" to match the mysterious properties of rubbed amber. They occasionally dug up heavy black stones that could attract pieces of iron. These stones were called "magnets," probably because they were particularly plentiful in the province of Magnesia in Asia Minor (now a part of modern Turkey).

1

Man is a curious animal. Other animals are content to accept their environment and attempt to adjust themselves to it. Not so man. He constantly seeks to learn the why and wherefore of nature. And with this knowledge he attempts to change his environment to his advantage.

Modern man studies his environment by means of an orderly process of observation and collection of facts and by experimentation. This process is called **science.** Since early man did not possess this technique, he tried to explain natural phenomena by means of legends and myths.

Thus he explained the attraction of straw to amber and of iron to the magnet by claiming that it was the "soul" of the amber and magnet that was responsible. Both the amber and the magnet were supposed to have the power of curing disease. Also, if a magnet were rubbed with garlic, it was supposed to lose its power of attraction. But if it then were smeared with goat's blood, this power would be restored.

Of course, we now know that these myths are false. But keep in mind that they were attempts by man to understand and explain his environment. And it was from these efforts to discover why amber and magnets behave as they do that the science of electricity came into existence.

There is another matter which should be explained at this point. Throughout this book we shall encounter numbers that are too large, or fractions that are too small, to be handled conveniently by ordinary means. Consider, for example, an electrical unit which is equal to the combined charges of over 6 million, million, million (6 followed by 18 zeros) electrons. Or consider an atomic particle so small that its diameter is a fraction of an inch represented by 22 preceded by 12 zeros up to the decimal point. Obviously, these are awkward ways for presenting such quantities.

Accordingly, mathematicians have worked out a sort of shorthand for dealing with such numbers and fractions. It would be useful to understand this method before proceeding further in the book.

If you multiply 10 by 10, you get 100. Since 100 is formed by **two** tens multiplied together, mathematicians express 100 as 10^2. The 10 is called the **base** and the superscript (2) is called the **exponent.**

Similarly, 1000 is formed by **three** tens multiplied together. Thus the exponent is 3, and 1000 may appear as 10^3. And so forth, adding one to the exponent for every multiplication by 10. Thus:

$$10^1 = 1 \times 10, \text{ or } 10$$
$$10^2 = 10 \times 10, \text{ or } 100$$

$10^3 = 10 \times 10 \times 10$, or 1000
$10^4 = 10 \times 10 \times 10 \times 10$, or 10,000
$10^5 = 10 \times 10 \times 10 \times 10 \times 10$, or 100,000
$10^6 = 10 \times 10 \times 10 \times 10 \times 10 \times 10$, or 1,000,000.

And so forth.

Any number may be broken down to an appropriate decimal number and some suitable multiplier. For example, consider the number 5,123,000. This number can be converted to:

Decimal number	Multiplier
512,300	$\times$ 10^1, or
51,230	$\times$ 10^2, or
5,123	$\times$ 10^3, or
512.3	$\times$ 10^4, or
51.23	$\times$ 10^5, or
5.123	$\times$ 10^6.

And so forth. Each of the above is equal to the original number, 5,123,000. Generally, a decimal number is chosen which has only one **significant** digit (that is, a digit other than 0) to the left of the decimal point.

The larger the number, the more apparent is the advantage of using this shorthand method. For example, consider the number previously mentioned — 6 million, million, million. Ordinarily, this enormous number would be represented by 6 followed by 18 zeros. Using our shorthand system, it appears as 6×10^{18}.

Fractions can be handled in a similar manner. Thus the decimal fraction 0.1 ($\frac{1}{10}$) becomes $\frac{1}{10^1}$. Another method for writing this fraction is 10^{-1}. The superscript (-1) is called the **negative exponent.** Thus:

$10^{-1} = 0.1$, or $1/10^1$, or $1/10$
$10^{-2} = 0.01$, or $1/10^2$, or $1/100$
$10^{-3} = 0.001$, or $1/10^3$, or $1/1000$
$10^{-4} = 0.0001$, or $1/10^4$, or $1/10,000$
$10^{-5} = 0.00001$, or $1/10^5$, or $1/100,000$
$10^{-6} = 0.000001$, or $1/10^6$, or $1/1,000,000$.

And so forth.

As an example, you will recall that we mentioned a particle whose diameter is a fraction of an inch represented by 22 preceded by 12 zeros up to the decimal point. Using our shorthand method, this fraction appears as 22×10^{-14} or, preferably, 2.2×10^{-13}.

In addition to the problem of handling very large numbers and very small fractions, some of the units used to measure the various electrical effects may present another sort of problem. We may find some of these units either many times too large or many times too small for certain applications. Accordingly, mathematicians have drawn up a series of prefixes, each of which stands for a specific value, to be used as a multiplication factor for these units. Thus:

Prefix	Symbol	Numerical value	
atto	a	0.000,000,000,000,000,001,	or 10^{-18}
femto	f	0.000,000,000,000,001,	or 10^{-15}
pico	p	0.000,000,000,001,	or 10^{-12}
nano	n	0.000,000,001,	or 10^{-9}
micro	μ	0.000,001,	or 10^{-6}
milli	m	0.001,	or 10^{-3}
centi	c	0.01,	or 10^{-2}
deci	d	0.1,	or 10^{-1}
deka	dk	10,	or 10^{1}
hecto	h	100,	or 10^{2}
kilo	k	1000,	or 10^{3}
mega	M	1,000,000,	or 10^{6}
giga	G	1,000,000,000,	or 10^{9}
tera	T	1,000,000,000,000,	or 10^{12}.

Thus, for example, **millivolt** is 0.001 volt, whereas **kilovolt** is 1000 volts.

INTRODUCTION
I

INTRODUCTION

WHAT IS ELECTRICITY?

1

A. EARLY THEORIES

For many years it was known that if a piece of amber was rubbed it acquired the ability to attract small bits of straw, paper, and such. About 1600 A.D. William Gilbert, physicist and doctor to the English Queen Elizabeth I, discovered that other substances, such as sulfur, sealing wax, and glass, had properties similar to that of amber. The scientists of his time believed that such substances, when rubbed, or **charged,** exuded a sort of fluid which attracted light objects. They called this "fluid" **electricity.**

Early in the eighteenth century a French scientist, Charles du Fay, came to the conclusion that there were **two** types of "fluids," or electricity. One type he called "vitreous," the electricity present in charged glasslike substances. The other type he called "resinous," the electricity present when objects such as amber, wax, and rubber are charged.

About 1747 Benjamin Franklin, the American statesman and scientist, decided that there was only one type of "fluid," or electricity. The "vitreous" and "resinous" charges, he believed, were only two oppo-

7

site phases of the same phenomenon. He arbitrarily called the "vitreous" charge **positive** (+) and the "resinous" **negative** (−).

For the next one hundred and fifty years, though dissatisfied with the "fluid" theory of electricity, scientists could find no better explanation. But at the beginning of the twentieth century, while they were investigating the nature of matter in general, the outlines of a better and more rational theory began to take shape.

B. THE STRUCTURE OF MATTER

1. The Atom

Our world is full of a great many things that we call **matter,** by which we mean anything that has weight and takes up space. There are three forms, or **states,** of matter. These are the **solid** state, such as stones, the **liquid** state, such as water, and the **gaseous** state, such as air.

But what is this matter made of? Well, suppose we start with a drop of water. It has certain properties which we recognize as peculiar to water. Divide this drop and you have two smaller drops. Each droplet still has the properties of water.

Continue dividing the droplets until you come (theoretically, at least) to a particle so small that it can be divided no further. This particle still is water and exhibits all its properties. The smallest particle of matter that retains the properties of that matter is called a **molecule.** Matter, then, is made up of countless molecules, each of which exhibits the properties of that particular type of matter. You can see, accordingly, that there are as many different types of molecules as there are types of matter.

It was found that it is possible to break down the water molecule into two gases, **oxygen** and **hydrogen.** Note, however, that these gases do not resemble the original water. The molecule seems to be made up of simpler substances.

A substance that can be broken down into two or more simpler substances is called a **compound.** If the substance cannot be broken down any further, it is called an **element.** Water is a compound; oxygen and hydrogen are elements. The smallest particle of a compound is a **molecule;** the smallest particle of an element is called an **atom.**

We know of only 92 naturally-occurring types of atoms.* And just as

---

*When scientists succeeded in unlocking the secret of nuclear energy, they also succeeded in creating a number of new types of atoms such as we do not normally find on our earth.

a few different kinds of bricks may be used to build a great many types of buildings, so these relatively few kinds of atoms may be combined to form the enormous number of different molecules known to man.

The atom was considered to be the smallest particle of matter until in 1897 J. J. Thomson, the famous English scientist, announced he had definite proof that atoms, under certain conditions, shot out still smaller particles of matter, which we now call **electrons.** The amazing thing about these electrons is that they are all alike, regardless of what substance emits them.

Once it was shown that the atom could be broken up, scientists delved deeper into its secrets. As a result, the **electron theory*** of the structure of matter was set forth. Scientists today generally believe this theory to be true. But keep in mind that it is only a theory. It may be modified from time to time, and even, if proved to be false, discarded.

According to the electron theory, all matter is composed mainly of three types of particles. These are (1) the **electron,** a particle carrying a negative electrical charge; (2) the **proton,** a particle carrying a positive electrical charge; and (3) the **neutron,** a particle that carries no electrical charge. All atoms are composed of these particles; the atoms differ from one another in the number of particles they contain and in the arrangement of these particles.

In the early part of the twentieth century Niels Bohr, a Danish scientist, gave us a theoretical picture of atomic structure which, at that time, was widely accepted.† According to Bohr, the atom is composed of a central **nucleus,** which is surrounded by revolving **electrons,** somewhat as our sun is surrounded by revolving planets. In fact, the electrons that

*When this theory was first stated it was believed that all atoms were made up of electrons and protons. Hence the name "electron theory." Today we know that the atom contains a number of other particles such as neutrons, mesons, positrons, neutrinos, antiprotons, and several more. Thus it is no longer strictly accurate to talk about the "electron theory of atomic structure." However, to avoid confusion between our present-day theory and the old atomistic theories, we will retain the term **electron theory.**

Except for the electrons, protons, and neutrons, the particles that make up the atom do not normally appear before us. They seem to be created and exist for an extremely short period of time as atoms break up. Accordingly, we will ignore these short-lived particles in this book. Also, the theories presented here must, of necessity, be in greatly simplified form.

†Although the picture of atomic structure originally presented by Niels Bohr has been considerably modified by the discovery of new facts, for the purpose of this book it may be better if we consider the earlier, and simpler, Bohr atomic structure.

revolve around the nucleus are called **planetary** electrons. (See Figure 1-1.)

The nucleus contains all the protons and neutrons. An atom of one element differs from an atom of any other in the number of protons contained in the nucleus. The number of protons in the nucleus is called the **atomic number** of the element and varies from 1, for the element **hydrogen,** to 92 for the element **uranium.** (The number is even higher for the new man-made elements.) The atomic number of **helium,** whose structure is shown in Figure 1-1, is 2.

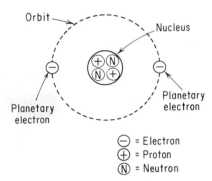

⊖ = Electron
⊕ = Proton
Ⓝ = Neutron

Fig. 1-1. Theoretical structure of an atom. The nucleus here contains two protons and two neutrons. The two planetary electrons revolve around the nucleus in the orbit indicated. This is the helium atom.

The negative charge of the planetary electron is equal and opposite to the positive charge of the proton. Since the nucleus contains all the protons, it carries a total positive charge that is equivalent to the number of protons present. Inasmuch as a normal atom is **neutral** — that is, it has no external electrical charge — the positive charges of the nucleus are exactly neutralized by the negative charges of the planetary electrons revolving about it.

It follows that the neutral atom has as many planetary electrons as there are protons in its nucleus. Consequently, the number of electrons revolving around the nucleus varies from 1 for hydrogen to 92 for uranium (and higher, for the man-made elements). Note that in the helium atom shown in Figure 1-1 there are two protons and two planetary electrons.

All atoms, except those of ordinary hydrogen, contain one or more neutrons in the nucleus. The helium atom (Figure 1-1) contains two neutrons; the uranium atom may contain 146 neutrons. The neutron carries no electrical charge and in some respects acts as though it were composed of a proton and electron combined, with the positive charge of the proton neutralized by the negative charge of the electron.

Although the opposite electrical charges carried by an electron and a proton are equal in magnitude, the **mass,** or weight, of the proton is about 1840 times as great as that of the electron. The mass of the neutron is about equal to that of the proton. We can readily see that practically the entire mass, or weight, of the atom is contained in its nucleus.

The number of protons and neutrons in the nucleus of an atom determines its weight, or **mass number.** The mass number of atoms varies from 1 for ordinary hydrogen, which is the lightest of the elements (one proton and no neutrons in its nucleus), to 238 for uranium which, until recently, was the heaviest element (92 protons and 146 neutrons in its nucleus). The new and heavier man-made elements have even greater mass numbers. The mass number for the helium atom shown in Figure 1-1 is 4 (two protons and two neutrons).

All of the atoms of an element contain the same number of protons in their nuclei, and the atoms of one element differ from those of all other elements in the number of protons so contained. For example, each hydrogen atom contains one proton in its nucleus and each uranium atom has 92 protons. But the atoms of the same element may differ in mass number, owing to different numbers of neutrons in their nuclei.

For example, three different types of hydrogen atoms have been found. All of these atoms have the same atomic number of 1 (one proton in the nucleus). However, one of these atoms (the type most commonly found in nature) has a mass number of 1 — that is, one proton and no neutron in its nucleus. A second type has a mass number of 2 — one proton and one neutron in the nucleus. A third type has a mass number of 3 — one proton and two neutrons. Except for the differences in mass, all three types of atoms have identical properties.

We call these different atoms of the same element **isotopes.** Most elements are known to have two or more isotopes. It is interesting to note that scientists have been able to produce artificial isotopes by bombarding the nuclei of atoms with neutrons. In this way, for example, an atom of uranium with a mass number of 238 (92 protons and 146 neutrons in its nucleus) sometimes captures a neutron, raising the mass number to 239 (92 protons, 147 neutrons).

So far, we have concentrated on the nucleus of the atom. Now let us turn our attention to the electrons revolving around the nucleus. As previously stated, the normal atom has one planetary electron for each proton in the nucleus. Thus, the number of such electrons will vary from 1 for hydrogen to 92 for uranium (and higher for the new elements).

These electrons do not revolve around the nucleus in a disorderly fashion. They follow concentric paths, called **orbits,** about the nucleus in a manner somewhat similar to the orbiting of the planets around the sun of our solar system. The planets are held in their orbits by the **gravitational** attraction between them and the sun. Similarly, the electrons of an atom are held in their orbits by the **electrostatic** attraction between the positive electrical charges carried by the protons of the nucleus and the negative electrical charges carried by the electrons.

However, if the electron is to hold its orbit and not fall into the nucleus, it must move at a constant speed around the nucleus in order to develop the centrifugal force required to counterbalance this attraction. Since it has a definite mass, the moving electron must possess a certain amount of energy which is determined by its mass and velocity.

The greater the radius of the orbit (that is, the farther the electron is from the nucleus), the greater must be the velocity and, hence, the energy of the electron. Thus the orbits correspond to the energy levels of the electrons that occupy them. Electrons with lower energy levels occupy orbits closer to the nucleus. Those with higher energy levels occupy orbits farther away from the nucleus. Should an electron absorb additional energy from some external source, it would jump to an orbit farther from the nucleus. Should it lose some of its energy, it would fall to an orbit closer to the nucleus.

The electron possesses energy in discrete amounts, and the larger amounts are whole-number multiples of the smaller. Since it may occupy only an orbit that corresponds to its energy level, only certain orbits are permissible. These permissible orbits are bunched in groups, called **shells,** arranged in concentric layers about the nucleus, somewhat as the layers of an onion. There is a certain maximum number of electrons that each shell may contain. The number of shells, up to seven, depends upon the atomic number of the atom — that is, the number of electrons in orbit around its nucleus. (See Figure 1-2.)

The shell nearest the nucleus (shell #1) may contain up to two electrons. If there are more than two electrons, the excess form a second shell (shell #2) around the first. This shell may hold up to eight electrons.

After shell #2 is completely filled, a third shell (shell #3) is formed. This shell may hold up to eighteen electrons. However, the outermost

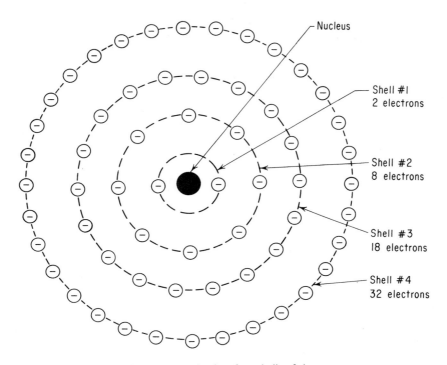

Fig. 1-2. Arrangement of electrons in the first four shells of the atom.

shell of any atom may not hold more than eight electrons. Thus, after shell #3 has eight electrons, shell #4 is formed. Only after this shell has one or two electrons does shell #3 receive its remaining ten electrons.

With shell #3 completely filled, additional electrons are added to shell #4. When the latter shell has eight electrons, electrons begin to appear in shell #5. Then electrons are added to shell #4 until its quota of thirty-two is completed. In a somewhat similar manner shells #6 and #7 are created.

Only those electrons in the outermost shell of a particular atom (that is, those electrons with the highest energy levels) are involved in chemical and electrical phenomena. These are called **valence electrons.**

The particles we have been discussing are fantastically small. For example, it has been estimated that if 250 million (2.5×10^8) hydrogen atoms were placed side by side, they would extend about one inch. And if 100,000 (1×10^5) electrons were placed side by side, they would be as large across as a single hydrogen atom. The proton and neutron are even smaller than the electron.

The mass of the proton has been estimated to be about 1.66×10^{-24} gram. (There are 453.6 grams to the pound.) The mass of the neutron is approximately the same as that of the proton. The electron has a mass of about 9.1×10^{-28} gram.

What is even more amazing is the fact that the major portion of the atom consists of — empty space! It has been estimated that if a copper one-cent piece were enlarged so that its circumference were equal to that of the earth's orbit around the sun (over 58 billion miles), the electrons would be the size of baseballs and would be about three miles apart.

2. How Atoms Combine to Form Molecules

Since the outermost shell of any atom may not contain more than eight electrons, an atom may have no more than eight valence electrons. Atoms are considered to be **stable** if their outermost shell is completely filled or, in the alternative, if this outermost shell contains eight valence electrons. Because such atoms are stable, they ordinarily do not combine chemically with other atoms. Examples of stable atoms are those of the so-called inert gases — **helium, neon, argon, krypton, xenon,** and **radon.**

Helium (illustrated in Figure 1-1) has an atomic number of 2. Hence its two electrons completely fill its outermost, and only, shell. Neon has an atomic number of 10. The first two electrons of its atom fill shell #1. The remaining eight electrons completely fill its outermost shell, shell #2. Argon, with an atomic number of 18, has its first two electrons filling shell #1. Its next eight electrons fill shell #2. This leaves the remaining eight electrons for the outermost shell, #3. In a similar manner, the atoms of the other inert gases have eight valence electrons in their outermost shells.

Atoms that contain between one and seven valence electrons are **unstable.** Because unstable atoms seek to achieve a stable state, they will combine with other atoms, thus forming molecules. In so doing, an atom containing more than four valence electrons may seize electrons from its neighbor or neighbors. On the other hand, if an atom has less than four valence electrons it may readily surrender them to other atoms.

The process whereby atoms combine to form molecules is called **atomic bonding.** There are three types of bonding—**ionic** bonding, **covalent** bonding, and **metallic** bonding.

As an example of ionic bonding, consider the combination of a sodium atom with a chlorine atom. The sodium atom has an atomic number of 11, that is, it has eleven protons in its nucleus and eleven planetary electrons. These electrons are arranged in three shells around the nucleus. The first shell is completely filled with two electrons. The second shell, too, is completely filled with eight electrons. The third shell contains the remaining electron. Thus the sodium atom has one valence electron.

The chlorine atom has an atomic number of 17 — seventeen protons in its nucleus surrounded by seventeen electrons in three shells. Shell #1 is filled with two electrons, shell #2 is filled with eight electrons, and shell #3 contains the remaining seven electrons. Thus, the chlorine atom has seven valence electrons.

The chlorine atom seizes the outermost electron from the sodium atom, thus becoming stable with eight valence electrons. But, in the process, it upsets its neutrality since it now has one more electron than the number of its protons. Hence, it now has a net negative electrical charge of 1. An atom that is not neutral but has an electrical charge is called an **ion.** Thus the chlorine atom becomes a **negative ion.**

The sodium atom readily surrenders its single valence electron to the chlorine atom, thus achieving stability because its outer shell now is completely filled. However, it, too, loses its neutrality since it now has one less electron than the number of its protons. Thus it has a net positive electrical charge of 1, and it becomes a **positive ion.**

Ions containing opposite electrical charges attract each other. Hence the sodium and chlorine ions are bound together to form a molecule of **sodium chloride** — ordinary table salt.

The hydrogen atom contains one electron in its one and only shell. Since this shell has a capability of two electrons, the hydrogen atom is unstable. As such, it readily surrenders its electron to become a positive ion and combine with a negative ion to form a molecule by ionic bonding.

The hydrogen atom also may achieve a stable state by a sharing process whereby two hydrogen atoms share their electrons. Thus each atom has the equivalent of two electrons to complete its shell. (See Figure

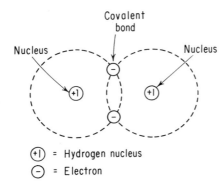

Covalent
bond

Nucleus

Nucleus

(+1) = Hydrogen nucleus
(−) = Electron

Fig. 1-3. How two hydrogen atoms combine to form a hydrogen molecule.

1-3.) The two atoms form a hydrogen molecule, held together by their shared electrons. The pair of shared electrons is called a **covalent bond** and the process of forming molecules in this manner is known as **covalent bonding.**

As for **metallic bonding,** as its name implies, it is the type of bonding that holds together the atoms of metals such as copper, silver, and so forth. Consider the copper atom. It has an atomic number of 29 — twenty-nine protons in its nucleus surrounded by twenty-nine electrons. Twenty-eight of these electrons fill the first three shells. The twenty-ninth electron lies in the fourth shell. Hence the copper atom has one valence electron.

Since the first twenty-eight electrons completely fill the first three shells, they are held firmly by the attraction of the nucleus. The single valence electron in the fourth shell, on the other hand, is held very loosely. As a matter of fact, it frequently leaves its parent nucleus and wanders off as a **free electron.** The free electron, together with the many other free electrons from the great number of atoms that constitute a piece of copper, form a sort of electron cloud which drifts about between the atoms.

As a copper atom loses its valence electron, it loses a negative charge and thus becomes a positive ion. Since like charges repel, the positive ions would tend to fly apart. However, the attraction between these positive ions and the negative electron cloud around them holds the ions in place; thus the piece of copper retains its shape.

Some molecules are made up of hundreds, and even thousands of atoms. However, even with the help of the electron microscope, only a few of the very largest molecules have ever been seen by the human eye.

3.   Conductors, Insulators, and Semiconductors

Since the free electrons are free to move, they will flow readily through a metal in a sort of **current** if subjected to an **electrical pressure,** somewhat as a liquid will flow through a pipe if subjected to a mechanical pressure. (Electrical pressure will be considered a little later in the book.) For a given pressure, the greater the number of free electrons, the greater will be the electron current. Or, put another way, the greater the number of free electrons, the smaller will be the resistance to the flow of current.

Substances that offer very little resistance to the flow of the electron current are called **conductors.** Metals, generally, are good conductors. It has been estimated that there are approximately 8.5×10^{22} free electrons in a cubic centimeter of copper. Or, if we consider resistance, a cubic centimeter of copper will offer a resistance of about 1.7×10^{-6} **ohm.** (The ohm is a unit of resistance and will be explained later; here it is used only for purposes of comparison.)

Other substances, which have very few free electrons and, hence, offer very large resistance to electron flow, are called **insulators.** One such substance, ordinary glass, offers a resistance of about 9×10^{13} ohms per cubic centimeter. Some other examples of insulators are hard rubber, wax, amber, and sulfur.

There is a third type of material, called **semiconductor,** whose resistance lies between that of conductors and insulators. An example is **germanium,** whose resistance is about 60 ohms per cubic centimeter.

C.   THE ELECTRIC FIELD

We have seen that in the normal atom the positive charges of the protons in the nucleus are exactly neutralized by the negative charges of the planetary electrons. Hence the normal atom is neutral — that is, it has no external electrical charge.

When a neutral atom loses one of its electrons, it now has more positive charges than negative. Hence the overall effect is to give the atom a positive charge. In other words, it becomes a positive ion.

Should an extra electron attach itself to a neutral atom, there would be more negative charges than positive. The atom thus would become a negative ion.

When two charged particles are brought near each other, an interaction is set up between them. If the particles have similar charges —

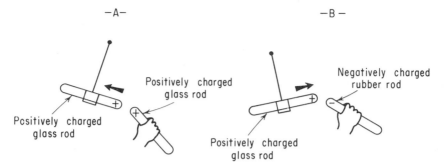

Fig. 1-4. A. Bodies having like charges repel each other.
B. Bodies having unlike charges attract each other.

that is, if they both are charged positive or both negative — the particles tend to repel each other. If, on the other hand, the two particles have opposite charges — that is, one is positive and the other negative — they attract each other.

This can be demonstrated by a simple experiment. Rub a small glass rod with a piece of silk cloth. As the rod is rubbed, some of the electrons in the outer shells of its atoms are torn away by the cloth. The cloth, then, has an excess of electrons, or a negative charge. The rod, having lost some of its electrons, has a positive charge.

Repeat with a second glass rod. We now have two rods each containing a positive charge. Suspend one of the rods so that it may swing freely. As the second positively charged glass rod is brought near it, the first rod will swing away from it. (See Figure 1-4A. The positive charge is indicated by $+$ and the negative charge by $-$.)

Rub a hard-rubber rod with a piece of fur. In this case the atoms of the rubber rod will tear electrons away from the atoms of the fur. Thus the fur will acquire a positive charge and the rubber rod a negative charge. Now bring the negatively charged rubber rod near the suspended positively charged glass rod. The glass rod will be attracted to the rubber rod. (See Figure 1-4B.)

Further, it will be found that the charged rods are able to attract certain light objects such as a small scrap of paper. How can the charged rod attract the uncharged paper?

The answer is shown in Figure 1-5. By rubbing the rubber rod, we charge it negatively (excess of electrons). As this negatively charged rod is placed near the uncharged paper, some of the electrons along the

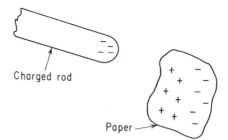

Fig. 1-5. How a charged body attracts
an uncharged one.

side of the paper nearest the rod are repelled and tend to move to the opposite side. This leaves the atoms of the paper nearest the rod with a deficiency of electrons (positive charge). We say that this nearer side is charged by **electrostatic induction.** Because the attraction between the positively charged nearer side of the paper and the negatively charged rod is greater than the repulsion between the latter and the negatively charged farther side, the paper will be attracted to the rod.

What determines which substance will seize electrons and which will lose them? The answer lies in the nature of the substance — that is, the number and arrangement of its outermost electrons. Some substances, such as glass and fur, have their outermost electrons so arranged that they can lose them quite easily (and so obtain a positive charge). On the other hand, some substances, such as silk and rubber, have their outermost electrons so arranged that, when they are rubbed, they will seize electrons from the material with which they are stroked, and so will accumulate an excess of electrons (negative charge).

When two bodies are able to affect each other through space, we say that a **field of force** exists between them. For example, the earth and the sun attract each other even though they are separated by more than 92,000,000 miles of space. In this case, the field of force is the **gravitational field** that surrounds all matter.

In the case of charged particles, the field of force between them is called a **dielectric, electrostatic,** or **electric field.** We may visualize this field as consisting of imaginary **electric lines of force** radiating in all directions from the charged particles. (See Figure 1-6.)

Figure 1-6A shows the lines of force about a positively charged particle. Arrowheads on these lines of force are used to indicate the **direction** of the field. The direction of the field is the one in which a charged particle moves or tends to move when acted on by the force of

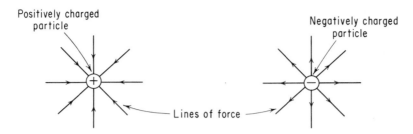

Fig. 1-6. A. Lines of force around a positively charged particle.
   B. Lines of force around a negatively charged particle.

the field, just as a cork floating in a stream shows which way the stream is flowing.

In testing the direction of the electric field, we conventionally use the small negative charge carried by the electron as an indicator. Thus the electron will be attracted to the positively charged particle, as indicated by the arrowheads pointing into the particle.

The lines of force about a negatively charged particle are illustrated in Figure 1-6B. Here, if an electron is placed near the negatively charged particle, it will be repelled away from the particle along one of the lines of force. The arrowheads pointing away from the negatively charged particle indicate the direction of motion.

If two oppositely charged particles are placed near each other, they will be linked by the electric lines of force radiating from each particle. (See Figure 1-7.) These lines of force, though imaginary, act somewhat as stretched rubber bands. Hence they tend to contract and, at the same time, push each other aside. It is this contractive force that pulls the two oppositely charged particles towards each other.

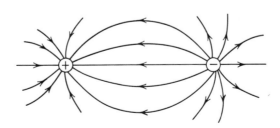

Fig. 1-7. Electric field of force between two oppositely charged particles.

−A−

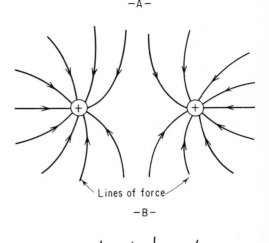

−B−

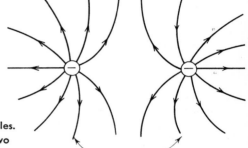

Fig. 1-8. **A.** Electric field of force between two positively charged particles.
B. Electric field of force between two
negatively charged particles.

If, as in Figure 1-8A and B, two similarly charged particles are placed near each other, the electric lines of force are not linked. Instead, they push each other aside and cause the particles to repel each other.

The electrons of the atoms of an insulator, such as the glass or rubber rod, are tightly held by their nuclei. Thus, when these rods are charged by losing or acquiring electrons, the action is local. That is, only the portion of the rod being rubbed is affected. If you take away or add electrons to one end of such a rod, the atoms at the other end remain neutral and thus the rod at that end has no charge.

On the other hand, the outer electrons of the atoms of conductors are loosely held. If an excess of electrons (negative charge) is placed at one end of a conductor, the repulsion between like charges will cause the loosely held electrons of neighboring atoms to move toward the other end. The movement of these electrons will cause a disturbance among electrons farther away and, as a result, the excess of electrons quickly distributes itself throughout the entire conductor.

Similarly, if electrons are removed from one end of a conductor (positive charge), electrons from neighboring atoms are attracted to compensate for the deficiency. Again, all the atoms are affected and soon the deficiency is spread throughout the entire conductor. Thus a charge placed upon any portion of a conductor quickly spreads itself.

QUESTIONS

In this and following chapters, wherever possible, diagrams should be used to clarify the answers to these questions. These diagrams need not be elaborate, but they should be drawn neatly with the significant portions clearly labeled.

1. Describe Bohr's theory of the structure of the atom.

2. What is meant by the **atomic number** of an atom? By its **mass number?**

3. In terms of the electron theory, how do atoms of one element differ from those of another?

4. What is meant by **isotopes?** How do isotopes of the same element differ from one another?

5. What is the relationship between the energy possessed by an electron and the distance of its orbit from its nucleus?

6. What is meant by the **valence** electron? What is the maximum number of valence electrons an atom may contain?

7. What is meant by a **stable** atom; an **unstable** atom?

8. Draw the theoretical picture of an atom of **calcium** whose atomic number is 20 and whose mass number is 40.

9. What is meant when we say an atom is **neutral?** What is meant when we say an object has a **positive charge**; a **negative charge?**

10. What is an **ion**; a **positive ion**; a **negative ion?**

11. Using an example for each, explain what is meant by **ionic bonding, covalent bonding,** and **metallic bonding.**

12. What is meant by the **electric field of force?** State the law of attraction and repulsion between charged particles.

13. Explain how a charged hard-rubber rod can attract small scraps of paper.

14. In terms of electron theory, explain what is meant by an **insulator**; a **conductor.**

STATIC
ELECTRICITY

2

A. ELECTRICAL PRESSURE

We now know that a field of force exists between two dissimilarly charged objects. Another way of looking at this is to picture the excess electrons (negative charge) as straining to reach a point where there is a deficiency of electrons (positive charge).

If the two oppositely charged bodies are connected by a conductor, the excess electrons will have an easy path to the point of deficiency. The loosely held and free electrons of the conductor are repelled from the end that has the negative charge (excess of electrons). At the same time they are attracted to the end that has the positive charge (deficiency of electrons). As a result, a stream of electrons flows through the conductor from the end that has the excess of electrons to the end that has the deficiency. This flow continues until the charges on both bodies are equalized.

But if the two oppositely charged bodies are separated by an insulator, the picture is different. The insulator has few, if any, free electrons and the outermost electrons are tightly held to their orbits. Accordingly, the excess electrons cannot move from the negative end of the insulator to satisfy the deficiency at the positive end.

If the two oppositely charged bodies are separated by air (which is an insulator) and we continue to pile up an excess of electrons on the negatively charged body and remove electrons from the positively charged one, the electrostatic field of force between the two bodies increases. The stress, or **electrical pressure**, may become quite great and, after a certain limit is reached, the air no longer can restrain the excess electrons and they rush across to the point of deficiency. This rush of electrons produces light and we see it in the form of a spark. At the same time, the air through which the electrons rush is heated and expands. We may hear the inrush of air that follows cooling in the form of a sharp, crackling sound, which accompanies the spark.

This rush of electrons, or spark, takes place in a small fraction of a second, and it might seem that the electrons merely jump from the point of excess to the point of deficiency. But closer examination indicates that the electrons jump back and forth between the two points many times (even millions of times) per second.

We may understand this better, perhaps, if we consider the behavior of a pendulum. Suspend a small weight from a string tied to a nail. The weight hangs straight down. Now lift the weight several inches to the right. The force of gravity tends to pull the weight back to its original position (that is, straight down). Release the weight. It swings toward its original position, but keeps right on going and now swings to the left. Again the force of gravity pulls it back, and once again it swings to the right. This process continues, each swing being a little less than before, until the weight comes to rest in its original position.

Similarly, electrons at a point that has a negative charge seek to satisfy the deficiency at the opposing point. When they rush over, more go than are needed to satisfy the deficiency. The two charges change places. The electrical pressure now is in the opposite direction and the electrons rush back. This to-and-fro surge of electrons continues until both points have equal charges and the electrons are at rest once more.

B. ELECTRICITY AND LIGHTNING

The similarity between the electric spark and the flash of lightning led to the conclusion that they were identical. This identity was proved by Benjamin Franklin in 1752. Reasoning that lightning was the passage of electricity from the clouds to the earth, he set out to draw some of that electricity by means of a kite raised during a thunderstorm. Some of the electricity flowed down the wet string of the kite to a metal key attached at its end. By placing his knuckle near the key, he was able to draw a spark as the electricity jumped from the key to the ground through his body. He probably was unaware that he was risking his life by that experiment!

We now know that the lightning flash occurs because of a tremendous electrical pressure arising from opposite charges on the clouds and on the earth. How do these charges get on the clouds?

We are not sure, but we do know that thunder clouds are accompanied by a tremendous upward rush of warm air. We believe this upward rush of air tears apart the droplets of water of the cloud. The friction between the rushing air and the fragments of water-drops generates an electrical charge. The positively charged particles generally are carried to the upper portion of the cloud, giving this upper portion a positive charge and leaving a negative charge on the lower portion of the cloud (Figure 2-1). By induction, the ground beneath the cloud receives

Fig. 2-1. Distribution of electrical charges on a cloud and on the earth beneath it. Sometimes the charges are reversed so that the underside of the cloud is positive and the ground is negative. Yet the result is the same whether the excess electrons start flowing from the cloud to the ground, or from the ground to the cloud.

a positive charge. Since trees and houses in contact with the ground acquire the same charge, they, too, are charged positively.

When the electrical pressure becomes great enough, the excess electrons on the under portion of the cloud surge downward toward the earth in the form of a lightning flash. As was true of the spark, so in the lightning flash the electrons surge back and forth many times between the earth and the cloud. All this may take place in a few millionths of a second; that is why a lightning flash appears to be a single bolt.

Sometimes the charges are reversed so that the lower portion of the cloud becomes charged positively. The ground beneath the cloud then receives a negative charge by induction and the first rush of electrons may be up to the cloud. The effect is the same whether the lightning stroke first starts moving upwards or downwards.

So great are the electrical charges involved, that a lightning flash may extend five miles from cloud to earth. Most lightning flashes, however, are from cloud to cloud and such discharges may be up to ten miles long though, generally, they are much shorter. The clap of thunder that accompanies the flash is caused by inrushing air. The air that had been expanded by the passage of the hot lightning bolt has cooled and contracted, creating a partial vacuum. The surrounding air rushes into the low-pressure zone.

You see now why you should not stand under a tree during a lightning storm. Since the tree has the same charge as the ground and is closer to the cloud than the ground beneath it, there is a better chance that the lightning will strike the tree. So great is the force when this occurs, that the tree may be splintered and the heat of the flash may set it on fire.

It was the fact that lightning tends to strike the highest point that led Benjamin Franklin to invent the **lightning rod,** a device that protects buildings from lightning. It is a long, slender, metal rod extending above the roof of the building and connected to the earth by a heavy metal cable. Hence it presents an easier path to ground than the building, and any electrical charge it receives is carried away harmlessly through the cable directly to the ground.

The lightning rod protects the building in another way, also. If a charge is placed upon a spherical conductor, the mutual repulsion between like charges will cause the charge to be distributed uniformly over the sphere's surface. But if the conductor is eggshaped, the pointed end will become more highly charged than the rounded end. The greater charge causes a greater electrical pressure and, as a result, the charge will leak off into the air more readily from the pointed end than from the rounded one. By the same token, a pointed conductor will receive a charge from a nearby object more readily than will a rounded conductor.

For this reason the upper end of the lightning rod is pointed and, consequently, its charge leaks off readily. Since the rod is attached to the building, the charge on the latter, too, will leak off through the sharp point of the lightning rod. Thus, the electrostatic field between the building and the cloud above it is lessened and the chances of the buildings being struck by lightning are reduced.

C. PRACTICAL APPLICATIONS OF STATIC ELECTRICITY

There are two ways in which we may put electricity to work. We may use the energy of the electrostatic field existing between two opposite charges. This is called **static** electricity. Or else we may use the energy of the movement of the electrons as they travel from a point of excess to a point of deficiency. This is called **current** electricity. For the remainder of this chapter we shall confine ourselves to practical applications of static electricity.

To further their studies of lightning, scientists sought to produce miniature flashes in the laboratory. Even on a smaller scale, a tremendous electrical pressure is required to cause electrons to jump, say, twenty feet of air from a negatively charged sphere to a positively charged one.

To obtain these tremendous electrical pressures, an American scientist, Robert Van de Graaff, in 1931 designed a machine that continuously placed small electrical charges into a large "container," thus building up the charge. Its action is somewhat similar to the manner in which an endless chain of buckets will bring water up from a well and fill a large tank on top of a building. A simplified version of the Van de Graaff generator is shown in Figure 2-2.

In essence, it consists of a large, hollow, metal ball set on a long tube. This metal ball, which may be ten or more feet in diameter, is the "container" for the electrical charge. The tube, which may be thirty or forty feet long, supports the metal ball, and is made of some insulating material in order to prevent the charge stored in the ball from flowing to the ground. Inside this tube a long silk belt is made

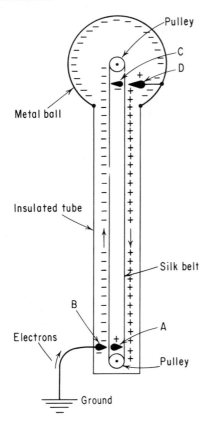

Pulley

C

D

Metal ball

Insulated tube

Silk belt

B

A

Electrons

Pulley

Ground

Fig. 2-2. Van de Graaff generator

to pass rapidly over a set of pulley wheels. These wheels are insulated from the ground. Note the positions of the four eggshaped metal electrodes marked A and B (near the bottom of the belt) and C and D (near the top).

Let us suppose that a small positive charge is deposited on the righthand (descending) side of the silk belt. (This charge may be placed there by friction or by some other means.) As this charged portion of the belt comes opposite the point of electrode A, electrons are drawn from this electrode to compensate for the deficiency on the belt. This leaves electrode A with a positive charge.

Electrode B is connected to the ground. As electrode A becomes positive, it induces a negative charge on electrode B, whose excess electrons flow up from the ground. Some of these electrons leak off the pointed end of electrode B and are deposited on the silk belt, which rapidly carries them up the structure.

Near the top, the belt passes electrode C. Some of its excess electrons leak over, making this electrode negative. As a result, electrode D is made positive by induction. Hence, it attracts the excess electrons still remaining on the silk belt as the latter passes by its pointed end, and it deposits them on the metal ball to which it is connected. Having lost its electrons, the descending silk belt has a positive charge and is ready to repeat the cycle when it passes by electrode A.

The belt revolves at high speed and each revolution adds to the negative charge on the surface of the metal ball. Of course, the metal ball may be charged positively by starting the process with a negative charge on the descending side of the belt.

Tremendous charges may be built up in this way. When the electrical pressure becomes great enough, the sphere discharges to ground in the form of a miniature stroke of lightning. Sometimes two such generators,

Westinghouse Electric Corp.

Fig. 2-3. External view of Van de Graaff generator.

one creating a positive charge and the other a negative, are operated simultaneously. When the electrical pressure becomes great enough, a tremendous discharge, many feet in length, takes place between the two metal balls.

In addition to producing miniature lightning flashes, the Van de Graaff generator is used for other purposes where tremendous electrical pressures are required. For example, in atomic research, it is used to generate the large electrical pressure needed to bombard the nuclei of atoms.

Electrical charges can be used for other purposes. For example, in Figure 2-4 you see a device that is used to filter dust and soot that otherwise would go up a factory smokestack together with the hot gases. Not only is the surrounding air kept purer, but often valuable chemicals are recovered.

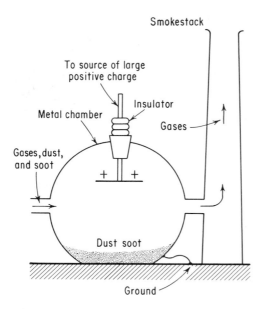

Fig. 2-4. How soot and dust are removed from factory exhaust gases.

The operation of this device is extremely simple. The hot gases, dust, and soot pass through a metal chamber before entering the smokestack. A metal plate is mounted from an insulator near the top of the chamber. A large positive charge is placed upon this plate. As the particles of dust and soot pass beneath the charged plate, they are attracted to it. Once they touch the plate, the particles, too, acquire

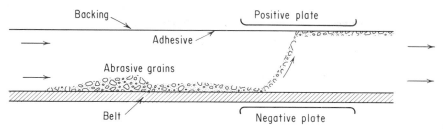

Fig. 2-5. How the electrostatic field is used in the manufacture of a better sandpaper.

positive charges. As such, they are repelled from the charged plate and accumulate at the bottom of the chamber, which is grounded. The hot gases, cleaned of dust and soot, continue up the smokestack.

The electrostatic field is used, too, in the manufacture of sandpaper and similar abrasives. Ordinarily, a backing of paper or cloth is covered with an adhesive and the abrasive grains are spread over this adhesive and permitted to harden in place. In an improved method, however, the abrasive grains are dropped onto a belt, which carries them between two oppositely charged plates. The backing, coated with adhesive, is passed beneath the upper plate (Figure 2-5). As the abrasive grains enter the electrostatic field between the two charged plates, they acquire the same charge as the lower plate. They are repelled from this plate and fly up to the adhesive backing where, since all the grains have similar charges, they repel each other, spacing themselves uniformly. When the adhesive hardens, therefore, the abrasive grains are found to be standing on end and uniformly spaced. The result is a great improvement in the cutting quality of the abrasive.

A similar method is used in the manufacture of certain fabrics, such as simulated velvet or carpeting. The cloth backing is coated with an adhesive and passed beneath the upper of two charged plates. A belt carries tiny textile fibers over the lower plate. These fibers are repelled from the lower plate and are shot into the adhesive where they stand on end and are evenly spaced, often packed in at 250,000 fibers per square inch. Sometimes, instead of being spread uniformly, the adhesive is applied in the form of a design. The result, then, is a raised design where the tiny fibers stick to the adhesive.

So far we have dealt with stationary charges. The excess electrons deposited on a body have remained there, except for the brief interval when they have distributed themselves over a conductor or during an electrical discharge. It is for this reason that we have headed this chapter Static, or **stationary,** Electricity. In Chapter 4 and succeeding chapters, however, we shall study the electrons as they move from point to point under the influence of an electrical pressure. We shall, therefore, be considering **current,** or **flowing,** electricity.

QUESTIONS

1. In terms of the electron theory, what is meant by **electrical pressure?**

2. Explain how lightning is produced.

3. Why is it unsafe to stand beneath a tree in an open field during a thunderstorm?

4. Explain how the lightning rod operates to protect a building from lightning.

5. Explain the operation of the Van de Graaff generator.

6. Explain how static electricity may be used to prevent dust and soot from polluting the air near a factory chimney.

7. Explain how static electricity may be used in the manufacture of sandpaper.

8. Explain how static electricity may be used in the manufacture of simulated velvet.

9. Explain the difference between **static** and **current** electricity.

MAGNETISM

3

The ancient people knew about that mysterious, heavy black stone (which we now know is a mineral containing iron) called a **magnet,** which could attract pieces of iron. They also knew that if a bar of iron were stroked with this stone, the iron, too, would be able to attract other pieces of iron.

(We know today that a magnet can attract not only pieces of iron, but certain other metals, such as nickel and cobalt, although with less force. We call substances that can be attracted by a magnet **magnetic,** and the ability of a magnet to attract magnetic substances we call **magnetism.**)

Further, legend has it that thousands of years ago, the Chinese discovered that if a magnet were suspended so that it could swing freely, one end would always point in the general direction of north and the other end in the general direction of south. The north-seeking end is called a **north pole** (N) and the south-seeking end a **south pole** (S). Such a device is called a **compass,** and for centuries navigators have used the compass to guide their ships. It is for this reason that the magnet was also known as **lodestone** (leading stone).

In addition, it was found that if the north-seeking end of the magnet were brought near the north-seeking end of another, the two magnets would be repelled. Similarly, if two south-seeking ends were brought together, they also would repel each other. On the other hand, if a north-seeking end were brought near a south-seeking end, they would attract each other.

About 1600 A.D. William Gilbert, the same man who investigated the properties of amber, undertook the first serious study of magnets. He found, among other things, that the attractive property of the magnet was concentrated at its two opposite ends, called **poles,** with very little attractive force between the poles.

He theorized that the earth is a huge magnet with its magnetic north pole near its geographic north pole and its magnetic south pole near its geographic south pole. It is the attraction between the north-seeking pole of the compass (which is, in reality, a south pole) and the north pole of the earth that causes the compass to act as it does.

A. THE MAGNETIC FIELD

This matter of attraction and repulsion between poles of magnets warrants close attention. It was found that the poles need not touch each other. Even if they are a distance apart, like poles will repel each other and unlike poles will attract each other. If a nonmagnetic substance is placed between the poles, their attraction or repulsion is unchanged. Thus, if a sheet of glass or copper is placed between two unlike magnetic poles, they continue to attract each other as though the glass or copper were not there.

We may understand this phenomenon a little more clearly if we consider the electric field existing between charged bodies (see Chapter 1, subdivision C). Here, under the influence of the field of force existing between them, bodies containing similar charges repel each other. If the bodies have unlike charges they attract each other.

It would appear, then, that a **magnetic force,** or **field,** exists between the two opposite poles of a magnet. We may find out more about this magnetic field by means of a simple experiment. Place a magnet on a wooden table. Over it, place a sheet of glass. Sprinkle iron filings on the glass and tap the glass lightly. The iron filings will assume a definite pattern on the glass sheet (see Figure 3-1).

The iron filings are attracted to the magnet through the glass sheet. Although the glass prevents these iron filings from touching the magnet,

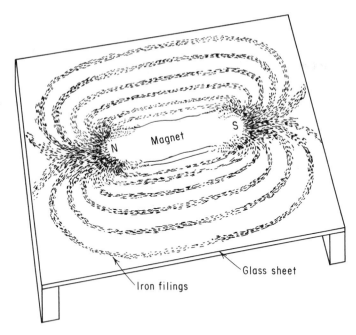

Fig. 3-1. How iron filings show the form of the magnetic field around a magnet.

nevertheless the filings will form a pattern which will show the form of the magnetic field. Note that the iron filings arrange themselves outside the magnet in a series of closed loops that extend from pole to pole.

It is in this way that we visualize the magnetic field around a magnet. We say that the magnet acts as though **magnetic lines of force** surround it in the pattern formed by the iron filings. Of course, the pattern formed in Figure 3-1 represents the magnetic field only in the horizontal plane. To obtain a true picture, we should consider the magnetic field surrounding the magnet as existing in three dimensions — that is, over and under the magnet as well as on either side of it.

Note that the magnetic lines of force, like the electric lines of force discussed in Chapter 1, are imaginary. Nevertheless, the field acts as though the lines of force were present. It would seem that these lines try to follow the shortest distance from pole to pole, at the same time repelling each other. As with the electric lines of force, it might help if we think of the magnetic lines of force as a bundle of stretched rubber bands that tend to shorten and, simultaneously, push each other away sidewise.

We arbitrarily assume the force acts from the north pole to the south pole. It is as if lines of force "flowed" from the north pole to the south pole. If, theoretically, we were to place a small north pole in the magnetic field, it would be repelled from the north pole of the magnet and move along the path of a line of force until it reached the south pole.

We can see now why like poles repel and unlike poles attract. If we place two unlike poles near each other, as in Figure 3-2A, the lines of force flow from the north pole to the south pole. Since these lines of force tend to shorten, the two magnets are pulled to each other. If, on the other hand, we place two like poles near each other, as in Figure 3-2B, the lines of force tend to repel each other and the two magnets are pushed apart.

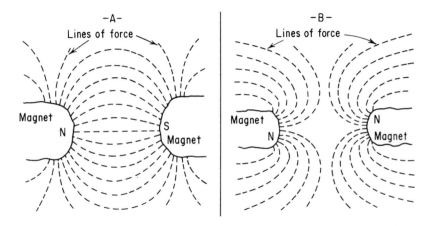

Fig. 3-2. A. Pattern of the resulting magnetic field when two unlike poles are placed near each other.

B. Pattern of the resulting magnetic field when two like poles are placed near each other.

The magnetic lines of force are known as the **magnetic flux,** which is considered to flow in closed loops. This is illustrated in Figure 3-3, which shows the flux around and through a magnet. The lines of force flow out of the north pole, through the air, and back into the south pole of the magnet. Within the magnet, they flow back to the north pole, thus completing the loop. (In the previous figures the flow of flux within the magnets was omitted for the sake of simplicity.)

We find that the magnetic flux encounters less opposition flowing

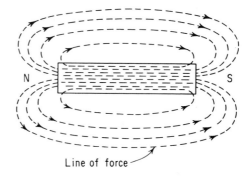

Fig. 3-3. Magnetic flux around and through a magnet.

Line of force

through magnetic substances, such as iron, than through nonmagnetic substances, such as air, glass, or copper. We call the opposition to the flow of magnetic flux, **reluctance.** The reluctances of all nonmagnetic substances are the same. Thus the flux will flow with the same ease (or difficulty) through air, glass, copper, and so on.

On the other hand, it will flow much more readily through a piece of iron placed in its path. Thus, if a piece of iron is placed within the magnetic field, the lines of force will distort their pattern to take the easier path through the iron, as shown in Figure 3-4A. If, as in Figure 3-4B, an iron washer is placed within the magnetic field, the flux will flow through the iron of the washer, and no lines of force will be found in the air space within the washer.

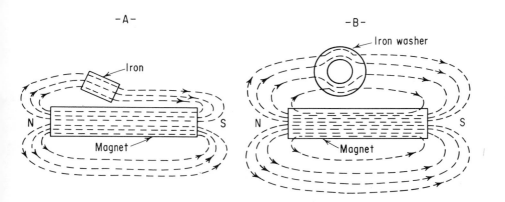

Fig. 3-4. A. Distortion of the magnetic field around a magnet by a piece of iron.
B. Distortion of the magnetic field around a magnet by an iron washer. Note that there are no lines of force in the air space within the washer.

B. THEORIES OF MAGNETISM

1. The Molecular Theory of Magnetism

A nineteenth-century German physicist, Wilhelm Weber, proposed a molecular theory of magnetism. According to this theory, the molecules of magnetic material, such as iron, are tiny magnets, each with a north and south magnetic pole and with a surrounding magnetic field.

In the unmagnetized piece of iron these magnetic molecules are arranged helter-skelter (see Figure 3-5A). As a result of this arrangement, the magnetic fields around the molecules cancel each other out and there is no external magnetic field. But if the iron is magnetized, the molecules line up in an orderly array, with the north pole of one molecule facing the south pole of another (see Figure 3-5B). The result, then, is that all the magnetic fields aid each other and we have a magnet with an external magnetic field.

This theory explains a number of things we know about magnets. For example, note that in the magnetized iron all the north poles of the molecules are facing one way and all the south poles are facing the

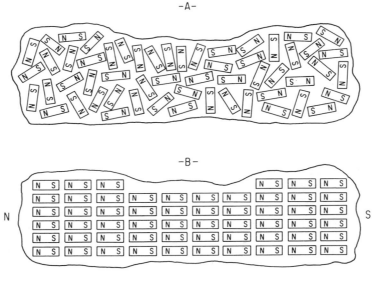

Fig. 3-5. Arrangements of molecules in a piece of magnetic material.
  A. Unmagnetized.
  B. Magnetized.

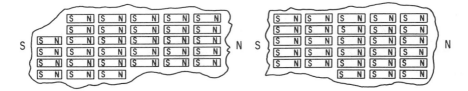

Fig. 3-6. Arrangement of molecules showing why two magnets are formed when a magnet is broken in two.

opposite way. The result, then, is that we have a magnet whose magnetism is concentrated at the two opposite poles.

Further, it was found that if you break a magnet in two, you obtain two magnets, each with a set of poles, as illustrated in Figure 3-6. Also, you can destroy the magnetism of a magnet by any means that will disarrange the orderly array of the molecules, such as by heating or jarring the magnet.

If a piece of unmagnetized magnetic material is placed in the magnetic field of a magnet (as, for example, a bar of iron is stroked by a magnet), the attraction between the external magnet and the molecules of the magnetic substance causes these molecules to line up in the necessary orderly array (see Figure 3-7). Magnetism produced in this way is called **induced magnetism.**

We now can explain why a magnet can attract an unmagnetized piece of iron. Under the influence of the magnetic field of the magnet, the

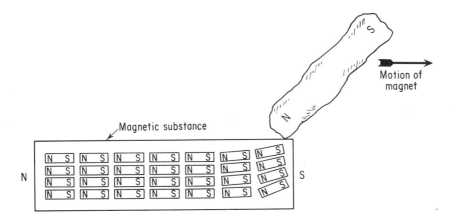

Fig. 3-7. How a piece of magnetic material may be magnetized by stroking with a magnet.

piece of iron becomes magnetized by induction, two poles being produced at the ends of the iron. The end nearest the magnet is an unlike pole. The attraction between unlike poles draws the piece of iron to the magnet.

A nonmagnetic substance, such as glass or copper, resists all attempts to align its molecules in orderly fashion. Nor do all magnetic substances submit to this lining-up process to the same degree. In some substances, such as soft iron, the molecules are easily moved and will line up readily under the influence of the magnetic field of another magnet. However, once the external magnetic field is removed, the molecules of the soft iron revert to their original, disorderly condition. The soft iron forms a **temporary magnet,** which is magnetized only so long as it is acted on by an external magnetic field.

On the other hand, the molecules of some substances, such as steel, require a much greater magnetic force to produce an orderly arrangement. However, when the external magnetic field is removed, these molecules will retain their positions, and consequently these substances form **permanent magnets.** As we have seen, however, heating or jarring the magnet will disarrange its molecules and thus destroy its magnetic properties.

The greater the number of aligned molecules of a magnetic substance, the stronger will be its magnetic field. When all the molecules are aligned, the substance is said to be **saturated.** Then, increasing the strength of the aligning force will produce no increase in the magnetic field of the substance.

2. Modern Theories of Magnetism

Armed with a better understanding of the structure of matter, modern scientists have modified Weber's molecular theory of magnetism. They believe that a magnetic field is produced by a moving electric field. And, conversely, that an electric field is produced by a moving magnetic field.

We have learned that the electrons of an atom rotate in concentric shells around the nucleus. It was found that, not only do these electrons rotate around the nucleus, but, as they do so, they spin on their own axis, somewhat as the earth spins on its axis as it rotates around the sun. And it was further discovered that the phenomenon of magnetism seems to be associated with the spin of the electrons within the third shell of the atoms of magnetic materials.

The electron carries a negative electrical charge. Hence it is surrounded by an electric field. As the electron spins, so does its electric field. It is this moving electric field that creates a magnetic field. The polarity of this magnetic field — that is, the direction of its lines of force — depends upon whether the electron is spinning clockwise or counterclockwise.

Since the electrons of all atoms spin, why, then, are not all atoms magnetic? The answer lies in the fact that, in the atoms of most materials, half of the electrons spin in one direction and half in the other. Thus the opposing polarities of the resulting magnetic fields cancel out and there remains no external magnetic field.

In the atom of a magnetic material, however, more electrons spin in one direction than in the other. Hence the magnetic fields do not cancel completely, and a net external magnetic field is produced.

Consider, for example, the magnetic material iron. Its atom has an atomic number of 26 — that is, its nucleus contains twenty-six protons surrounded by twenty-six electrons arranged in four shells. The first shell contains two electrons, one spinning in a clockwise direction and the other in a counterclockwise direction. The resulting magnetic fields thus cancel out. The second shell contains eight electrons, four of which are spinning in one direction and four in the opposite direction. Again, the magnetic fields cancel out.

The third shell is incomplete, containing fourteen electrons (instead of the eighteen it can hold). Nine of these electrons spin in one direction and five spin in the opposite direction. The net result, then, is an external magnetic field due to the four electrons whose magnetic fields are not canceled.

The fourth shell contains the two valence electrons, one spinning in one direction and the other in the opposite direction. Hence their magnetic fields cancel. The overall effect, then, is that the atom has an external magnetic field due to the four uncanceled electrons in the third shell. That is, the atom becomes a small magnet.

In magnetic material the interaction between the atoms causes a number of neighboring atoms to line up parallel to each other in such a way that their magnetic fields aid each other. Such an array (which, it has been estimated, contains approximately 10^{15} atoms and occupies a volume of about 10^{-8} cubic centimeter) is called a **domain.** The magnetic material, then, contains a large number of domains, each of which is a tiny magnet.

In the unmagnetized material these domains are arranged in a random pattern. Hence their magnetic fields cancel out and there is no external magnetic field around the material. But if this material is placed in an external magnetic field, the domains are forced to line up in the direction of the lines of force of that field. Hence the domains' fields aid each other and, as a result, the magnetic material has an external magnetic field. That is, it becomes magnetized.

Except that we now talk of magnetic domains instead of magnetic molecules, the explanations offered by Weber's theories for the various magnetic phenomena discussed in the previous subdivision are essentially unchanged.

C. MAGNETIC UNITS AND TERMS

Throughout the ages man has been compelled to find means of expressing the concepts of "how big," "how heavy," and "how much time." Practically every nation established its own units of length, mass, and time. However, to facilitate the exchange of goods and ideas, there arose the need for a uniform system of units, in which a certain unit would mean the same thing to all men.

In the area of science there are three such systems. One is the **English** system based upon the **inch, pound,** and **second.** Where the metric system of measurement is employed, there are two systems of units. One is the **cgs** system based upon the **centimeter, gram,** and **second.** The other is the **mks** system based upon the **meter, kilogram,** and **second.** To avoid confusion, the magnetic units described here will be in the cgs system.

MAGNETIC CIRCUIT. We have seen that the magnetic field seems to "flow" along magnetic lines of force. We also have seen that these lines of force form closed loops around and through the magnetic material. We call the path we can trace around these loops a **magnetic circuit.**

As an example of what is meant by a magnetic circuit, consider Figure 3-8. Here you see the magnetic circuit of an electric generator (which will be discussed later in the book). An iron frame contains a set of poles made of the same material. Coils of wire are wound on these poles and, as current flows through these coils, a magnetic field is set up. (This, too, will be explained later.) The polarity of these poles is as indicated in the diagram. An iron armature core is suspended between the two poles.

Note that there are two magnetic paths. One starts from the north pole, passes across the right-hand air gap between the right-hand pole

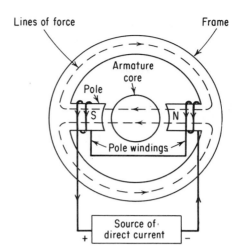

Fig. 3-8. Magnetic circuit of a generator.

and the armature core, through the armature core, across the left-hand air gap, through the left-hand pole, through the upper half of the frame, through the right-hand pole, and back to the north pole. The other path is similar to the first, except that it passes through the lower half of the frame instead of the upper half. Both paths are closed loops. Note, too, that the magnetic path may be through both magnetic materials (such as the iron) and nonmagnetic materials (such as the air).

MAGNETOMOTIVE FORCE (**mmf**). In order to establish the magnetic lines of force in a magnetic circuit, a **magnetomotive force** must be applied. This force may come from the magnetic field around another magnet or (as we shall see later in the book) from the flow of an electric current through a coil of wire. The symbol for the magnetomotive force is the script letter ℱ and its unit is the **gilbert.** The value of this unit generally is expressed in electrical terms, and it will be discussed further when such terms are encountered.

MAGNETIC FLUX. The magnetic field is represented by lines of force. The total number of lines of force of the field is called the **magnetic flux.** The symbol for magnetic flux is the Greek letter ϕ (**phi**) and the name of its unit is the **maxwell.** A maxwell is equal to one magnetic line of force.

RELUCTANCE. The opposition that a magnetic path offers to magnetic flux when a magnetomotive force is applied is called **reluctance,** whose symbol is the script letter ℛ. The relationship between magnetomotive force, magnetic flux, and reluctance may be expressed by the following formula:

$$\text{Reluctance} = \frac{\mathfrak{F}}{\phi},$$

where $\mathfrak{F}$ is the magnetomotive force in gilberts, and
ϕ is the magnetic flux in maxwells.

MAGNETIZING FORCE. Whereas the magnetomotive force produces the magnetic flux, the **magnetizing force** (designated by H) is the magnetomotive force that produces the flux per unit length of the magnetic circuit. Its unit is called the **oersted**, which is the magnetomotive force per unit length. Thus:

$$H = \frac{\mathfrak{F}}{L},$$

where H is the magnetizing force in oersteds,
$\mathfrak{F}$ is the magnetomotive force in gilberts, and
L is the length in centimeters.

MAGNETIC FLUX DENSITY (or MAGNETIC INDUCTION). Whereas the magnetic flux indicates the total number of lines of force in the magnetic field, the **flux density** (designated by B) is a measure of the number of lines of force per unit area taken at right angles to the direction of flux. Its unit is the **gauss**, whose value is **one maxwell per square centimeter.** Thus:

$$B = \frac{\phi}{A},$$

where B is the flux density in gauss,
ϕ is the magnetic flux in maxwells, and
A is the area in square centimeters.

PERMEABILITY. The measure of how much magnetic flux will be produced in a certain material by a given magnetizing force is called the **permeability** of that material. Thus

$$\text{Permeability} = \frac{B}{H},$$

where B is the flux density in gauss, and
H is the magnetizing force in oersteds.

The symbol for permeability is the Greek letter μ (**mu**).

The permeability of a vacuum — and, for practical purposes, air and all other nonmagnetic substances — is unity. Magnetic substances, on the other hand, have greater permeabilities. Thus permeability is the ratio between the magnetic flux produced in a magnetic material and the flux that would be produced if air were used instead, the magnetizing force remaining constant.

Not all magnetic substances, however, have the same permeability. The permeability of soft iron is much greater than that of steel. And there are some magnetic alloys that have a greater permeability than soft iron.

The relationship between flux density (*B*) and the magnetizing force (*H*) for a magnetic material can be shown by the graph illustrated in Figure 3-9. As the magnetizing force is increased, so is the flux density. But only up to a point. At the **saturation point** the material has acquired practically all the magnetism of which it is capable. Increasing the magnetizing force beyond this point results in very little more magnetic flux in the material.

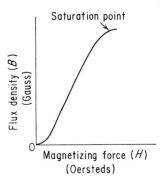

Fig. 3-9. Typical magnetization curve for steel.

In certain applications magnetic material is located within a coil of wire through which an **alternating electric current** flows. Such a current (which will be discussed later in the book) starts from zero, flows in one direction, increasing gradually to a maximum. Then it decreases gradually back to zero. At this point the current reverses its direction of flow, increasing gradually to a maximum (but this time in the opposite direction). Then the current decreases gradually back to zero. After the current completes this series of changes it repeats the entire cycle.

As you have learned, a current flowing through a coil of wire produces a magnetic field. Since this magnetic field corresponds to the current flow, the magnetic material finds itself in a magnetic field that is

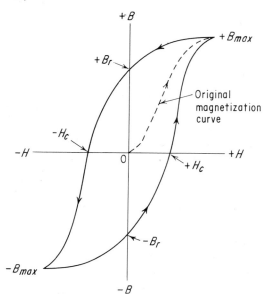

Fig. 3-10. Typical hysteresis curve for steel.

changing in step with the current. How the magnetic material is affected by this changing magnetic field is shown by the graph illustrated in Figure 3-10.

We start with the material unmagnetized. When the current is at 0, the magnetization curve for the magnetic material, too, is at 0. As the current flow increases in one direction, the flux density $(+B)$ also increases until the point of maximum flux density $(+B_{max})$ is reached. The material is saturated and beyond this point the flux density will not increase, regardless of any increase in current or magnetizing force. (Note that this curve is, in effect, the magnetization curve shown in Figure 3-9.)

As the current starts to fall back to zero, the flux density also starts to decrease. But its decrease is not as great as was its rise. When the current reaches zero, the magnetizing force (H) produced by it, too, is zero. However, a certain amount of flux, called the **residual flux density,** or **residual inductance,** remains in the material (as shown by point $+B_r$ of the curve).

As the current reverses and flows in the opposite direction, it produces a magnetizing force once more. But this force, too, now is in the opposite direction $(-H)$. It now is called a **demagnetizing force.** The flux produced by this demagnetizing force acts to neutralize the residual flux remaining in the material. When the demagnetizing force becomes great enough, the flux it produces completely cancels the residual flux (point $-H_c$). The value of the demagnetizing force at this point is known as the **coercive force.**

As the demagnetizing force increases still further, it produces a magnetic flux in the material again, but this time in the direction opposite to the original flux. That is, it produces a negative flux $(-B)$. This process continues until the material is saturated once more and can hold no more flux (point $-B_{max}$). Now the current, and the magnetizing force it produces, starts to fall back to zero. The flux density, too, decreases slowly. At point $-B_r$ the current and the magnetizing force reach zero. The material, however, still retains some residual negative flux.

At this point the current and the magnetizing force change direction again. The positive flux they now produce tends to neutralize the residual negative flux in the material. When the magnetizing force becomes great enough, the positive flux it produces completely cancels the residual negative flux (point $+H_c$). As the current and magnetizing force continue to increase, a positive flux is produced once more in the material. This flux continues to increase with the increase in current and magnetizing force until saturation is reached (point $+B_{max}$, again). As the current repeats the cycle, the entire process is repeated (except, of course, for the portion indicated by the original magnetization curve).

A curve of this type is known as a **hysteresis curve.** The word "hysteresis" means "a lagging behind." The curve gets its name from the fact that it shows how the magnetizing (and demagnetizing) effect lags behind the magnetizing (and demagnetizing) force.

D. PRACTICAL APPLICATIONS OF PERMANENT MAGNETS

Under certain circumstances (as we shall see later in the book) it is desirable that we be able to turn the magnetic effect on or off at will. For this purpose we employ a temporary magnet, which uses a material that will become magnetized only so long as it is influenced by an external magnetic field. Then, when this field is removed, the material loses its magnetism.

Such a material (as, for example, pure iron) is called a **magnetically soft** material. It has a relatively high permeability — that is, its domains are easily aligned by the magnetizing force. On the other hand, its residual flux density (B_r) is relatively low and it requires very little coercive force (H_c) to overcome this residual flux.

Permanent magnets, on the other hand, are made of **magnetically**

hard material. Such a material has a relatively low permeability. To align its domains requires a relatively large magnetizing force. But once aligned, the domains tend to retain their aligned positions and resist any force that tends to disturb this alignment. Hence it has a relatively large residual flux density, and a relatively large coercive force is required to overcome this residual flux. Thus a permanent magnet can resist the demagnetizing effect of any small stray magnetic field in which it may find itself.

The earliest permanent magnets were made of steel (an alloy of iron and carbon). At the beginning of this century it was found that the magnetic qualities of steel could be improved by adding small amounts of tungsten and cobalt. About 1933 Alnico, an improved type of permanent-magnet material, was developed. This substance is an alloy of iron, aluminum, nickel, and cobalt, Small amounts of copper and titanium, too, may be added. In recent years ceramic magnetic material, consisting of combinations of iron and barium oxides, was developed.

Incidentally, the earth's magnetic field amounts to about half a gauss. The magnetic fields that may be obtained from natural magnets (lodestone) amount to several hundred gauss. Permanent magnets made from modern alloys are able to produce fields of as much as 10,000 gauss.

The permanent magnet has many uses, both in the home and in industry. Its earliest application, in the magnetic compass, is still being used. In the home it is used in such devices as the telephone, the radio and television receiver, the magnetic can opener, magnetic door latches,

General Electric Company.

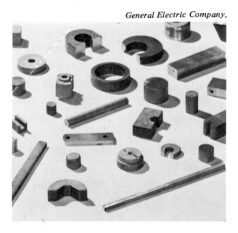

Fig. 3-11. Permanent magnets can be constructed in many different shapes.

and toys. In industry it is used with certain types of electron tubes, in small motors and generators, in measuring instruments, in tool holders, for separating magnetic from nonmagnetic materials, and for handling magnetic materials on conveyers. These are but a few of the many applications of permanent magnets.

Common shapes of manufactured permanent magnets are the **bar magnet,** which has a pole at each end, and the **horseshoe magnet,** which really is a bar magnet that has been bent into a U or horseshoe shape so that the poles are closer together. Thus, practically all its magnetism is concentrated into a smaller space. Magnets are also made in a great many other shapes, such as disks, rings, cylinders, and balls. (See Figure 3-11.)

QUESTIONS

1. List three magnetic substances; three nonmagnetic substances.

2. Draw a diagram showing the magnetic lines of force around a bar magnet.

3. State the law of magnetic attraction and repulsion.

4. Explain why two north poles will repel each other; why a north pole will attract a south pole.

5. A watch contains steel gear wheels. Should these gear wheels become magnetized, the time-keeping quality of the watch will be impaired. How would you protect such a watch from an external magnetic field? Explain.

6. In terms of the electron theory, explain why some substances are magnetic and others nonmagnetic.

7. Explain the difference between a magnetized and an unmagnetized piece of iron.

8. Explain how a magnet can attract an unmagnetized piece of iron. Why will it not attract a piece of wood?

9. Explain what is meant by the **magnetic flux.**

10. Explain what is meant by the **magnetic flux density.**

11. Explain what is meant by **magnetomotive force.**

12. Explain what is meant by **magnetic reluctance.**

13. Explain what is meant by **permeability.**

14. What is the difference between a temporary and a permanent magnet?

15. Give two methods for destroying the magnetic property of a permanent magnet. Explain.

DIRECT CURRENT ELECTRICITY

II

THE ELECTRIC CURRENT

4

A. CURRENT CARRIERS

If an electron is removed from a neutral atom, the atom becomes a **positively** charged ion. If an electron is added, the neutral atom becomes a **negatively** charged ion. In the space between any two charged particles there exists a stress, or **field of force.** We call this field of force the **electric,** or **electrostatic, field.** Particles bearing like charges tend to repel one another, whereas particles bearing unlike charges tend to attract one another. The movement of charged particles arising from the presence of this field is called the **electric current.**

Moving particles that carry an electrical charge are called **current carriers.** There are three types of carriers. In solid conductors, such as a copper wire, as an electron escapes from its parent nucleus, the remaining positive ion is held in place. It is only the free electron that is able to move about. Hence the current carriers of such a conductor consist of electrons, which are particles carrying a negative electrical charge.

In liquids and gases the charged ions, too, are free to move. Thus we have a second type of carrier — the ion.

In semiconductors, such as **germanium** or **silicon,** we encounter a third type of carrier. The atoms of semiconductors are held together by **covalent bonds** (see Chapter 1, subdivision B, 2). This bond, you will recall, consists of two valence electrons, one each from two neighboring atoms. If sufficient energy (such as heat or light energy) is applied, the bond may be broken. Then one of the bonded electrons escapes, becoming a free electron which is free to move about. In its place in the bond is left a deficiency, called a **hole.** Since this hole represents a deficiency of a negative charge, it may be considered as the equivalent of a particle with a positive charge.

The attraction of the positive charge of the hole may then cause an electron from a nearby covalent bond to leave its bond and drop into the hole, thus repairing the original broken bond. As it does so, however, the electron leaves a hole in its bond. The attraction of this hole then attracts an electron from a bond farther away.

The result, then, is the progressive movement of the hole and its accompanying positive charge from bond to bond. Thus, in a semiconductor, we may have the movement of negative carriers (electrons) and positive carriers (holes). Note, however, that these carriers move in opposite directions.

B. THREE FACTORS OF AN ELECTRIC CURRENT

1. Electromotive Force (emf)

Whenever an excess of electrons (negative charge) occurs at one end of a conductor and a deficiency (positive charge) at the other end, the electric field between the two ends will set up an electrical pressure that will cause the loosely held electrons from the outer shells of the atoms of the conductor to stream from the point of excess to the point of deficiency. Thus, an electric current will flow through the conductor from the negative end to the positive one.

To get an idea of **electrical pressure,** let us consider a simple analogy. Assume that we have a U-shaped tube with a valve or stopcock at the center (Figure 4-1). Assume that the valve is closed. We now pour water into arm A to a height represented by X. We pour water into arm B to a height represented by Y. If the valve now is opened, the water will flow from arm A to arm B until X and Y are equal.

What caused the water to flow? It was not **pressure** in arm A, because

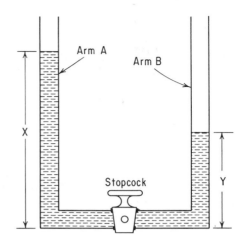

Fig. 4-1. Diagram illustrating the fact that it is the *difference* in pressure which causes a fluid to flow through a tube.

when X and Y are of equal length, no water flows even though the water in arm A still exerts pressure. It was the **difference in pressure** between the two arms that caused the water to flow. The flow continued until the pressures in the two arms were equal and the **difference in pressures** was zero.

So it is with electrons. If, at one end of a conductor, electrons are piled up, and at the other end, electrons are fewer in number or are being taken away, the excess electrons will flow toward the point of deficiency.

In Figure 4-1 the water in arm A can do no work until the valve is opened. Nevertheless, it represents a **potential** source of energy — that is, energy due to position. However, the actual work is not done by the potential energy of the water in arm A, but by the **difference in potential energy** between the water in A and the water in B.

Similarly, in Figure 4-2, it is not simply the potential energy of the excess electrons at one end of the conductor that causes the electrons to flow. It is the **difference** between the amounts of potential energy at the two ends of the conductor that does the work. We may say that electrons

Fig. 4.2 Diagram illustrating the fact that it is the *difference* in electrical pressure which causes electrons to move through a conductor. (The electrons are indicated by ⓔ→ and the arrow shows their direction of movement.)

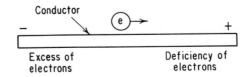

flow through a conductor (that is to say, an electric current flows) because of the difference in potential energy between the ends of the conductor. The force that moves the electrons from one point to another is known as the **potential difference,** or **electromotive (electron-moving) force.**

2. Resistance

There is a factor other than electromotive force, or potential difference, that affects electrical flow. Suppose that we were to suspend two metallic balls in air several inches apart, and place a negative charge on one and a positive charge on the other (Figure 4-3).

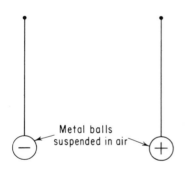

Metal balls
suspended in air

Fig. 4-3. Two oppositely charged balls suspended apart in air. Although there is a potential difference between them, no current flows because the resistance of the path between them is too great.

Here we have a potential difference, and yet no current flows. The reason is that the air between the two balls offers too great a **resistance** to the flow of current. If you connect the two balls by a piece of metal, however, the electric current will flow from the negatively charged ball to the other one. The resistance of the metal strip is low enough so that the potential difference may send the electric current flowing through it.

But it is not necessary to connect the two balls with a metal strip to cause the electrons to flow from one to the other. All we need do is increase the charges. When the potential difference becomes great enough, the electrons will jump across through the air in the form of an electric spark. We conclude, therefore, that, for electric current to flow, **the potential difference must be great enough to overcome the resistance of the path.**

Different substances offer different resistances to the flow of electric current. Metals, generally, offer little resistance and are good conductors. Silver is the best conductor known, and copper is almost as good. Insulators, such as glass, rubber, sulfur, and the like, offer a very high resistance. But all substances will permit the passage of some electric current, provided the potential difference is high enough.

3. Current Flow

For the third factor that affects electrical flow, refer back to Figure 4-1. We measure the flow of water from one arm to the other in terms of quantity per unit of time. We say so many gallons flow past a certain point in a minute. Similarly, we measure the flow of electricity by the number of electrons that flow past a point on a conductor in one second.

What determines the amount of water per unit of time that flows through the valve in Figure 4-1? Obviously, it is the difference between the amounts of potential energy of the water in the two arms of the tube, and it is also the size of the opening in the valve (that is, the resistance the valve offers to the flow of water).

In the case of the electric current, the quantity of electricity (the number of electrons per second) that flows in a conductor depends upon the potential difference and the resistance of the conductor. **The greater the potential difference, the larger the quantity of electricity that will flow; the greater the resistance, the smaller the quantity of electricity.**

C. UNITS OF MEASUREMENT

1. Quantity of Electric Charge

Adding an electron to a neutral atom gives it a negative charge; taking an electron away from a neutral atom gives it a positive charge. Thus the electrical charge of the electron is our basic unit.

Since the charge of one electron is very small, a **coulomb** is used as a practical unit for measuring the quantity of electric charge. A coulomb is equal to the combined charges of 6.25×10^{18} electrons.

Of course, a figure this large is meaningless to the average person. It is stated here merely to emphasize that the electrical charge on the electron is extremely small and that the combined charges of many electrons are employed to make up the coulomb.

2. Electric Current

When we talk of electric current, we mean electrons in motion. When the electrons flow in one direction only, the current is called a **direct**

current (abbreviated **dc**). In the discussion of electricity in this portion of the book we are speaking only of direct currents. (As we shall see later, the electrons may flow, alternately, first in one direction and then in the other. Such a current is known as an **alternating current.**)

It is important to know the number of electrons that flow past a given point on a conductor in a certain length of time. If a coulomb flows past a given point in one second, we call this amount one **ampere** of electric current. Hence, the unit of electric current is the ampere.

Aside from the fact that electrons are too small to be seen, we would find it impossible to count them as they flowed by. Fortunately, we have an electrical instrument, called the **ammeter** (to be described later) that indicates directly the amount of current flowing through it.

Where the ampere is too large a unit to be used, we may employ the **milliampere,** which is a thousandth (1/1000) of an ampere, or the **micro-ampere,** which is a millionth (1/1,000,000) of an ampere. In an electrical formula, the capital letter *I* stands for current.

3. Resistance

A number of factors determine the resistance that a substance offers to the flow of electric current. First of all, there is the nature of the substance itself. The greater the number of free electrons present in the substance, the lower is its resistance.

Resistance is also affected by the length of the substance. The longer an object is, the greater its resistance. Another factor is the cross-sectional area of the substance, which is the area of the end exposed if we slice through the substance at right angles to its length. The greater the cross-sectional area, the less the resistance to current flow.

Resistance is also affected by the temperature of the substance. Metals generally offer higher resistance at higher temperatures. Certain nonmetallic substances, such as carbon, on the other hand, offer lower resistance at higher temperatures.

The unit of resistance is the **ohm.** By international agreement, the ohm is the resistance to the flow of electric current offered by a uniform column of mercury, 106.3 centimeters long, having a cross-sectional area of one square millimeter, at 0°C. Where the ohm is too small a unit, we may employ the **kilohm** (1000 ohms) and the **megohm** (1,000,000 ohms). The symbol for the ohm is the Greek capital letter Ω **(omega).** In an electrical formula, the capital letter *R* stands for resistance.

All the substances offer a certain amount of resistance to the flow of

−A−

−B−

International Resistance Co.

Fig. 4-4. Fixed resistors.
    A. Wire-wound type.
    B. Composition type.

current. There are times, however, when we wish deliberately to introduce definite amounts of resistance into the current path. To do so we
use devices called **resistors.** Where the resistance required is not too
great, the resistor may consist of **nichrome** wire, whose resistance is more
than 50 times that of copper, of suitable length and thickness wound
upon a ceramic tube. An insulating ceramic coating usually is applied
over the winding. This is called a **fixed** resistor. (See Figure 4-4A.) If the
resistance is to be large, the resistor may consist of a thin coat of a carbon composition deposited on a ceramic tube and covered with some
insulating material (Figure 4-4B). Connections are made by means of
wires attached to the ends of the resistor.

    Where the resistor is to be variable, the wire is wound on a fiber strip
which may be bent into a circular form. A metal arm or slider, manipu-
lated by a knob, is made to move over the wire, thus making contact
with any desired point. In this way the length of wire, and hence the resistance, between one end of the wire and the slider may be varied. Such
variable resistors are called **rheostats** or **potentiometers.** (See Figure 4-5.)

−A−

−B−

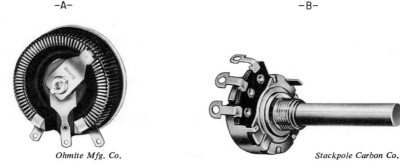

Ohmite Mfg. Co.

Stackpole Carbon Co.

Fig. 4-5. Variable resistors.
    A. Wire-wound type.
    B. Composition type.

Where the resistance is to be large, a carbon composition deposited on the fiber strip may replace the coiled wire.

4. Electromotive Force

The electromotive force creates the electric pressure that causes the current to flow through a conductor. Another name for this force is **voltage.** We can measure the voltage between any two selected points on a conductor by means of an electrical instrument, called the **voltmeter** (to be described later).

The unit of measurement of electromotive force, or voltage, is called the **volt.** The volt is defined as **that electromotive force that is necessary to cause one ampere of current to flow through a resistance of one ohm.** Where the volt is too large a unit, we may use the **millivolt** (1/1000 of a volt) or the **microvolt** (1/1,000,000 of a volt). Where the volt is too small a unit, we may use the **kilovolt** (1000 volts). In an electrical formula, the capital letter E stands for voltage.

5. Electric Power and Energy

A body at rest tends to remain in this condition or, if it is in motion, it tends to continue this motion in a straight line. This property of the body is known as **inertia.** To overcome the effect of inertia, **force** is required. Thus force may be considered as a push or pull, and, in mechanics, it commonly is measured in units of **pounds.**

Work is the product of the force and the distance through which it acts and is measured in **foot-pounds.** If a 1-pound weight is lifted one foot, the work done is 1 foot-pound. If the weight is lifted five feet, the work done is 5 foot-pounds.

Power is the rate at which work is done. A child might lift the weight five feet in five seconds. A more powerful adult might lift it the same distance in one second. The formula then may be

$$\text{Power} = \frac{\text{work done}}{\text{time}}.$$

The unit of power (for mechanics) is **foot-pound per second.** Thus, in the case of the child,

$$\text{Power} = \frac{1 \text{ pound} \times 5 \text{ feet}}{5 \text{ seconds}} = 1 \text{ foot-pound per second.}$$

In the case of the adult,

$$\text{Power} = \frac{1 \text{ pound} \times 5 \text{ feet}}{1 \text{ second}} = 5 \text{ foot-pounds per second.}$$

Five hundred and fifty foot-pounds per second, or 33,000 foot-pounds per minute, equals one **horsepower.**

In electricity, work is done as the electromotive force causes electrons to move through a conductor. The **electric power** (the rate of doing work) is the product of the electromotive force and the number of electrons set flowing per unit of time. But the number of electrons flowing per unit of time is the current (measured in units of ampere). Hence, the power, whose symbol is P, is the product of the electromotive force, in volts, and the current, in amperes. The unit of electric power is the **watt.** The formula for power thus becomes

$$P \text{ (watts)} = I \text{ (amperes)} \times E \text{ (volts).}$$

EXAMPLE.   What is the power required to enable a dry cell whose electromotive force is 1.5 volts to cause a current of 2 amperes to flow through a conductor?

ANSWER. $P = I \times E = 2$ amperes $\times 1.5$ volts $= 3$ watts.

The dry cell generates the power; the power is consumed as the current flows through the conductor.

Most electrical appliances carry labels stating the voltage at which the appliance is to operate and the power it consumes. Thus an electric lamp may bear a label indicating "60 watts, 120 volts." From this you can calculate that the filament of this lamp has 0.5 ampere of current flowing through it.

As previously stated, the unit of electric power is the **watt,** which is defined as the power used when an electromotive force of one volt causes a current of one ampere to flow through a conductor. Where the watt is too large a unit to be used conveniently, we employ the **milliwatt** (1/1000 of a watt). Where the watt is too small a unit, we may use the **kilowatt** (1000 watts).

The horsepower (33,000 foot-pounds per minute) is the unit of mechanical power. The kilowatt is the unit of electric power. Since mechanical energy can be converted to electric energy, and vice versa, the two units of power may be equated. Thus:

$$1 \text{ kilowatt} = 1.34 \text{ horsepower,}$$
$$1 \text{ horsepower} = 0.746 \text{ kilowatt.}$$

To measure the total electrical energy consumed by an appliance we must know how much power it uses (in watts) and the length of time (in seconds) it continues to consume this power. Thus, the unit of energy is the **wattsecond,** or **joule** — that is, one watt of power applied for one second. Since the wattsecond is a small unit, we frequently use the **watthour** (one watt applied for one hour) or the **kilowatthour** (1000 watts applied for one hour).

It is on the basis of kilowatthours that we pay our electric bill. For example, how much would it cost to run five 60-watt lamps four hours a day for 30 days if we have to pay at the rate of $.05 per kilowatthour?

Each day we consume

$$5 \times 60 \text{ watts} \times 4 \text{ hours} = 1200 \text{ watthours.}$$

In 30 days we consume

1200 watthours $\times$ 30 = 36,000 watthours or 36 kilowatthours.

At $.05 per kilowatthour, our bill is $1.80.

D. OHM'S LAW

The relationship between the electromotive force, the current, and the resistance was discovered by a German scientist, George Simon Ohm, at the beginning of the nineteenth century. The unit of resistance was named in his honor.

This relationship, which is called **Ohm's law,** can be expressed mathematically by means of the following formula:

$$\text{Current} = \frac{\text{electromotive force}}{\text{resistance}} \quad \text{or} \quad I = \frac{E}{R},$$

where I is measured in amperes, E in volts, and R in ohms. This means that the greater the electromotive force is, the greater will be the current; and the greater the resistance, the smaller the current.

Let us try a problem. How much current will flow through a conductor whose resistance is 10 ohms, when the electromotive force is 100 volts? Using our formula $I = E/R$, we get

ANSWER. $\qquad I = \dfrac{100 \text{ volts}}{10 \text{ ohms}},$ or 10 amperes.

Our Ohm's-law formula can be transposed as follows:

$$E = I \times R \quad \text{and} \quad R = \frac{E}{I}.$$

EXAMPLE. How many volts are required to make a current of 5 amperes to flow through a conductor whose resistance is 2 ohms?

Substituting our known values in the formula $E = I \times R$, we get

ANSWER. $E = 5$ amperes $\times 2$ ohms, or 10 volts.

EXAMPLE. What is the resistance of a conductor if 100 volts are required to force 2 amperes of current through it?

Since $R = E/I$,

ANSWER. $R = \dfrac{100 \text{ volts}}{2 \text{ amperes}}$, or 50 ohms.

We also can use these formulas to show the relationship between power (in watts), current (in amperes), and resistance (in ohms). You will recall that

Power = current $\times$ voltage   or $P = I \times E$.

If we substitute for E its equivalent ($I \times R$), we get

$$P = I \times I \times R \quad \text{or} \quad P = I^2 \times R.$$

EXAMPLE. How much power will be required to force a current of 2 amperes to flow through a conductor whose resistance is 5 ohms?

Since $P = I^2 \times R$, substituting our values we get

ANSWER. $P = (2)^2 \times 5 = 4 \times 5 = 20$ watts.

In a similar manner, we can show the relationship between power (in watts), electromotive force (in volts), and resistance (in ohms). Since $P = I \times E$, substituting for I its equivalent (E/R) we get

$$P = \frac{E}{R} \times E \quad \text{or} \quad P = \frac{E^2}{R}.$$

EXAMPLE. If the electrical pressure is 10 volts, how much power will be dissipated by a resistance of 20 ohms?

Since $P = E^2/R$, then

ANSWER. $P = \dfrac{(10)^2}{20} = \dfrac{100}{20} = 5$ watts.

QUESTIONS

1. List and explain three types of current carriers.

2. Explain what is meant by **electric current**; **direct current.** In what unit do we measure current?

3. Explain what is meant by **electromotive force.** What is its unit of measurement?

4. Explain what is meant by **resistance.** What is its unit of measurement?

5. Explain four factors affecting the resistance of a substance.

6. Explain Ohm's law.

7. How many volts are required to make a current of 5 milliamperes flow through a 1-megohm resistor? [*Ans.* 5000 volts.]

8. If a voltage of 100 volts be applied to a 50-ohm resistor, how much current will flow through it? [*Ans.* 2 amperes.]

9. What must be the resistance of a resistor if 60 volts will cause 3 amperes of current to flow through it? [*Ans.* 20 ohms.]

10. Explain what is meant by **electric power.** What is its unit of measurement?

11. What is the power in an electrical circuit whose resistance is 4 ohms if the applied voltage is 10 volts? [*Ans.* 25 watts.]

12. How much power is consumed in a circuit where a current of 50 milliamperes flows through a conductor whose resistance is 1000 ohms?
[*Ans.* 2.5 watts.]

13. Explain what is meant by electric energy. What is its unit of measurement?

14. The resistance of the heating element of an electric stove operating on a 120-volt line is 12 ohms. How much would it cost to operate this stove for 5 hours if the power company charged $.04 per kilowatthour? [*Ans.* $.24 .

HOW CURRENT FLOWS
THROUGH A CIRCUIT

5

A. THE ELECTRIC CURRENT IN SOLIDS

If electrons are added to one end of a solid conductor, such as a piece of copper wire, and some are taken away from the other end, a difference of potential is set up between the ends of the wire. This potential difference (voltage) tends to cause free electrons in the wire to move from the negative end to the positive one. As previously stated, this movement of electrons is an electric current.

The free electron moves comparatively slowly through the wire and travels but a short distance before it collides with an atom. This collision generally knocks an electron free from the atom, and this new free electron travels a short distance toward the positive end of the wire before it collides with another atom. Thus, there is a slow drift of electrons from the negative to the positive end of the wire.

But although the drift of electrons is comparatively slow, the disturbance that causes this drift travels through the wire at a very high speed (thousands of miles per second). This action may be understood

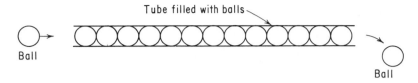

Tube filled with balls

Ball

Ball

Fig. 5-1. Diagram illustrating the motion of electrons through a solid conductor.

by visualizing a long, hollow tube completely filled with balls. (See Figure 5-1.) If a ball is added to one end of the tube, a ball at the other end is forced out immediately. Thus, although each ball moves slowly and for only a short distance, the disturbance is transmitted almost instantaneously through the entire tube.

Note that the electric current consists, essentially, of the movement of the electrons of the conductor itself. The generator that furnishes the electromotive force may be considered as a sort of "pump" that removes electrons from one end of the conductor (thus creating a deficiency or positive charge at that end) and piles them up at the other end (creating an excess or negative charge).

B. THE ELECTRIC CURRENT IN LIQUIDS

A molecule of ordinary table salt is made up of an atom of sodium and an atom of chlorine. When these two atoms are combined to form a molecule of salt, the chlorine atom seizes a valence electron from the sodium atom and thus becomes a negative ion. The sodium atom, having lost an electron, becomes a positive ion. The attraction between oppositely charged ions forms the ionic bond which holds the salt molecule together. When the molecule of salt is dissolved in water it **disassociates,** or separates into its components, the negative chlorine ion (Cl^-) and the positive sodium ion (Na^+). (See Figure 5-2.)

If two metal plates (called **electrodes**) are set at opposite ends of the solution and a source of electromotive force is connected to these plates so that one becomes a positive (electron-deficient) electrode and the other a negative (electron-excess) electrode, an electric field is created between these two electrodes. Since opposite charges attract, the negative chlorine ion is attracted to the positive electrode and the positive sodium ion is attracted to the negative electrode. Upon reaching the positive electrode, the chlorine ion surrenders its extra electron to the electrode and becomes a neutral chlorine atom. As the sodium ion reaches the

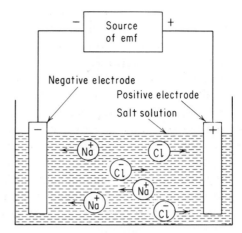

Fig. 5-2. Movement of ions through a liquid. The symbol Na⁺ represents the positive sodium ion, Cl⁻ the negative chlorine ion.

negative electrode, it obtains an electron from the electrode and becomes a neutral sodium atom.

The effect of the electromotive force, then, is to cause a movement of negative ions through the solution toward the positive electrode and of an equal number of positive ions toward the negative electrode. This movement of charged particles constitutes an electric current, and in this way the electric current flows through a liquid.

An **electrolyte** is a liquid containing substances that are able to disassociate into free ions. The resistance of an electrolyte depends upon the degree to which these ions are released **(ionization)**. The greater the amount of ionization, the less is the resistance of the liquid. Thus, for a given electromotive force, the greater the ionization the greater is the current transmitted by ions through the liquid.

C. THE ELECTRIC CURRENT IN GASES

The molecules of a gas are in a perpetual state of motion, constantly colliding with one another. These collisions knock off electrons, producing free electrons and converting the atoms that have lost electrons into positive ions. Since, by definition, a charged particle is called an ion, we may consider the free electrons as negative ions. Thus the gas contains positive and negative ions, just as an electrolyte does. If positive and negative electrodes are placed in the gas, the free electrons tend to travel to the positive electrode, and the positive ions to the negative electrode, thereby producing an electric current.

Normally, a positive or negative ion cannot travel very far in a gas before meeting an ion of opposite charge. This meeting would tend to produce neutralization and would result in neutral molecules. Since neutral molecules are not affected by the electric field between the two electrodes, the current would tend to cease flowing.

But if the gas is placed in a sealed container (such as a glass tube or bulb, with the two electrodes sealed in), and if most of the gas is pumped out, then the ions can travel considerable distances without being obstructed. The effect of the electric field is to accelerate, or speed up, the motion of the ions, so the farther they travel, the more velocity they attain. If a fast-moving ion collides with a neutral molecule, the ion tends to knock electrons off the neutral molecule, thereby creating more ions. This process is cumulative and tends to keep a constant stream of ions moving toward the electrodes. In this manner an electric current flows through a gas.

When the positive ions reach the negative electrode, they acquire electrons to become neutral molecules once more. The negative ions (electrons) are attracted to the positive electrode. Then the entire process is repeated.

D. THE ELECTRIC CURRENT IN A VACUUM

If a free electron were in a vacuum within the electric field set up between positive and negative electrodes, the negatively charged electron would be attracted to the positive electrode. The movement of the electron would constitute a flow of electric current. It is on this principle that the electron tubes used in our radio and television receivers operate.

We can construct an electron tube by sealing a pair of metal electrodes into opposite ends of a glass bulb and evacuating the air from within the bulb, leaving a vacuum. Connecting the electrodes to a source of electromotive force makes them positive and negative, respectively. (See Figure 5-3.) A question now arises. How can we get the free electron into the tube?

As previously described, there always is a disorderly movement of free electrons within all substances, especially metals. If the difference of potential between the two sealed-in electrodes be made great enough, some of the free electrons of the negative electrode will be attracted so strongly to the positive electrode that they will leave the former and fly through the vacuum to the latter.

Or else, if a substance is heated, the movement of free electrons

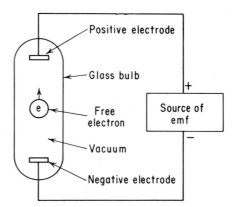

Fig. 5-3. Diagram illustrating the basic principle of the electron tube. The symbol ⓔ represents a free electron.

Positive electrode

Glass bulb

Free electron

Vacuum

Negative electrode

Source of emf

within that substance is increased. If the temperature is raised high enough, the movement of free electrons is increased to the point where some of the electrons actually fly off from the substance. We call this process **thermionic electron emission.**

In most electron tubes the negative electrode is heated to the point where it emits electrons. These electrons are attracted to the positive electrode and constitute a one-way flow of electric current through a vacuum from the negative to the positive electrode.

Further, certain substances, such as sodium, potassium, and cesium, will emit electrons if they are exposed to light. This phenomenon is known as **photoelectric emission.** If the negative electrode of the tube is made of such a material and light is permitted to fall on it, it will emit electrons that will be attracted to the positive electrode. Such a tube is called a **photoelectric cell.**

E. TYPES OF CIRCUITS

Just as water flows downhill from a point of high potential energy to a point of low potential energy, so the electric current flows from a high-potential point (excess of electrons) to a low-potential point (deficiency of electrons). The path or paths followed by the current flow is called the **electric circuit.***

*It will be noted that the movement of electrons (that is, the flow of current) is from a point of excess (negative) to a point of deficiency (positive). Before the acceptance of the electron theory it was believed that electric current flowed from positive to negative. Even after the old concept was found to be untrue, a number of authors still stated that current flowed from positive to negative, masking the fallacy by calling such a flow the **conventional current flow.** There is, however, an increasing tendency in modern books to correct this misconception. **In this book we will always consider current as flowing from negative to positive.**

All circuits must contain a source of electromotive force to establish the difference of potential that makes possible the current flow. This source may be a dry cell, a mechanical generator, or any of the other devices that will be discussed later in the book. All paths of the circuit lead in closed loops from the high-potential (negative) end of the electromotive-force source to its low-potential (positive) end.

The current in a circuit may flow through solid conductors, liquids, gases, vacuums, or any combinations of these. Its path may include lamps, toasters, motors, or any of the thousand-and-one electrical devices available in this electrical age. But, regardless of the type of circuit and the devices through which the current flows, all circuits offer some resistance to the current.

This resistance may be high or low, depending upon the type of circuit and the devices employed. Sometimes this resistance is undesirable, as, for example, the resistance of the wires connecting the various devices in a circuit. Accordingly, we keep this resistance at a low level by using wires made of copper having a large cross-sectional area, and by keeping their lengths as short as possible. Sometimes, however, it is desirable to introduce concentrated, or lumped, resistances into the circuit. Such a lumped resistance is called a **resistor.** In electrical diagrams the symbol for a **fixed** resistor — that is, one whose resistance is constant — is -⋀⋀⋀-. The symbol for a **variable** resistor is ⋀⋀⋀ or ⋀⋀⋀ .

In this chapter we will consider three general types of circuits. One is the **series** circuit, which offers a single, continuous, external path for current flow from the negative side of the electromotive-force source to the positive side. Another is the **parallel** circuit, which offers two or more parallel paths for current flow from negative to positive. The third type is the **series-parallel** circuit, a combination of the other two.

1. The Series Circuit

The series circuit, we have stated, offers a single, continuous, external path for current flow from the negative side of the electromotive-force source to the positive side. Such a circuit is illustrated in Figure 5-4. Electrons flow (as indicated by the symbol e →) from the negative side of the source, through resistor R, and back to the positive side of the source.

As these electrons encounter the top of the resistor they are slowed down by its resistance and piled up, thus making the top more negative (as indicated by −) than the bottom. Hence a difference of potential

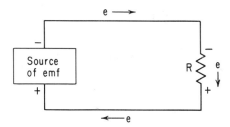

Fig. 5-4. Series circuit. The symbol e→ indicates current flow.

exists between the ends of the resistor. This difference of potential is called the **voltage drop** across the resistor. You will note that the source of electromotive force is connected across the resistor. Hence the voltage of the source is equal to the voltage drop.

Since there is only one path over which the current may flow, the current is the same in all parts of the series circuit. This can be proven by inserting a current-measuring device, called an **ammeter**, in different portions of the circuit and noting that the reading is always the same. We may determine what this current will be if we know the electromotive force of the source (in volts) and the resistance of the resistor (in ohms). Then by applying Ohm's law ($I = E/R$) we can find the current (in amperes).

EXAMPLE. In Figure 5-4, if the electromotive force is 100 volts and the resistance of R is 20 ohms, what current will flow through the circuit (neglecting, for the sake of simplicity, the resistance of the source and the connecting wires)?

ANSWER.
$$I = \frac{E}{R} = \frac{100 \text{ volts}}{20 \text{ ohms}} = 5 \text{ amperes.}$$

The series circuit, though offering a single, continuous path for current flow, may contain more than one resistor. For example, examine the circuit illustrated in Figure 5-5. As before, the current flow is the same in all parts of the circuit.

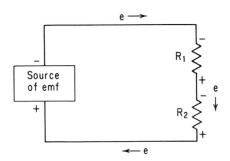

Fig. 5-5. Series circuit containing two resistors.

71

The total resistance of the series circuit is equal to the sum of all the individual resistances. Thus the total resistance of the series circuit shown in Figure 5-5 is equal to the sum of the resistance of R_1, the resistance of R_2, the resistance of the connecting wires, and the resistance of the electromotive-force source (all expressed in ohms). The total resistance of a series circuit may be determined by the following formula:

$$R_{total} = R_1 + R_2 + R_3 + R_4, \text{ and so forth.}$$

EXAMPLE. In Figure 5-5, if the resistance of R_1 is 30 ohms and R_2 is 20 ohms, what is the total resistance of the circuit (neglecting the resistance of the source and the connecting wires)?

ANSWER. $R_{total} = R_1 + R_2 = 30$ ohms $+ 20$ ohms $= 50$ ohms.

If the voltage of the source is 100 volts, what current will flow through the circuit?

By Ohm's law,

$$I = \frac{E}{R} = \frac{100 \text{ volts}}{50 \text{ ohms}} = 2 \text{ amperes.}$$

Although the current is the same in all portions of the series circuit, the electromotive force, or electrical pressure, is not. Starting with full pressure at the negative terminal of the source, the electrical pressure gradually diminishes as it is expended driving the electrons against the resistances encountered in the circuit until it reaches zero at the positive terminal of the source.

Let us see how this applies to the circuit illustrated in Figure 5-5. We again assume that R_1 is 30 ohms, R_2 is 20 ohms, and the voltage of the source is 100 volts. We also know that the current is equal to 2 amperes. For the sake of simplicity we again neglect the resistances of the connecting wires and of the source.

We start at the negative terminal of the source with the full voltage of 100 volts. A certain amount of this voltage is required to force 2 amperes of current to flow through R_1 (30 ohms). This can be determined by Ohm's law ($E = I \times R$).

$$E = I \times R = 2 \text{ amperes} \times 30 \text{ ohms} = 60 \text{ volts.}$$

Since 60 volts are required to force the current through R_1, only 40 volts are left to drive the electrons around the rest of the circuit.

At R_2 more of the voltage is lost forcing the current through that resistor. Again, by Ohm's law,

$$E = I \times R = 2 \text{ amperes} \times 20 \text{ ohms} = 40 \text{ volts}.$$

Thus the remaining 40 volts have been used up and the voltage has dropped to zero.

Note that the current flowing through it causes a difference of potential to appear between the ends of a resistor. This potential difference is equal to the voltage drop which, in turn, is equal to the product of the resistance (ohms) and the current (amperes). Hence the voltage drop across a resistor is called its *IR* **drop**. In a series circuit the sum of the *IR* drops equals the voltage of the source.

As current flows through R_1 (Figure 5-5), the top of the resistor becomes negative with respect to its bottom. Similarly, the top of R_2 becomes negative with respect to its bottom. This is indicated by the $+$ and $-$ signs at each resistor and shows the polarity of the voltage drop across each.

Note that the voltage drop across R_1 is 60 volts and the current flowing through it is 2 amperes. From the formula P (watts) $= E$ (volts) $\times I$ (amperes) we may determine the electric power dissipated by that resistor. Thus:

$$P = E \times I = 60 \text{ volts} \times 2 \text{ amperes} = 120 \text{ watts}.$$

This means that R_1 must be able to safely dissipate at least 120 watts. The voltage drop across R_2 is 40 volts and the current flowing through it is 2 amperes. Thus:

$$P = E \times I = 40 \text{ volts} \times 2 \text{ amperes} = 80 \text{ watts}$$

and R_2 need dissipate only 80 watts.

The total power consumed by the entire circuit may be found by multiplying the voltage of the source by the total current. Thus:

$$P = E \times I = 100 \text{ volts} \times 2 \text{ amperes} = 200 \text{ watts}.$$

Note that this is equal to the sum of the power dissipated by R_1 and R_2.

(There is an interesting sidelight we may consider. Suppose that R_2 becomes defective, opening the circuit. Since the circuit is broken, no current flows. Hence there are no *IR* drops. However, since the ends of R_2 are still connected to the source, the full 100 volts appears across those ends.)

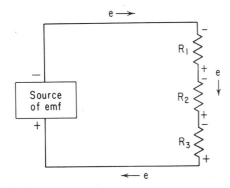

Fig. 5-6. Voltage divider.

Assume we have three resistors R_1, R_2, and R_3 connected in series across a voltage source, as illustrated in Figure 5-6. Further assume that the voltage of the source is 100 volts and that the resistances of R_1, R_2, and R_3 are 20 ohms, 30 ohms, and 50 ohms, respectively. The total resistance of the circuit (neglecting the resistances of the source and the connecting wires) is $20 + 30 + 50 = 100$ ohms. Accordingly, since by Ohm's law $I = E/R$, the current flowing through the circuit is 100 volts/100 ohms = 1 ampere.

The *IR* drop across R_1 is equal to 1 ampere $\times$ 20 ohms = 20 volts. Similarly, the *IR* drops across R_2 and R_3 are 30 volts and 50 volts, respectively. Thus the voltage of the source is divided across the network of R_1, R_2, and R_3 according to their resistances — that is, in a 2, 3, 5 ratio. Note the polarities of the voltage drops across these resistors.

Such a network is called a **voltage divider** and is used where we wish to obtain a portion of the source voltage. If, for example, we wish 20 volts, we may obtain it from across R_1. If we wish 30 volts, we may obtain it from across R_2, and we may obtain 50 volts from across R_3. If we desire 80 volts, we may obtain it from across R_2 and R_3 in series.

Of course, any combination of resistors may be used to obtain other voltage drops. In practice, the voltage divider may be a single resistor whose resistance is equal to the sum of all, with taps at suitable points.

You may have noticed we have been saying "neglecting the resistance of the source." Actually, the resistance of the source is quite important. Here is why.

In Figure 5-7A we see the circuit of a resistor (R) in series with a voltage source. Its equivalent circuit appears in Figure 5-7B, where the voltage source is divided into two parts — the voltage and the internal

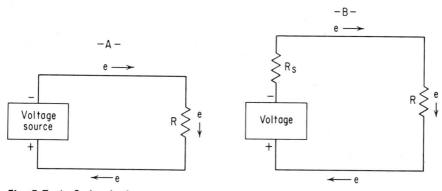

Fig. 5-7. A. Series circuit.
B. Its equivalent circuit.

resistance of the source as represented by R_s. Note that R_s is in series with the rest of the circuit.

As current flows through the circuit a voltage drop occurs across R_s which reduces the voltage left for resistor R. If the current drawn by R (which is called the **load**) is low, the voltage drop across R_s, too, is low. If, on the other hand, the current drawn by the load is high, the voltage drop across R_s is high. Thus it can be seen that the resistance of R_s (the internal resistance of the source) must be very low if extreme voltage fluctuations with variations in the load are to be avoided.

2. The Parallel Circuit

The parallel circuit is illustrated in Figure 5-8. Note that the electro-motive force across all components of such a circuit is the same. Current flows (as indicated by the symbol e→) from the negative side of the source to the upper junction of resistor R_1 and R_2. Here it divides, part flowing through R_1 and part through R_2. At the lower junction of R_1 and R_2 both currents reunite and flow back to the positive side of the source.

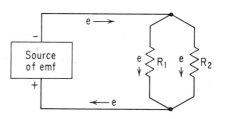

Fig. 5-8. Parallel circuit.

[At this point, it might be well to mention a few facts concerning electrical circuit diagrams. Often, we must show two wires crossing each other. There are several methods of designation, but in this book, if two wires cross and connect with each other, this connection will be indicated by a dot at the point of crossover (+). If there is no connection, a loop (‾⌒‾) will be used to indicate this fact.]

The total resistance of resistors connected in parallel may be expressed by the following formula:

$$\frac{1}{R_{total}} = \frac{1}{R_1} + \frac{1}{R_2} + \frac{1}{R_3} + \frac{1}{R_4} +, \text{ and so forth.}$$

EXAMPLE. Assume, as in Figure 5-8, that R_1 is 90 ohms, and R_2 is 45 ohms. What is the total resistance of R_1 and R_2 in parallel?

ANSWER. $\frac{1}{R_{total}} = \frac{1}{R_1} + \frac{1}{R_2} = \frac{1}{90 \text{ ohms}} + \frac{1}{45 \text{ ohms}} = \frac{3}{90},$

$$R_{total} = \frac{90}{3} = 30 \text{ ohms.}$$

Now let us carry our problem a step further. Assuming a voltage source of 90 volts, what will be the total current flowing in the circuit (neglecting the resistance of the source and connecting wires)?

Since the total resistance is 30 ohms and the voltage is 90 volts, by Ohm's law ($I = E/R$) we get

ANSWER. $I = \frac{E}{R} = \frac{90 \text{ volts}}{30 \text{ ohms}} = 3 \text{ amperes.}$

If a total current of 3 amperes is flowing through the circuit, how much current flows through R_1 and through R_2? Note that the same voltage is applied across R_1 as R_2 — that is, the *IR* drop across R_1 and R_2 is 90 volts. Again, by Ohm's law, we can determine the current flowing through each resistor. Thus, for R_1:

ANSWER. $I = \frac{E}{R} = \frac{90 \text{ volts}}{90 \text{ ohms}} = 1 \text{ ampere.}$

And for R_2:

ANSWER. $I = \frac{E}{R} = \frac{90 \text{ volts}}{45 \text{ ohms}} = 2 \text{ amperes.}$

Note that the greater current flows through the resistor with the lower resistance. And because R_2 has half the resistance of R_1, twice as much current flows through R_2 as through R_1.

Power dissipated by each resistor may be determined from the formula $P = E \times I$. For resistor R_1, E equals 90 volts and I equals 1 ampere. Hence the power is 90 watts. For resistor R_2, E equals 90 volts and I equals 2 amperes. Hence the power is 180 watts. The power consumed by the entire circuit may be found by multiplying the voltage of the source by the total current. Thus:

$$P = E \times I = 90 \text{ volts} \times 3 \text{ amperes} = 270 \text{ watts.}$$

Note that this is equal to the sum of the power dissipated by R_1 and R_2. Note, too, that in the series circuit the total power dissipated is also equal to the sum of the power dissipated by each of the individual resistors.

3.   The Series-Parallel Circuit

As its name implies, this circuit is a combination of the other two types. A simple example of such a circuit is illustrated in Figure 5-9. Note that R_2 and R_3 are connected in parallel with each other and that both together are connected in series with R_1. To find the total resistance of this circuit, we first find the joint resistance of R_2 and R_3 in parallel and then add this joint resistance to that of R_1, as in any other series circuit.

Let us assume that the voltage of the source is 150 volts. Further assume R_1 is 20 ohms, R_2 is 90 ohms, and R_3 is 45 ohms. What is the total resistance of the circuit (neglecting the resistance of the connecting wires and of the source)?

First we find the joint resistance of R_2 and R_3 in parallel.

$$\frac{1}{R_{\text{joint}}} = \frac{1}{R_2} + \frac{1}{R_3} = \frac{1}{90 \text{ ohms}} + \frac{1}{45 \text{ ohms}} = \frac{3}{90},$$

$$R_{\text{joint}} = \frac{90}{3} = 30 \text{ ohms.}$$

Next, we find the total resistance of R_{joint} and R_1 in series.

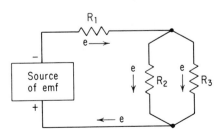

Fig. 5-9. Series-parallel circuit.

ANSWER. $R_{total} = R_{joint} + R_1 = 30$ ohms $+ 20$ ohms $= 50$ ohms.

To find the current flowing through R_1, R_2, and R_3, first find the total current of the circuit. Knowing the voltage of the source and the total resistance, we may find the total current by Ohm's law. Thus:

$$I = \frac{E}{R} = \frac{150 \text{ volts}}{50 \text{ ohms}} = 3 \text{ amperes.}$$

Since R_1 is in series with the rest of the circuit, the total current flows through it. Thus the current flowing through R_1 is 3 amperes.

We now can calculate the voltage drop across R_1.

$$E = I \times R = 3 \text{ amperes} \times 20 \text{ ohms} = 60 \text{ volts.}$$

Since 60 volts are expended forcing the current through R_1, 90 volts are left for the rest of the circuit. Because R_2 and R_3 are connected in parallel, the same voltage is applied to each. Hence the voltage drops across R_2 and R_3 are 90 volts each.

Since the resistance of R_2 is 90 ohms and its voltage drop is 90 volts, then:

$$I = \frac{E}{R} = \frac{90 \text{ volts}}{90 \text{ ohms}} = 1 \text{ ampere flowing through } R_2.$$

Similarly, since R_3 is 45 ohms and its voltage drop, too, is 90 volts, then:

$$I = \frac{E}{R} = \frac{90 \text{ volts}}{45 \text{ ohms}} = 2 \text{ amperes flowing through } R_3.$$

To find the total power consumed by the circuit, multiply the total voltage (150 volts) by the total current (3 amperes). This comes to 450 watts. The power dissipated by R_1 may be found by multiplying its voltage drop (60 volts) by its current (3 amperes). This comes to 180 watts. To find the power dissipated by R_2, multiply its voltage drop (90 volts) by its current (1 ampere). This comes to 90 watts. To find the power dissipated by R_3, multiply its voltage drop (90 volts) by its current (2 amperes). This comes to 180 watts. If we add up the power dissipated by R_1, R_2, and R_3, it equals the total of 450 watts consumed by the entire circuit. Thus, regardless of the method of connection, the total power consumed by the circuit is equal to the sum of the power consumed by all of its parts.

Gustav Kirchhoff, a German physicist, noted several interesting facts about circuits. At any point in the circuit, he found, the current flowing towards the point is equal to the current flowing away from that point. Let us see how this applies to the circuit we have just discussed.

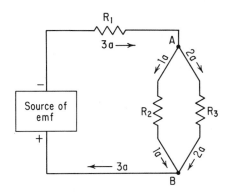

Fig. 5-10. Circuit diagram illustrating
Kirchhoff's First Law.

In Figure 5-10 you see this circuit upon which we have indicated the amount of current flowing through its various portions and the direction of current flow. Consider point A. Three amperes of current are flowing towards it. On the other hand, three amperes of current are flowing away from it — one ampere through R_2 and two amperes through R_3. Similarly, at point B, three amperes of current are flowing towards it, and three amperes are flowing away from it.

Kirchhoff set his findings in a set of laws that bear his name. **If current flowing towards a point is designated as plus (+) and current flowing away from the point is designated as minus (−), then the algebraic sum of all currents flowing to and away from a point in any type of circuit is equal to zero.**

Kirchhoff formulated a second law, which deals with electromotive forces and voltage drops (*IR* drops) in a circuit. He found that **the algebraic sum of all the voltage drops and electromotive forces in any closed circuit, when taken with their proper signs, is equal to zero.**

For example, look at Figure 5-11. Here are two sources of electromotive force (batteries B_1 and B_2) connected in a series circuit with

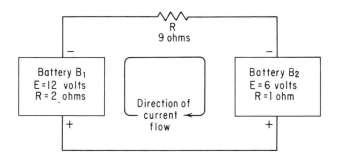

Fig. 5-11. Circuit diagram illustrating Kirchhoff's Second Law.

a resistor (R). Battery B_1 has an electromotive force of 12 volts and an internal resistance of 2 ohms. Battery B_2 has an electromotive force of 6 volts and an internal resistance of 1 ohm. Note that the batteries are connected so that their electromotive forces oppose each other. The resistance of R is 9 ohms.

To add the electromotive forces of the batteries algebraically, we may call the direction of the electromotive force of one of the batteries (in this case, the direction of the one with the larger voltage, B_1) plus (+). Since the direction of the electromotive force of B_2 is in the opposite direction, we call this direction minus (−). To add the two voltages algebraically, we add +12 volts and −6 volts, which gives a resultant of +6 volts. The total resistance of this series circuit (2 ohms + 9 ohms + 1 ohm) is 12 ohms. We may find the current flowing in this circuit by Ohm's law:

$$I = \frac{E}{R} = \frac{6 \text{ volts}}{12 \text{ ohms}} = 0.5 \text{ ampere.}$$

Now let us see how Kirchhoff's law applies. Start with the positive (+) terminal of battery B_1. Going from the positive to the negative pole of the battery, we encounter a voltage increase of 12 volts. At the same time, owing to the internal resistance of the battery, we suffer a voltage drop of 1 volt ($E = IR = 0.5$ amp $\times$ 2 ohms = 1 volt). Thus the electromotive force is +11 volts. Passing to resistor R we suffer another voltage drop of 4.5 volts (0.5 amp $\times$ 9 ohms = 4.5 volts), thus reducing the electromotive force to 6.5 volts. At battery B_2, because it is connected in opposition to battery B_1, we suffer a voltage decrease of 6 volts. At the same time, we have another voltage drop of 0.5 volt owing to the internal resistance of that battery (0.5 amp $\times$ 1 ohm = 0.5 volt). Thus the algebraic sum of the applied electromotive forces drops to zero (6.5 volts − 6.5 volts).

QUESTIONS

1. Explain how electric current flows through a solid conductor.
2. What is meant by an **ion**; an **electrolyte?**
3. Explain how electric current flows through a liquid.
4. Explain how electric current flows through a gas.
5. Explain how electric current flows through a vacuum.
6. What is meant by an **electric circuit**; a **series circuit**; a **parallel circuit?**
7. Two resistors, one 40 ohms and the other 20 ohms, are connected in series across a 120-volt line.

(*a*) What is their total resistance?   [*Ans.* 60 ohms.]

(*b*) What is the total current flowing through the entire circuit?

[*Ans.* 2 amperes.]

(*c*) How much current will flow through each resistor?

[*Ans.* 2 amperes.]

(*d*) What will be the voltage drop across each resistor?
[*Ans.* 80 volts across the 40-ohm resistor, 40 volts across the 20-ohm resistor.]

(*e*) What is the total power dissipated by the circuit?   [*Ans.* 240 watts.]

(*f*) How much power is dissipated by each resistor? Check this answer against (*e*) above.

[*Ans.* 160 watts by the 40-ohm resistor, 80 watts by the 20-ohm resistor.]

8. Two resistors, one 60 ohms and the other 30 ohms, are connected in parallel across a 120-volt line.

(*a*) What is their total resistance?   [*Ans.* 20 ohms.]

(*b*) What is the total current flowing through the entire circuit?

[*Ans.* 6 amperes.]

(*c*) How much current will flow through each resistor?
[*Ans.* 2 amperes through the 60-ohm resistor, 4 amperes through the 30-ohm resistor.]

(*d*) What will be the voltage drop across each resistor?

[*Ans.* 120 volts.]

(*e*) What is the total power dissipated by the circuit? [*Ans.* 720 watts.]

(*f*) How much power is dissipated by each resistor? Check this answer against (*e*) above.

[*Ans.* 240 watts by the 60-ohm resistor, 480 watts by the 30-ohm resistor.]

9. Two resistors, one 30 ohms and the other 60 ohms, are connected in parallel. They then are connected in series with a 40-ohm resistor across a 120-volt line.

(*a*) What is their total resistance?   [*Ans.* 60 ohms.]

(*b*) What is the total current flowing through the circuit?

[*Ans.* 2 amperes.]

(*c*) How much current will flow through each resistor?
[*Ans.* 2 amperes through the 40-ohm resistor, $1\frac{1}{3}$ amperes through the 30-ohm resistor, $\frac{2}{3}$ ampere through the 60-ohm resistor.]

(*d*) What will be the voltage drop across each resistor?
[*Ans.* 80 volts across the 40-ohm resistor, 40 volts across the 30-ohm resistor, 40 volts across the 60-ohm resistor.]

(*e*) What is the total power dissipated by the circuit?   [*Ans.* 240 watts.]

(*f*) How much power is dissipated by each resistor? Check this answer against (*e*) above.

[*Ans.* 160 watts by the 40-ohm resistor, $26\frac{2}{3}$ watts by the 60-ohm resistor, $53\frac{1}{3}$ watts by the 30-ohm resistor.]

10. State Kirchhoff's laws for current and voltage.

11. A battery whose electromotive force is 10 volts and whose internal resistance is 1 ohm, a battery with an electromotive force of 20 volts and an internal resistance of 2 ohms, and a resistor are connected in series. The batteries are connected so that their electromotive forces oppose each other. It is found that the current flowing in this circuit is 2 amperes. What is the resistance of the resistor? [*Ans.* 2 ohms.]

EFFECTS OF
ELECTRIC CURRENT

6

A. THERMAL EFFECT

When mechanical energy is applied to a machine, it meets a kind of resistance, called **friction.** Mechanical power is "lost" overcoming this friction. However, it is not really lost since it shows up as heat at the point or points of friction. It has merely been changed from mechanical energy to heat energy.

Similarly, when electrical energy is applied to a conductor, the resulting current flow must overcome the resistance of the conductor. Electrical power is "lost" overcoming this resistance. As is true of the mechanical power, the electrical power is not really lost, but is converted to heat in the conductor. That is, the electrical energy has been changed to heat energy.

You will recall that the electrical power consumed by a circuit is equal to the product of the current and the electromotive force. Thus:

P (in watts) $= E$ (in volts) $\times I$ (in amperes).

Since, by Ohm's law, $E = I \times R$, by substituting for E in the first equation its equivalent $(I \times R)$, we may indicate the power equation in terms of current (I) and resistance (R). Thus:

$$P = E \times I = (I \times R) \times I = I^2 \times R.$$

Similarly, since $I = E/R$, we may indicate the power equation in terms of electromotive force (E) and resistance (R). Thus:

$$P = E \times I = E \times \frac{E}{R} = \frac{E^2}{R}.$$

Thus, if we wish to determine the power consumed or "lost" as current flows through a resistor, we may multiply the square of that current by the resistance of the resistor. Still another method is to divide the square of the voltage drop by the resistance. Both these methods produce the same result.

Since the voltage of a circuit usually is kept at a constant value, the two variables generally are the resistance of the resistor and the current flowing through it. Hence the power loss most frequently is expressed in terms of $I^2 \times R$. Since this power produces heat, the heating effect of an electric current often is called the I^2R **loss.**

In many instances the heat so produced is undesirable, and steps are taken to keep it at a minimum. For example, where large currents are to flow, conductors may be made of heavy copper bars to keep the resistance low. Where these steps are not sufficient to keep the heat at safe levels, the heat itself may be conducted away. Thus many motors are constructed with a built-in fan to blow cool air over the wires heated by the current flowing through them.

However, there are instances where the heat is desirable. Certain devices, such as toasters, heaters, electron-tube filaments, and such, are constructed with special conductors made of alloys that offer a fairly high resistance to current flow. This resistance, in turn, produces heat. Some of these devices will be discussed later in this book.

B. LUMINOUS EFFECT

As you know, the electrons of an atom move in distinct orbits around the nucleus. If heat energy is applied to the atom, some of its electrons may acquire sufficient energy to jump from their normal orbits to orbits farther removed from the nucleus. An atom in this state is said to be "excited." Since this is an unstable state, these electrons soon fall back to their normal orbits. As they do so, they release the excess energy they

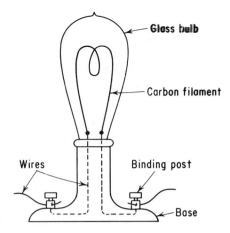

Fig. 6-1. Edison incandescent lamp.

had acquired in the form of light energy. Thus, if a substance such as a metal wire, for example, is heated sufficiently, some of its atoms may become excited and emit light as they return to their normal states.

You now know that as a current flows through a conductor, heat is produced as a result of the I^2R loss. If the current and resistance be large enough, the heat so produced may be great enough to make the conductor emit light. This is the principle of the **incandescent lamp,** invented by Thomas A. Edison in 1879.

To provide for a high enough resistance, Edison used a wire, or filament, made of carbon. However, if this filament be heated until it emits light — that is, to **incandescence** — it burns up in the air, which supports combustion. Accordingly, Edison sealed the carbon filament in a glass bulb from which he pumped out the air. (See Figure 6-1.)

The passage of an electric current may heat a gas, as well as a solid, to incandescence. This principle underlies the **carbon-arc light** which was used extensively for street lighting at the beginning of the twentieth century. In this type of street light, current is led to two carbon rods. (See Figure 6-2.) The tips of these rods are touched together and then

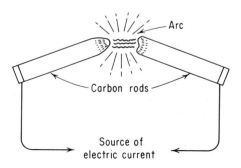

Fig. 6-2. Carbon-arc light.

slightly separated. As a result a hot electric spark, or **arc,** jumps from tip to tip. The heat of this arc vaporizes some of the carbon, and the passage of current through this carbon vapor heats it to burning and incandescence. (The tips of the rods, too, are heated to incandescence, thus furnishing an additional source of light.)

So far, we have discussed light produced when a solid or a gas is heated to incandescence. There are other electrical methods for producing light without the necessity for heating a substance. You will recall that electric current flows through a gas by the movement of ions to oppositely charged electrodes sealed into opposite ends of a tube from which most of the gas has been evacuated (Chapter 5, subdivision C). When the positively charged gas ion reaches the negative electrode, it receives an electron and becomes a neutral atom once more. As it does so, light is emitted (as a result of the rearrangement of the electrons of the gas atom).

It is not only at the negative electrode that the gas atom emits light. Anywhere in the tube, when a positive gas ion meets an electron, neutralization takes place and light is emitted. Hence the tube glows over its entire length. This process is continuous since the neutral atom soon is struck by a speeding ion and becomes an ion once more. Each type of gas produces light of characteristic color. Neon produces a reddish light. Helium produces a pinkish light; argon produces a bluish-white light; mercury vapor produces a greenish-blue light.

In addition to producing a visible greenish-blue light, mercury vapor produces an invisible ultraviolet light. When the ultraviolet light strikes the atoms of certain chemicals, called **phosphors,** some of these atoms absorb the light energy, become excited, and, as their electrons fall back to their normal states, emit the excess energy in the form of light energy. The interesting point here is that, although invisible light energy had been absorbed, the light energy emitted is of the visible type, its color depending upon the composition of the phosphor. This is the principle upon which the fluorescent lamps so widely used today operate.

A long glass tube is coated on its inner surface with one of these phosphors. Within the tube, mercury is heated until it vaporizes and forms a mercury vapor. Current is passed through this mercury vapor, producing ultraviolet light. As the ultraviolet light strikes the phosphor, visible light is produced by the phosphor glow. (A discussion of the practical applications of the various luminous effects of the electric current will be found later in this book.)

C. CHEMICAL EFFECT

We touched upon the transformation of electrical energy into chemical energy when we discussed how the electric current broke the salt molecule into its component sodium and chlorine atoms (Chapter 5, subdivision B). This type of energy transformation is used in a great number of industrial processes such as electroplating and the manufacture of aluminum. We shall consider this matter in greater detail later in this book.

D. MAGNETIC EFFECT

In 1819 Hans Christian Oersted, a Danish physicist, brought a small compass near a wire that was carrying an electric current. He noticed that the compass was deflected. When he turned the current off, the compass assumed its original position. This discovery started a chain of events that has helped shape our industrial civilization.

Let us examine the significance of Oersted's discovery. The deflection of the compass while current was flowing through the wire indicated that it was being acted upon by an external magnetic field. Where did this magnetic field come from?

Not from the copper wire, which we know is nonmagnetic. Obviously, it could come only from the electric current flowing through the wire. The compass was deflected only when the current flowed through the conductor and continued to be deflected only so long as the current continued to flow. When the flow of current ceased, so did the deflection.

We believe that a magnetic field always accompanies the motion of a charged particle. It is in this way that we explain the magnetic fields around the electrons revolving about the nucleus of the atom. There is no reason why a magnetic field should not surround an electron moving in a conductor. Since the flow of current consists of the movement of electrons, we should expect a magnetic field around a conductor through which a current is flowing. This Oersted found to be true.

As might be expected, the greater the current flow — that is, the greater the number of electrons flowing per second — the greater is the magnetic field. Experimentation has shown that the pattern of the mag-

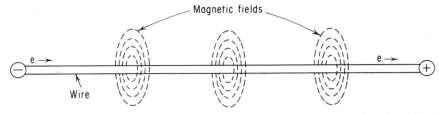

Fig. 6-3. Magnetic field formed around a conductor carrying a current. (For the sake of simplicity, the field is shown as a series of concentric circles instead of cylinders.)

netic field surrounding the conductor is in the shape of a series of concentric cylinders with the conductor at the center. (See Figure 6-3.)

Further experimentation produced a simple method for determining the direction of this magnetic field. If the conductor is grasped in the left hand with the extended thumb pointing in the direction of the current flow, the fingers then circle the conductor in the direction of the magnetic lines of force. This is illustrated in Figure 6-4.

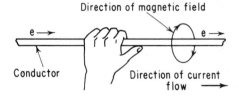

Fig. 6-4. Left-hand rule for finding the direction of the magnetic field around a conductor carrying a current.

Suppose we bend the current-carrying conductor into a loop. The magnetic field then would appear as illustrated in Figure 6-5. Note that the lines of force all pass through the enclosed portion of the loop. If we add more loops, each loop adds its magnetic field, thus producing a greater overall magnetic effect. The resulting magnetic field would appear as shown in Figure 6-6.

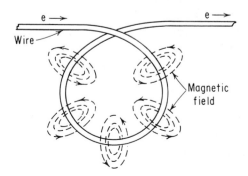

Fig. 6-5. Magnetic field around a loop of wire carrying a current.

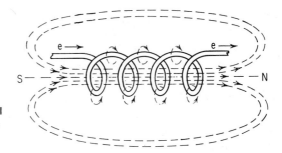

Fig. 6-6. Magnetic field around a coil of wire through which a current is flowing.

Note that the coil becomes a temporary magnet (that is, it is a magnet only while current flows through it) with a set of north and south poles. The greater the number of turns, the stronger the magnetic field will be. Also, the greater the current flowing through it, the stronger the field will be. Accordingly, we say that the strength of the magnetic field depends upon the **ampere-turns** of the coil. (The ampere-turns are designated by the symbol *NI*, where *N* stands for the number of turns and *I* for the current flowing through these turns.)

The polarity of the magnet formed by the coil may be determined by grasping it in the left hand so that the fingers follow around the coil in the direction in which the electrons are flowing. The extended thumb then will point toward the north pole. (See Figure 6-7.)

In Chapter 3, subdivision C, you will recall, we stated that the magnetomotive force (℥) generally is expressed in electrical terms. Accordingly, we may obtain the value of this force from the following formula:

$$\mathfrak{F} = K \times NI,$$

where ℥ is the magnetomotive force (in gilberts), *N* is the number of turns of the coil, and *I* is the current (in amperes) flowing through it. The letter K stands for a constant which is equal to 0.4π. (The Greek letter π, pronounced **pi,** stands for 3.14.)

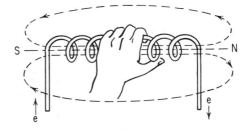

Fig. 6-7. Left-hand rule for finding the polarity of a coil through which a current is flowing.

We may increase the strength of the magnetic field by winding the coil on a core of magnetic material. As current flows through the coil, the magnetic field it produces magnetizes the core material. Thus the magnetism of the core is added to that of the coil, increasing the overall magnetic field. Since we generally desire the coil and its core to act as a temporary magnet, this core usually is made of some "soft" magnetic material, such as iron, which becomes a magnet only when under the influence of the magnetic field of the coil. The coil, of course, produces a magnetic field only as long as the current flows through it.

The coil with its core is called an **electromagnet.** Without its core the coil usually is called a **solenoid,** or **helix.** In electrical diagrams the symbol for a coil of wire is ⌇⌇⌇. If the coil has a magnetic core, the symbol becomes ⌇⌇⌇.

The main factors that determine the magnetic strength of an electromagnet are the number of turns of the coil, the current flowing through it, and the material and size of its core. Other determining factors are the shape and size of the windings and the mechanical arrangement of the core. There are a great many practical applications of the electromagnet. Some of these will be discussed later in this book.

There is another magnetic effect produced as current flows through a conductor. Since the conductor is surrounded by a magnetic field, if two current-carrying conductors are placed near each other the two magnetic fields will react with each other. Thus the conductors will be brought closer together or pushed apart, depending upon the directions of the two fields. This principle is employed in a number of applications, such as the electric motor, which, too, will be discussed later.

QUESTIONS

1. By means of an illustration, explain the thermal effect of the electric current.

2. By means of an illustration, explain the luminous effect of the electric current.

3. By means of an illustration, explain the chemical effect of the electric current.

4. State the left-hand rule for finding the direction of the magnetic field around a conductor carrying an electric current.

5. State the left-hand rule for finding the polarity of the magnetic field around a coil through which a current is flowing.

6. State three methods for increasing the strength of an electromagnet.

D-C
MEASURING
INSTRUMENTS

7

It is extremely important to have instruments by means of which we may measure directly the quantity of certain factors, such as current, voltage, and resistance, that may be present in various portions of the electrical circuit. These instruments are called **meters** and they operate by measuring the various effects produced by the electric current.

Generally, the meter consists of a **movement** which translates current flowing through it into a displacement of a **pointer**, which indicates this displacement on a **scale.** The whole is enclosed in a **case** for protection. Since the displacement is proportional to the current causing it, the excursion of the pointer over the scale indicates this amount of current. Depending upon the circuit in which the meter is connected, the scale may be calibrated in amperes, volts, watts, ohms, and so forth.

A. MEASUREMENT BASED UPON THE CHEMICAL EFFECT OF ELECTRIC CURRENT

We have seen that when a current is passed through a solution, such as of table salt, the sodium atoms will be deposited on the negative electrode and the chlorine atoms will accumulate at the positive electrode (Chapter 5, subdivision B). The stronger the current, the more sodium and chlorine will be accumulated at their respective electrodes. Here, then, is a method for determining the strength of the current by measuring the amount of sodium or chlorine deposited on one of the electrodes in a certain period of time.

In practice, it is difficult to weigh the amount of chlorine so deposited because chlorine is a gas and escapes readily. The sodium atoms, too, are difficult to weigh because they are very active chemically and react with the water of the solution the moment they are formed at the negative electrode. But if we use a silver salt, the silver atoms will be deposited on the negative electrode in the form of a coating or **plate.** Thus, we can weigh this negative electrode before the current flows, permit the current to flow, and then weigh the electrode with its coating of silver after the current has flowed for a definite length of time. The gain in weight represents the amount of silver that has been deposited out of the solution by the current in that length of time.

By careful control and measurement, and by using a silver solution of a certain definite composition, it has been found that one ampere of current will cause 0.001118 gram of silver to be deposited in one second. Of course, you can readily see that this method for measuring current is not very practical and is employed only as a laboratory experiment.

B. MEASUREMENT BASED UPON THE THERMAL EFFECT OF ELECTRIC CURRENT

When an electric current flows through a conductor, heat is created. Most metallic conductors expand upon heating. Thus, the more current that flows through such a conductor, the more it is heated, and the more it expands. We can, therefore, use the amount of expansion of such a conductor to indicate the amount of current flowing through it.

Figure 7-1 illustrates such a measuring instrument which, appropriately enough, is called a **hot-wire movement.** Current passes through a thin wire, usually composed of an alloy of platinum and silver. As the current flows through it, the wire is heated and expands. The heating effect and the resulting expansion are proportional to the current.

Attached to this wire is the **tension wire** and, in turn, a silk thread is attached to the tension wire. This thread passes over a small **roller,** and tension is maintained on the whole movement by means of the **spring.** As the hot wire expands, this spring pulls the thread to the left. The motion of the thread causes the roller to rotate. This, in turn, causes the

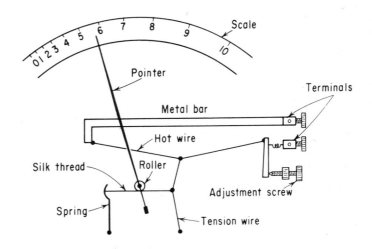

Fig. 7-1. Hot-wire movement.

pointer to move over the **scale.** Since the amount of expansion (and the consequent movement of the pointer) is proportional to the current flowing through the wire, the amount of current is thus indicated.

When the current ceases flowing, the wire cools and contracts, and the thread is pulled to the right, returning the pointer to zero on the scale. The **adjustment screw** is used to compensate for variations in the tension on the wire.

Note that the scale is not divided into uniform divisions. The heating effect of the current, you will recall (Chapter 6, subdivision A), is the result of the product of the square of the current and the resistance (I^2R). Thus, the heating effect is not **directly** proportional to the current but, rather, it is proportional to the **square** of the current. If the current is increased two times, the heating effect is increased four times (two squared). If the current is increased three times, the heating effect is increased nine times (three squared). And so on. Accordingly, since the scale indicates current (in amperes) and the pointer moves according to the heating effect, the divisions of the scale cannot be uniform, but resemble those illustrated in Figure 7-1. Such a scale is called a **square-law scale.**

C. MEASUREMENT BASED UPON THE MAGNETIC EFFECT OF ELECTRIC CURRENT

Most electrical measuring instruments make use of the magnetic effect of the electric current. As you recall, a conductor carrying an electric current is surrounded by a magnetic field. If this conductor is wound in the form of a coil, the magnetic field is concentrated. The strength of this field will depend upon the number of turns in the coil, and the amount of current flowing through it.

Armed with these facts, we now can construct a current-measuring instrument. Look at Figure 7-2. A soft-iron **vane** fastened to a delicately pivoted shaft is placed inside a coil of wire. (For the sake of simplicity, a cut-away view of the coil is shown.) As current flows through the coil, magnetic lines of force that are parallel to its axis are set up. The iron vane is magnetized by these lines of force and its lines of force tend to line up with those of the coil. The vane, therefore, tends to line up with the axis of the coil. In doing so, it rotates the shaft to which it is fastened.

The rotation of the shaft is opposed by the tension of the flat spiral spring that is attached to it. Hence there appear two opposing forces — (1) the rotation of the shaft due to the lining-up effect of the iron vane, and (2) the opposition of the spring. The shaft then will come to rest at a point where these two opposing forces are equal.

The stronger the current flowing through the coil, the greater will be the strength of the magnetic field and, hence, the greater will be the lining-up effect on the iron vane and the greater will be the tendency of the shaft to rotate. Thus the shaft will be able to rotate more before the increasing tension of the spring brings it to a halt. You can see, then, that we may measure the strength of the current flowing through the coil in terms of the amount of rotation of the shaft. This rotation is indicated by a pointer that is fastened to the shaft and that moves over a suitably calibrated scale.

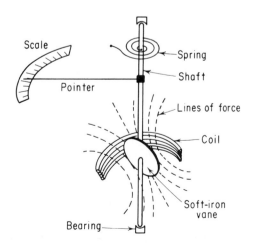

Fig. 7-2. Inclined-coil movement.

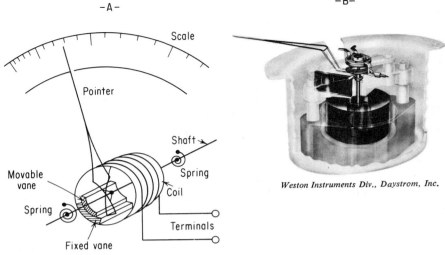

Fig. 7-3. A. Repulsion-vane movement.
　　　　　B. "Phantom" view of commercial repulsion-vane movement.

Note that the coil is mounted in an inclined position. Hence this type of instrument is known as an **inclined-coil movement.**

Another type of instrument depending upon the magnetic effect is the **repulsion-vane movement** illustrated in Figure 7-3. Two soft-iron pieces, or **vanes,** are placed inside a coil of wire. One of these vanes is fixed and the other is free to move. Attached to the movable vane is a shaft carrying a pointer. As current passes through the coil, the vanes become magnetized. Since they are magnetized in the same way (north pole to north pole and south pole to south pole), the two vanes repel each other. As a result, the movable vane is deflected around the center shaft, turning the shaft and carrying the pointer with it. The springs tend to restore the pointer to its original position. The greater the current flowing through the coil, the greater will be the deflection of the pointer.

Note that the scale is of the square-law type. Doubling the current will make the magnetism of each vane twice as great. Thus the repulsive force between them will be four times as large. Tripling the current will make the repulsive force nine times as great.

Still another instrument is the **solenoid-type movement** illustrated in Figure 7-4. Current passes through a coil of wire **(solenoid)** wound on a tube of some nonmagnetic material. As current flows through the coil, it becomes a magnet and its lines of force tend to pull the soft-iron plunger into the coil. The stronger the current, the greater is the magnetic field and the greater is the pull on the plunger.

The motion of the plunger pulls the pointer attached to it over the face of the scale. However, the motion of the plunger is opposed by the tension of the flat spiral spring. The pointer comes to rest at the point where the pulling effect of the coil and the opposing force of the spring are equal. Hence the amount of deflection of the pointer over the scale is an indication of the strength of the current flowing through the coil.

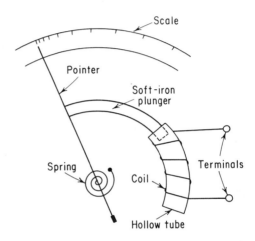

Fig. 7-4. Solenoid-type movement.

Note that here, too, the scale is nonuniform, though not quite of the square-law type. As the soft-iron plunger moves into the solenoid, it strengthens the magnetic field because a greater portion of the coil now has an iron core. Consequently, the force of attraction, and the resulting deflection of the pointer, increases more rapidly than the increase in current.

The types of movements we have described are fairly simple and rugged. However, they are not too accurate, they are inefficient, and they are not very sensitive. Accordingly, most electrical measuring instruments in use today are based upon a design invented by a French physicist, Arsène d'Arsonval.

In essence, this instrument consists of an electromagnet pivoted between the two opposite poles of a permanent horseshoe magnet (see Figure 7-5). As current flows through the turns of the electromagnet, the latter becomes a magnet. If the current flows as indicated in Figure 7-5A, the left-hand side of the electromagnet becomes a north pole, and the

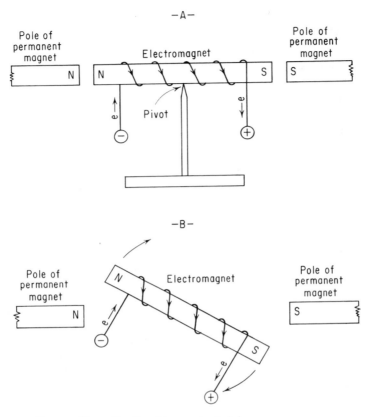

Fig. 7-5. Diagram illustrating the d'Arsonval principle.
A. The electromagnet is pivoted to rotate between two fixed poles of a magnet.
B. Looking down on the top of the electromagnet.

other side a south pole. Since like poles are facing each other, they repel. However, the permanent magnet is fixed and cannot move. Accordingly, the electromagnet rotates around its pivot, as indicated in Figure 7-5B.

The amount of repulsion between the like poles depends upon the relative strengths of the magnetic field around the permanent magnet and that around the electromagnet. Since the magnetic field around the permanent magnet is a fixed value, the stronger the magnetic field around the electromagnet, the greater will be the repulsion and rotating effect. But the strength of the magnetic field around the electromagnet will depend upon the current flowing through it. Thus, the amount of repulsion and the rotating effect will depend upon the strength of the electric current flowing through the coil of the electromagnet.

Here, then, is a convenient way to measure the current flowing through the coil of the electromagnet. All we need do is fix a pointer to this electromagnet and measure the amount of rotation on a suitable scale. Edward Weston, an American scientist, developed a practical instrument using this d'Arsonval principle. (See Figure 7-6.)

The electromagnet, called the **armature,** consists of a coil of very fine wire wound on a soft-iron cylinder as a core. The complete armature is delicately pivoted upon jewel bearings and is mounted between the poles of a permanent horseshoe magnet. Attached to these poles are two soft-iron **pole pieces** which concentrate the magnetic field.

As current flows through the armature coil, a magnetic field is set up around it in such a way that it opposes the field of the permanent magnet. As a result, the armature is rotated in a clockwise direction, carrying the pointer which is attached to it. The **spring** opposes the rotation and brings the pointer back to the no-current position when the current ceases. The greater the current, the greater will be the deflection of the pointer. This deflection is indicated on the scale over which the pointer passes.

Note that rotation will take place only when the poles of the permanent magnet face like poles on the armature. Thus the current must flow through the coil of the armature in such a way that this condition prevails. If the current flows in the opposite direction, unlike poles will face and attract each other, and there will be no rotation. Accordingly,

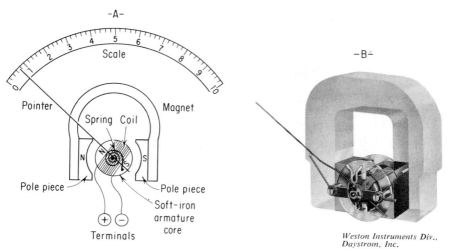

Weston Instruments Div.,
Daystrom, Inc.

Fig. 7-6. A. Moving-coil movement.

B. "Phantom" view of commercial permanent-magnet moving-coil movement.

the terminals of the instrument are suitably marked **plus** (+) or **minus** (−). If the minus terminal is connected to the high-potential (negative) side of the circuit under test and the plus terminal to the low-potential (positive) side, current will flow through the instrument in the proper direction.

This instrument is called, appropriately, a **moving-coil movement.** Such instruments can be constructed to be highly accurate and extremely sensitive. Its efficiency, too, is high as it requires very little power for its operation. Another advantage is its uniform scale, since the deflection of the pointer is directly proportional to the amount of current flowing through the coil of the armature.

Generally, the armature is suspended by hardened steel pivots set in jeweled sockets to keep friction to a minimum. A recent development is the **taut-band suspension** whereby the pivots, jeweled sockets, and spring are replaced by a set of metal ribbons or bands. (See Figure 7-7.) One end of the band is fastened securely to the armature. The band then is given a half-twist and is held taut by means of a tension spring at the other end. The rotation of the armature is opposed by the elasticity of the twisted, taut band. As a result of this construction, friction is eliminated and a more rugged instrument is produced.

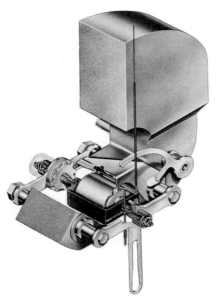

Fig. 7-7. Cutaway view of moving-coil movement showing the taut-band suspension.

Assembly Products, Inc.

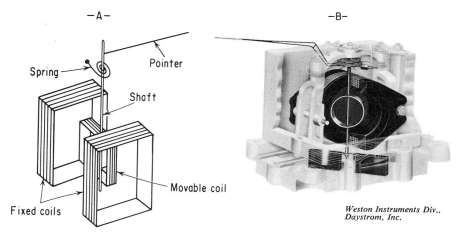

Fig. 7-8. A. Dynamometer movement.

B. "Phantom" view of commercial dynamometer movement.

A variation of the moving-coil movement is the **dynamometer movement** whose principle is illustrated in Figure 7-8. As in the previous type, a moving coil rotates within a magnetic field and the amount of rotation depends upon the strength of the current flowing through that coil. However, instead of using a permanent magnet to set up this magnetic field, two fixed coils placed on either side of the moving coil are employed. The three coils are connected in series and their magnetic fields are created by the current that flows through all.

The dynamometer-type instrument generally employs a square-law scale. This is because as the current increases, the magnetic fields of **both** the fixed and movable coils increase. Hence the deflection of the pointer, which results from the interaction between the magnetic fields of the fixed and movable coils, increases faster than the current. (But note that in the case of the **wattmeter,** which will be discussed later, the dynamometer movement employs a uniform scale.)

However, since the dynamometer movement does not employ a permanent magnet, it is not subject to changes in its calibration due to aging of the magnet or anything else that may affect its magnetism (such as heating, or jarring). In addition, it has a number of other advantages, as we shall learn a little later.

D. TYPES OF METERS

There are a great many types of electrical meters. Most of them employ movements that are mere variations of the ones we have discussed in the preceding subdivisions of this chapter. Since it is impossible to take up all types of electrical instruments in this book, we shall cover only those that are most common.

1. The Galvanometer

The **galvanometer** is an instrument used to indicate the presence, strength, and direction of very small currents in a circuit. Because we wish to show the presence of very small currents, only the moving-coil type of all those we have discussed is sensitive enough. Because we wish to show the direction of current flow, we cannot use the dynamometer type. Regardless of the direction in which it passes through this instrument, current flow causes a repulsion between the fixed and movable coils, and the pointer is deflected across the scale.

Of all the instruments we have considered, only the Weston-type movement with a fixed permanent magnet is suitable for use as a galvanometer. It has the required sensitivity and, you will recall, its pointer will be deflected only when current flows through its moving coil from its minus terminal to its plus terminal. Thus we can tell, by the manner in which the pointer is deflected, the direction of the current.

Actually, should the instrument be connected so that current flows through it from the plus terminal to the minus one, the pointer will be deflected backwards (to the left), as an examination of Figure 7-6 will show. This may bend the pointer and otherwise damage the instrument. Accordingly, this practice should be avoided.

However, it is possible to modify the instrument so that this danger is avoided. The coil and pointer are so mounted that, with no current flowing through the instrument, the pointer comes to rest (zero) at the center of the scale (see Figure 7-9). Then a deflection of the pointer to the right indicates a current flow from the right-hand terminal to the left-hand one. A deflection of the pointer to the left indicates a flow in the opposite direction. Such an instrument is known as a **zero-center** galvanometer.

The numbers on the scale generally are arbitrary ones and merely indicate the **relative** strength of the current. Thus, if the pointer is deflected two divisions, the current is twice as great as one that deflects the pointer one division. Some galvanometers are further calibrated so that the deflection of the pointer may be translated into an indication of the **absolute** strength of current. For example, a deflection from zero to the end of the scale of a certain galvanometer may indicate a current strength

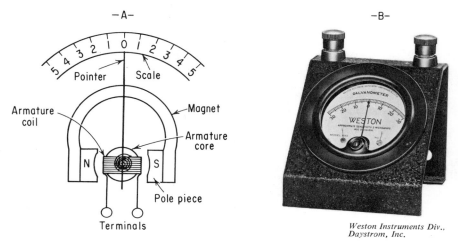

Fig. 7-9. A. Zero-center galvanometer.
   B. Commercial zero-center galvanometer.

of 0.0005 ampere (0.5 milliampere or 500 microamperes). Generally, however, galvanometers are used where only relative values of current are desired.

The moving coil must be light and, hence, is wound with very fine wire. Accordingly, care must be taken that only very small currents flow through it lest the wires burn up. The coil generally is wound with wire that can safely carry a maximum current of about 0.03 ampere (30 milliamperes).

The symbol for the galvanometer, when used in an electrical diagram, is ─ⓖ─ .

2. The Ammeter

The **ammeter** is used to measure the flow of current through a conductor somewhat as a **flowmeter** is used to measure the flow of water through a pipe. In both cases the circuit is broken and the meter is inserted in the break so that all the water (for the water circuit) or all the current (for the electrical circuit) flows through the meter. (See Figure 7-10.) In both cases the meter is inserted in **series** with the circuit under test.

If the current is small enough so that it does not burn up the coil, a

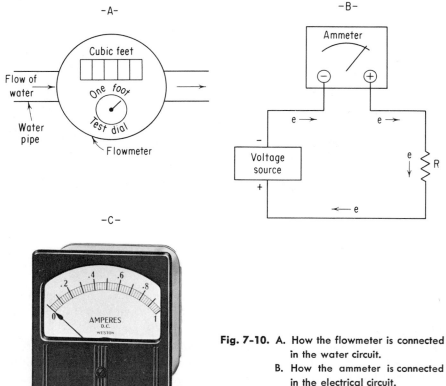

Fig. 7-10. A. How the flowmeter is connected in the water circuit.
B. How the ammeter is connected in the electrical circuit.
C. Commercial ammeter.

meter similar to the moving-coil galvanometer previously described may be used. The scale is calibrated to read milliamperes or microamperes, depending on the magnitude of the current. Generally, zero-center scales are not employed. Instead, a scale with the zero position at the extreme left is used. Values increase the farther the pointer is deflected to the right. Such an ammeter can be used up to about a 30-milliampere full-scale deflection.

As in the case of the flowmeter, which must be connected so that the water flows through it only in one direction, so the moving-coil ammeter must be connected so that the current flows through it only in one direction. Thus the minus (−) terminal must be connected to the high-potential (negative) side of the circuit and the plus (+) terminal must be connected to the low-potential (positive) side.

Where the current to be measured is greater than that which the moving coil can safely pass, a device called a **shunt** is used. This shunt consists of a metal wire or ribbon, usually of some alloy such as **manganin,** which is placed in parallel with the moving coil. Thus the current in the circuit divides, the greater portion flowing through the coil or the shunt, depending upon which has the lower resistance. If the shunt has the same resistance as the coil, half the current will flow through the shunt and half through the coil. If the shunt has half the resistance of the coil, two-thirds of the current will flow through the shunt and one-third through the coil. The scale of the meter, of course, must be calibrated accordingly.

By choosing the proper resistance ratio between the coil and the shunt, we can use a meter to measure any quantity of current. Assume, for example, that we have a meter whose maximum range is 0.001 ampere and that the resistance of the moving coil is 50 ohms. Suppose we wish to use this meter to measure currents up to 0.1 ampere. We must arrange for a shunt that will be able to carry 0.099 ampere and permit 0.001 ampere to flow through the coil. Hence the resistance of the shunt must be $\frac{1}{99}$ that of the coil. The resistance of this shunt then will be $\frac{50}{99}$, or 0.5 ohm (approximately). The scale of the meter must be recalibrated to indicate 0.1 ampere for full-scale deflection.

Note that when we use a shunt, the movement actually is serving as a **voltmeter,** measuring the voltage drop across the shunt. In the example we are considering, the movement has a resistance of 50 ohms, and a current of 0.001 ampere flowing through its coil will produce a full-scale deflection. Since $E = I \times R$, a voltage of 0.050 volt (50 millivolts) applied across the movement will produce a full-scale deflection.

Accordingly, the resistance of any shunt used with this movement must be such as to produce a voltage drop of 0.050 volt when the full-load current flows through the ammeter (movement and shunt). Hence, shunts used with this movement are marked to indicate the full-load current for the ammeter (in amperes) and the voltage required to produce a full-scale deflection of the movement (50 mv in this case, which is most commonly used). In this way we are sure that the shunt and movement are properly matched.

Where ammeters are used to measure currents of the order of milliamperes, they are usually called **milliammeters.** Where the full-scale deflection is less than one milliampere, the meter usually is called a **microammeter.** Moving-coil ammeters may be obtained in ranges of from about 10 microamperes (0.00001 ampere) to thousands of am-

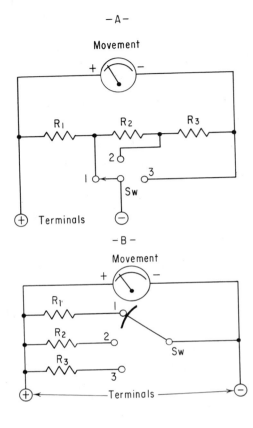

Fig. 7-11. Multirange ammeters.
A. Universal shunt.
B. Individual shunts.

peres. Where shunts are used, they generally are enclosed in the meter case for ammeters up to about 30 amperes. Beyond that, the shunts usually are external to permit adequate heat dissipation. Alloys such as manganin are used for shunts because their resistances vary little with the temperature changes caused by the passage of current.

A single movement may be used with several shunts to make a **multirange** ammeter. These shunts may be connected in parallel with the moving coil by means of switches or external connections. In Figure 7-11A is shown the circuit of a multirange instrument using a tapped resistor (consisting of R_1, R_2, and R_3) for that purpose. This is called a **universal,** or **Ayrton, shunt.** Different ranges may be obtained by means of the switch Sw (whose symbol is ⌐⌐⌐) which connects the coil of the movement across various portions of the resistor.

With the switch at position 1, the resistances of R_2 and R_3 are added to the resistance of the coil, and R_1 acts as the shunt for the movement. With the switch at position 2, the resistance of R_3 is added to the resistance of the coil. R_1 and R_2 in series now act as the shunt. With the switch at position 3, R_1, R_2, and R_3 in series act as the shunt for the coil.

Another type of multirange ammeter, using individual shunts, is shown in Figure 7-11B. Switch Sw selects which shunt (R_1, R_2, or R_3) is to be used across the coil of the movement. Of course, the various shunts used in multirange instruments must have the proper resistance ratios to that of the coil for the range employed.

Note the shape of the moving contact of the switch. This is to insure that when this moving contact is shifted from one position to another, it makes contact with its new fixed contact before it breaks contact with its previous fixed contact. This is to insure that, at all times, there is a shunt across the coil of the movement. Otherwise, when the switch is changed from one shunt to another, there might be an instant when there would be no shunt across the coil. Under such circumstances the full current flowing through the coil might burn it out.

The symbol for the ammeter, when used in an electrical diagram, is ⏦Ⓐ⏦ . If the instrument is a milliammeter, its symbol is ⏦ⓂⒶ⏦ . If it is a microammeter, its symbol is ⏦Ⓐ⏦ .

The dynamometer movement, also, may be used for the ammeter. However, since the pointer will be deflected the same way regardless of the direction of current flow through the movement, we need not worry about how the terminals are connected to the circuit, provided, of course, that the meter is in series. Where the current is great enough to operate the movement and where great accuracy is not required, any of the other movements previously described may be employed. As in the dynamometer type, their terminals may be connected either way. All the meters may employ shunts to extend their ranges.

3. The Voltmeter

The **voltmeter** is used to measure the difference of potential (electrical pressure or voltage drop) between two points in a circuit somewhat as the **pressure gage** is used to measure the water pressure in a pipe (see

Figure 7-12). In Figure 7-12B the voltmeter is used to measure the voltage drop across resistor R. In both cases the measuring instrument is connected in **parallel** with the circuit under test.

Note that in the electrical circuit the meter actually measures the **current** flowing through it. But since this current depends upon the voltage drop across R, we may calibrate the scale in volts rather than amperes. Since the moving-coil movement will operate only if the current flows through it in the proper direction, care must be taken to connect its minus terminal to the negative side of the resistor and the plus terminal to the positive side. If any of the other types of movements are employed, their terminals may be connected either way.

The resistance of the voltmeter must be large for two reasons. First, since the current flowing through the parallel circuit divides inversely

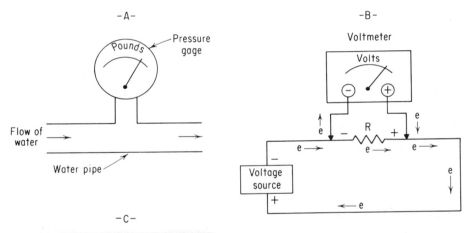

Weston Instruments Div..
Daystrom, Inc.

Fig. 7-12. A. How the pressure gage is connected in the water circuit.
B. How the voltmeter is connected in the electrical circuit.
C. Commercial voltmeter.

according to the ratio between the resistances of the meter and R, the meter resistance must be large so that very little current will flow through the instrument. Otherwise it will take too much current from the circuit and thus give a false reading of the voltage drop across R. Second, the meter resistance must be large to prevent too much current from flowing through its movement and burning up its movable coil. Since the resistance of the moving coil is low, high resistances, called **multipliers,** are connected in series with it. Thus, the voltmeter consists of the movement and the series multiplier (generally enclosed in the same case).

Assume, for example, that we wish to convert the 1-milliampere meter whose coil has a resistance of 50 ohms to a voltmeter whose full-scale deflection would indicate 100 volts. This means that with 100 volts across the voltmeter, one milliampere (0.001 ampere) of current must flow through it. From Ohm's law, $R = E/I$, or $R = 100/0.001$. Thus the total resistance of the voltmeter is $100/0.001$, or 100,000 ohms. Since the resistance of the coil is 50 ohms, the resistance of the series multiplier must be 99,950 ohms. (Generally, the resistance of the coil is neglected and a series multiplier of 100,000 ohms is employed.)

The multipliers usually are constructed of high-resistance wire, such as manganin, wound on wooden spools. Where the resistance is so high as to make the windings prohibitively large and costly, composition resistors are employed. Some of the less expensive types of voltmeters use composition resistors throughout.

Multirange voltmeters use a number of multipliers which may be placed in series with the moving coil by means of switches or external connections. In Figure 7-13A is shown the circuit of a multirange instrument using a tapped resistor (R_m) for that purpose. Different ranges may be obtained by means of the switch, which connects various portions of the resistor in series with the moving coil. As shown in the diagram, the voltmeter is at its lowest range.

Another type of multirange voltmeter using individual multipliers (R_1, R_2, and R_3) is shown in Figure 7-13B. Of course, the various multipliers used in multirange instruments must have the proper resistance for the range employed.

The symbol for the voltmeter, when used in an electrical diagram, is $-\text{Ⓥ}-$. If the instrument is a millivoltmeter, its symbol is $-\text{ⓂⓋ}-$.

Any of the other movements may also be employed as a voltmeter, though, except for the dynamometer type, they seldom are used in this manner. Where the resistance of the movement is too low, suitable multipliers may be employed.

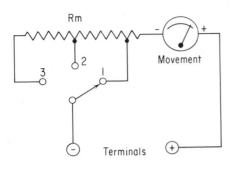

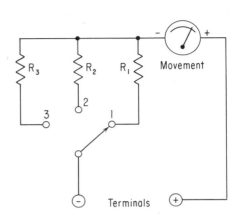

Weston Instruments Div.,
Daystrom, Inc.

Fig. 7-13. Multirange voltmeters.
A. Using a tapped multiplier resistor.
B. Using separate multiplier resistors.
C. Commercial multirange d-c voltmeter using external posts for connections to series multipliers.

4. The Wattmeter

The electric power consumed in a circuit is equal to the product of the current and voltage (watts = amperes × volts). A voltmeter and ammeter may be connected to indicate these values and the watts calculated by multiplying the two readings. For example, if we wish to determine the power consumed by a resistor (R) in a circuit, the meters may be connected as indicated in Figure 7-14.

The same result may be obtained by using a single instrument of the

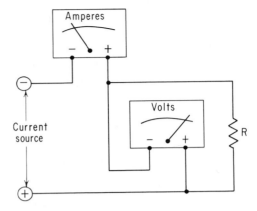

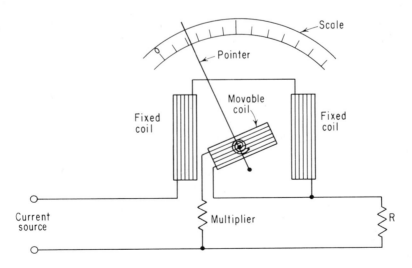

Fig. 7-14. Circuit used to determine the power consumed by resistor R.

dynamometer type connected in the circuit as shown in Figure 7-15. The two fixed coils are wound with wire that is heavy enough to pass the current and are connected in series with the circuit under test. Thus, the magnetic fields around them will be proportional to the current flowing in the circuit. The movable coil is connected in series with a multiplier, and both are connected across the circuit. Thus, they act as a voltmeter, and the magnetic field around the movable coil will be proportional to the voltage. Since the movement of the coil is the result of the interaction of the magnetic fields around the fixed and movable coils, the movement

Fig. 7-15. How the wattmeter is connected into the circuit.

of the pointer across the scale will indicate the watts being consumed by the circuit.

The scale, you will notice, is of the uniform type since it indicates, directly, the product of the current and voltage. Current flows through the fixed and movable coils in the same direction, regardless of how the meter's terminals are connected to the circuit. The symbol for the **wattmeter,** when used in an electrical diagram, is -ⓦ- .

5. The Watthour Meter

The **watthour** is the unit for electrical energy (see Chapter 4, subdivision C, 5). The **watthour meter,** which is used to measure electrical energy, combines the basic principles of the dynamometer movement and the electric motor. (The electric motor will be described in greater detail later in this book. Essentially, a magnetic field around a fixed coil, or coils, called the **field coil,** or **stator,** interacts with the magnetic field around a movable coil wound on an iron core, called the **rotor,** causing the latter to rotate. Current generally is brought to the rotor by means of a set of **brushes.**)

The watthour meter, of a type commonly used to measure the amount of electrical energy consumed by the various circuits in a house, is illustrated in Figure 7-16. The field coils of the motor are wound with heavy wire and are connected in series with the power line. Thus the magnetic field around these coils is proportional to the amount of current drawn from the power line by the house circuits. The rotor coil is wound with fine wire and is connected across the power lines through a series multiplier resistor (R_m). Thus its magnetic field is proportional to the voltage of the line.

The speed of the motor is dependent upon the magnetic fields of the field coils and rotor. Since these magnetic fields are proportional to the line current and voltage, respectively, the speed of the motor depends upon the power (in watts or kilowatts) consumed by the house circuits.

The rotor of the motor is connected through a set of gears to a series of dials that indicate the number of its revolutions. The faster the motor revolves and the longer it continues to do so, the greater will be the dial indications. Accordingly, the dials indicate the power being consumed by the house circuits and the length of time such power is being consumed. But power × time = energy (in watthours or kilowatthours). Accordingly, these dials indicate the kilowatthours of electrical energy consumed by the house circuits. The right-hand dial indicates kilowatt-

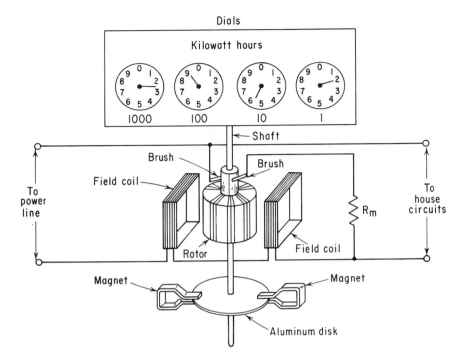

Fig. 7-16. The watthour meter.

hours in units of 1; the next one in units of 10; the next in units of 100; and the left-hand dial in units of 1000.

Note the aluminum disk attached to the lower portion of the rotor shaft. This disk rotates between the poles of two permanent magnets. As this disk rotates and cuts across the magnetic fields of the magnets, a current, called an **eddy current,** is set flowing in it (as you will learn later in this book). This current sets up a magnetic field around the disk that interacts with the magnetic fields of the permanent magnets. As a result of this interaction, the disk, and the rotor which is attached to the same shaft, are braked down. This prevents the rotor from rotating too fast and from continuing to rotate after the current is turned off.

6.   The Ohmmeter

The measurement of resistance is based upon current flowing through the circuit under test and on the voltage drop across the circuit produced

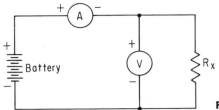

Fig. 7-17. Basic circuit for the resistance test.

by that current. If the current and voltage drop are known, the resistance can be calculated from Ohm's law ($R = E/I$). The basic circuit for this resistance measurement is shown in Figure 7-17.

Assume that a voltage source, such as a **battery,** of 100 volts is connected in series with an ammeter (A) and an unknown resistor (R_x) whose resistance we wish to measure. (Later in this book we shall discuss electric cells and batteries. The symbol for the cell is ⊣⊢, and for a battery of cells the symbol is ⊣⊦⊢⊢.) The voltmeter (V) will indicate the voltage drop across R_x. Since we will neglect here the slight voltage drop across the ammeter and connecting wires and the slight amount of current flowing through the voltmeter, the reading on the voltmeter will be 100 volts, the voltage of the battery. If the ammeter reads, say, 2 amperes, then, since $R = E/I$, $R_x = 100/2 = 50$ ohms.

You can readily see that if the voltage is known and remains constant, we may remove the voltmeter and can calibrate the ammeter to read resistance directly. The greater the resistance, the less the current flowing in the circuit and the less will be the reading on the ammeter. We call the recalibrated ammeter with its voltage supply an **ohmmeter.**

The circuit of a commercial ohmmeter is shown in Figure 7-18. A 400-microampere, 250-ohm microammeter is connected in series with a fixed resistor of 3000 ohms, a variable resistor of 6565 ohms, and a 1.5-volt dry cell. When the test terminals are short-circuited, current will flow through the circuit. To obtain full-scale deflection of the meter (0.0004 ampere) the resistance of the circuit must be 3750 ohms ($R = 1.5/0.0004$). The variable resistor is adjusted to obtain this full-scale deflection.

If any resistance now is introduced into this circuit, less current will flow and the deflection no longer will be full-scale. The greater this new resistance, the less will be the reading on the meter. Thus, the meter can be calibrated directly in terms of the amount of resistance introduced.

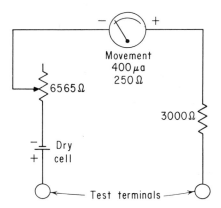

Fig. 7-18. Typical ohmmeter circuit.

If no resistance is introduced (test terminals short-circuited), the reading will be full-scale. If a resistance equal to the resistance of the circuit (3750 ohms) is introduced, 200 microamperes will flow and the reading will be half-scale. If twice the resistance of the circuit (7500 ohms) is introduced, the deflection will be one third of full-scale. With three times the resistance (11,250 ohms), the deflection will be quarter-scale. Thus, resistances up to about 250,000 ohms may be measured.

Note that the scale runs backward; zero resistance produces full-scale deflection and the larger the unknown resistance, the smaller the deflection. Note, too, that the scale is not uniform, but crowds together toward the left-hand side of the scale. When discussing ohmmeter ranges we generally speak of the resistance that will produce half-scale deflection — in this case, 3750 ohms. Since the divisions crowd up at the left-hand side of the scale, we generally can obtain readings with a fair degree of accuracy up to about ten times the half-scale value.

The variable resistor is used to compensate for variations in the dry cell produced by aging. Before each test, the test leads connected to the terminals should be short-circuited by touching them together and the variable resistor adjusted to produce a full-scale deflection. When this can no longer be obtained, the dry cell should be replaced.

To obtain different ohmmeter ranges, the values of the calibrating resistors, the voltage of the battery, or both, may be changed. The circuit of a commercial multirange ohmmeter is shown in Figure 7-19A. The unknown resistor is connected across the terminal marked COMMON and one of the others, depending upon the range desired. At the $R \times 1$ range, the center-scale reading is 25 ohms. At $R \times 10$, the center-scale reading is 250 ohms and all scale readings should be multiplied by 10. At

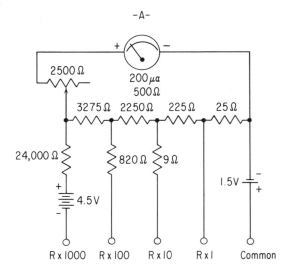

Weston Instruments Div.,
Daystrom, Inc.

Fig. 7-19. A. Typical multirange ohmmeter circuit.
B. Commercial ohmmeter.

$R \times 100$, the center-scale reading is 2500 ohms and all scale readings are multiplied by 100. All these three ranges use the 1.5-volt dry cell. At the $R \times 1000$ range, the 4.5-volt battery is connected in series. The center-scale reading then is 25,000 ohms and all scale readings should be multiplied by 1000. In all cases, the zero adjustment is made for full-scale deflection with the test terminals shorted before attempting to measure the unknown resistance.

E. MISCELLANEOUS NOTES

The meter is a delicate instrument and should be handled with care. If it is of the magnetic type it should not be placed within a strong magnetic field. The position of the meter, too, is important, since the effect of gravity on the movement may throw off the calibration. If the meter is placed in an improper position, excessive friction in the bearings may cause inaccurate readings. A meter should be operated in the position for which it is designed — namely, vertical, horizontal, or inclined.

The pointer is a little distance from the scale. Thus, if it is viewed from the side, an error of several divisions may occur. This phenomenon is called **parallax.** To avoid this error, the meter should be viewed from a position directly in front of the pointer. Some meters have a mirror incorporated in the scale so that the error may be discerned more readily. If the meter is viewed properly, the pointer and its image coincide. If the meter is viewed from the side, the image in the mirror appears to one side of the pointer.

Most meters are equipped with a screw adjustment to set the pointer at zero before it is connected in the circuit. This is called the **zero-set** adjustment and care must be taken that the pointer is properly adjusted before using the meter.

The ammeter must be connected in series with the circuit under test. If it were connected in parallel, its low resistance would act as a short-circuit and the large current set flowing through it would burn up or otherwise damage the instrument. The voltmeter, on the other hand, must be connected in parallel with the circuit under test. If it were connected in series, its high resistance would tend to restrict the flow of current through the circuit and the meter might be damaged.

If the meter is of the permanent-magnet moving-coil type, its terminals must be connected properly to the circuit under test. Thus the minus terminal must be connected to the negative side of the circuit and the plus terminal to the positive side. If the terminal connections are reversed, damage may result to the movement.

Care must be taken, too, that the current flowing through the meter is not great enough to burn out the movement. Thus the proper range should be employed. When testing a circuit whose current or voltage is unknown, start with the highest range and work down to the proper one. In this respect, it should be noted that the proper range is one where a significant deflection is evident. It would be extremely difficult, for example, to obtain an accurate reading of a current in the neighborhood of one milliampere on a 1000-milliampere scale. Generally, it is best to use a range where the reading appears at about mid-scale. Where the scale is nonuniform, try to use a range where the reading is obtained on the most spread-out portion of the scale.

The full-scale range of an ammeter is a measure of its **sensitivity.** Thus a 0–500 microampere meter is more sensitive than a 0–1 milliampere one. If 500 microamperes flow through the former, it produces a full-scale deflection. If the same current flows through the latter, it will produce only a half-scale deflection.

The sensitivity of a voltmeter, on the other hand, is determined by dividing the resistance of the meter (including the multiplier) by the reading of a full-scale deflection. Thus, if a 0–100 volt voltmeter has a resistance of 100,000 ohms, its sensitivity is 100,000/100, or **1000 ohms per volt.** The greater the number of ohms per volt, the greater is the sensitivity of the voltmeter. Since the voltmeter is connected in parallel with the circuit it measures, the higher the sensitivity of the meter (the greater the number of ohms per volt) the less power the meter draws from that circuit.

QUESTIONS

1. By means of a simple diagram, explain the hot-wire meter movement.

2. By means of a simple diagram, explain the solenoid-type meter movement.

3. By means of a simple diagram, explain the d'Arsonval moving-coil meter movement.

4. Explain how a d'Arsonval-type galvanometer may be converted into an ammeter; into a voltmeter.

5. Assume you have a milliammeter whose resistance is 100 ohms and whose full-scale deflection is 1 milliampere. What must be the resistance of the shunt needed to convert the meter to read a full-scale deflection of 1 ampere?

[*Ans.* 0.1 ohm.]

6. Assume you have a milliammeter whose full-scale deflection is 1 milliampere and whose resistance is 100 ohms. What must be the resistance of a series multiplier to convert this meter to a voltmeter whose full-scale deflection reads 500 volts? [*Ans.* 499,900 ohms.]

7. Explain how you would find the resistance of a resistor using an ammeter and voltmeter.

8. What is meant by the sensitivity of an ammeter; of a voltmeter?

ALTERNATING CURRENT ELECTRICITY

INDUCED EMF
AND THE A-C CYCLE

8

A. INDUCED ELECTROMOTIVE FORCE

Oersted discovered that an electric current flowing through a conductor can create a magnetic field. Can a magnetic field create an electric current in a conductor? In 1831 Michael Faraday, the famous English scientist, discovered that this could be done. Strangely enough, at about the same time, Joseph Henry, an American scientist and teacher, independently discovered the same thing.

Let us try an experiment to illustrate this point. Connect the terminals of a zero-center galvanometer to a coil of about 50 turns of wire wound in the shape of a cylinder about two inches in diameter. Now plunge the north end of a permanent magnet into the center of the coil (Figure 8-1A). You will observe that the pointer is deflected to the right, showing that an electric current was set flowing for a moment in the coil and galvanometer.

When the magnet comes to rest inside the coil (Figure 8-1B), the pointer swings back to zero, showing that the current has ceased flowing.

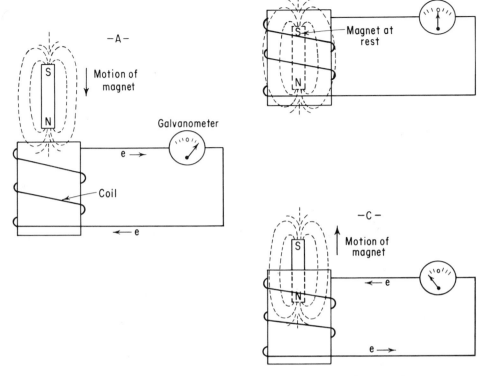

Fig. 8-1. Demonstration of how a current is induced in a conductor as it cuts across a magnetic field.

Now remove the magnet from the coil (Figure 8-1C). As you do so, the pointer swings to the left, showing that once more an electric current is set flowing, but this time in the opposite direction. The same effect may be obtained if the magnet is held stationary and the coil moved.

How can we explain this? You know that the permanent magnet is surrounded by a magnetic field. As the magnet is moved into or out of the coil, this magnetic field cuts across the wire of the coil. **When a conductor cuts through a magnetic field, an electromotive force is set up between the ends of the conductor.** If an external circuit is connected to the ends of the conductor, this electromotive force will cause current to flow through the conductor and the external circuit. If there is no external circuit, no current will flow even though the electromotive force is present, just as water under pressure will not flow when the faucet is closed.

It makes no difference whether the conductor is stationary and the magnetic field is moving across it, or the magnetic field is stationary and the conductor is moving through it. But if both the conductor and magnetic field are stationary (as in Figure 8-1B), no electromotive force will be set up and no current will flow.

We call an electromotive force set up in a conductor in this way an **induced electromotive force.** And the current set flowing as a result is an **induced current.**

Experimentation has evolved a rule to determine in which way an induced current will flow. Examine Figure 8-2, where a conductor is moving across a magnetic field set up between two poles of a horseshoe magnet.

Assume that the conductor is moving down between the poles of the magnet. Extend the thumb, the forefinger, and the middle finger of the left hand so that they are at right angles to one another. Let the thumb point in the direction in which the conductor is moving (down). Now let the forefinger point in the direction of the magnetic lines of force (you will recall that we assume they go from the north to the south pole). The

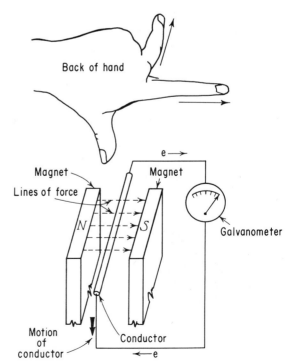

Fig. 8-2. Left-hand rule for determining the direction of induced current flowing in a conductor.

middle finger then will indicate the direction in which electrons will be set flowing by the induced electromotive force (away from the observer).

There is another important principle in connection with induced currents. Turn back to Figure 8-1. As the north pole of the magnet approaches the coil, the magnetic field, cutting across the turns of the coil, induces an electromotive force in those turns. This electromotive force causes an induced current to flow through the coil and the meter connected to it. You already know that when a current flows through the conductor, it sets up a magnetic field around this conductor. Thus, the coil becomes a magnet. The induced current in this coil is set flowing in such a direction that the end of the coil facing the north pole of the magnet becomes a north pole, too. Since like poles repel, this arrangement of magnets tends to prevent the insertion of the north pole of the magnet into the coil. Work must be done to overcome the force of repulsion.

When you try to remove the magnet from the coil, the induced current is reversed. The top of the coil becomes a south pole and, by attraction to the north pole of the magnet, tends to prevent you from removing it. Thus, once again, work must be done — this time to overcome the force of attraction. You see, you must perform work to create the induced electric current. Or, to put it another way, mechanical energy is converted to electrical energy. Of course, the same holds true if the magnet is stationary and the coil is moved.

These results may be summarized in a law formulated by Heinrich Lenz, a German scientist who investigated this phenomenon. According to Lenz's law:

> **An induced current set up by the relative motion of a conductor and a magnetic field always flows in such a direction that it forms a magnetic field which opposes the motion.**

Let us return again to the coil-and-magnet arrangement illustrated in Figure 8-1. The stronger the field around the permanent magnet, the greater will be the induced electromotive force in the coil as it cuts through this magnetic field. Also, the more turns in the coil, the greater the number of conductors cutting the magnetic field. Since each conductor adds its induced electromotive force to the total, the greater the number of turns in the coil, the greater will be the total induced electromotive force.

Notice, too, inserting the magnet slowly into the coil causes a slight deflection of the galvanometer pointer. Rapid insertion causes greater deflection. Thus, the faster the conductors cut across the magnetic field, the greater will be the induced electromotive force and the induced current set flowing by it.

From all of this we may conclude:

> 1. The stronger the magnetic field, the greater the induced electromotive force.
>
> 2. The greater the number of conductors cutting the magnetic field, the greater the induced electromotive force.
>
> 3. The greater the speed of relative motion between the magnetic field and conductors, the greater the induced electromotive force.

It has been found that if a conductor cuts across 10^8 lines of force in a second, an electromotive force of one volt will be induced between its ends.

B. THE SIMPLE GENERATOR

Here, then, is the beginning of our electrical age. We could convert the mechanical energy of a steam engine or water turbine to electric current for light, heat, and power to operate the marvelous electrical machines that soon were invented.

From a mechanical point of view, it is not practical to move our magnet in or out of a stationary coil of wire, or to move the coil over a stationary magnet. The same thing can be accomplished more simply by rotating a loop of wire between the poles of a magnet, thereby inducing a current in the wire as the magnetic field is cut. Let us examine such an arrangement, as illustrated in Figure 8-3.

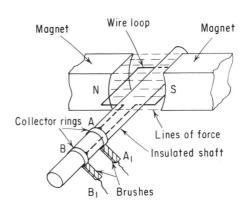

Fig. 8-3. Simple generator consisting of a loop of wire revolving between the poles of a magnet.

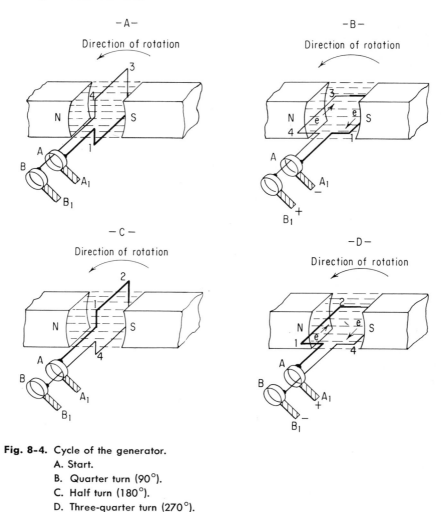

Fig. 8-4. Cycle of the generator.
  A. Start.
  B. Quarter turn (90°).
  C. Half turn (180°).
  D. Three-quarter turn (270°).

A simple loop of wire (called an **armature coil**) is mounted so that it may be rotated mechanically on a shaft between the north and south poles of a magnet. The two ends of the loop are connected to two brass or copper rings, A and B respectively, called **collector rings,** which are insulated from each other and from the shaft on which they are fastened. These collector rings rotate with the loop. Two stationary **brushes** (A_1 and B_1) make a wiping contact with these rotating collector rings and lead the current that has been induced in the loop to the external circuit. These brushes usually are made of copper or carbon. This arrangement of loop, magnetic field, collector rings, and brushes constitutes a simple **generator.**

Let us assume that the loop starts from the position shown in Figure 8-4A, and rotates at a uniform speed in a counterclockwise direction. In its initial position no lines of force are being cut because conductors 1-2 and 3-4 (the arms of the loop) are moving parallel to the lines of force, not across them.

As the loop revolves, however, the conductors begin to cut across the lines of force at an increasing rate and, therefore, the induced electromotive force becomes larger and larger. At the position shown in Figure 8-4B, the loop has the maximum electromotive force induced in it because conductors 1-2 and 3-4 cut across the maximum number of lines of force per second, since the conductors are moving at right angles to the magnetic field.

As the loop rotates to the position of Figure 8-4C, the electromotive force is still in the same direction, but is diminishing in value, until it is zero again. The loop now has made one-half turn, during which the induced electromotive force increased to a maximum and then gradually fell off to zero. Since conductors 1-2 and 3-4 are now in reversed positions, the induced electromotive force changes direction in both conductors. The electromotive force, however, again increases in strength and becomes maximum when the loop is again cutting the lines of force at right angles (Figure 8-4D).

Finally, the last quarter of rotation brings the loop back to its original position (Figure 8-4A), at which point the electromotive force is zero again. As the rotation is continued, the cycle is repeated.

This, then, is how the generator operates. Of course, practical generators are not constructed as simply as the one illustrated here. We shall discuss them further in a subsequent chapter.

C. THE ALTERNATING-CURRENT CYCLE

The term **cycle** really means "circle" — a circle or series of events which recur in the same order. A complete revolution of the loop of the generator is a cycle. So, also, is the series of changes in the induced electromotive force and the current set flowing by it. As the loop of the generator makes one complete revolution, every point in the conductors describe a circle. Since the circle has 360 degrees (360°), a quarter turn is equal to 90°; a half turn to 180°; a three-quarter turn, 270°; and a full turn, 360°. The number of degrees, measured from the starting point, is called the **angle of rotation.**

Thus, Figure 8-4A represents the starting point, or zero-degree (0°) position; Figure 8-4B, the 90° position; Figure 8-4C, the 180° position; Figure 8-4D, the 270° position; and Figure 8-4A again (after a complete revolution), the 360° position. Of course, positions in between these points may be designated by the corresponding degrees. However, it is customary to use the **quadrants** (that is, the four quarters of a circle) as the angle of rotation for reference.

We now are ready to examine more closely the induced electromotive force in the loop of the generator as it goes through a complete cycle or revolution. Note that during half the cycle the direction of the induced electromotive force is such as to cause electrons to move onto brush A_1. During the next half-cycle the direction of the induced electromotive force is reversed so that the electrons move onto brush B_1. To avoid confusion, let us designate the induced electromotive force in one direction by a plus ($+$) and in the other direction by a minus ($-$).

Let us assume that the armature loop makes a complete revolution (360°) in one second. Then, at $\frac{1}{4}$ of a second the loop will be at the 90° position, at $\frac{1}{2}$ of a second the loop will be at the 180° position, and so on. Assume, too, that the maximum electromotive force generated by this machine is 10 volts. Now we are able to make a table showing the electromotive force being generated during each angle of rotation.

Time in seconds	0	$\frac{1}{4}$	$\frac{1}{2}$	$\frac{3}{4}$	1
Angle of rotation	0°	90°	180°	270°	360°
Induced EMF (volts)	0	$+10$	0	-10	0

You will note that in one complete revolution of the loop there are two positions (Figures 8-4A and C) at which there is no induced voltage and, therefore, no current flowing to the brushes (or to the external circuit that is connected to them). There are also two positions (Figures 8-4B and D) at which the induced voltage is at its maximum value, although in opposite directions. At intermediate positions the voltage has intermediate values. Note that as the loop rotates, there are two factors

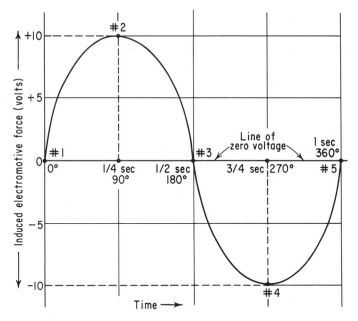

Fig. 8-5. Graph showing the sinusoidal waveform of the generator's alternating-voltage output.

that are continuously changing — the position of the loop and the value of the induced electromotive force, or voltage.

Now let us return to our table above, which shows the relationship between the induced electromotive force in the loop of the generator and the degrees of rotation of the loop (or, what amounts to the same thing, the time in seconds which the loop rotated). Let us try to show this relationship by means of a graph.

First draw the horizontal line of zero voltage (Figure 8-5) and divide it into four equal sections of a quarter-second each. Since we assume that the loop makes one revolution (360°) in one second, each quarter-second will correspond to 90°. Accordingly, these sections may be marked in degrees as well.

Next, draw the vertical line showing the induced electromotive force in volts. You will recall that this voltage is in one direction for half the cycle and then reverses and is in the other direction during the next half-cycle. You will recall, too, that we decided to indicate the voltage in one direction by a plus (+) sign and that in the other direction by a minus ($-$) sign. All the plus values of voltage will appear above the line of zero voltage, and all the minus values below it. Accordingly, all the numbers indicating induced voltage that appear above the line of zero voltage bear plus signs, and all those below bear minus signs.

We now are ready to transpose the values of our table to the graph. At 0° rotation, the table shows the time to be zero and the induced electromotive force to be zero as well. This point is located on the graph where the horizontal line meets the vertical (#1). At the 90° or ¼-second mark, the induced electromotive force has risen to a value of $+10$ volts. We find this point on the graph by drawing a vertical line up from the ¼-second (90°) mark and a horizontal line to the right from the $+10$-volts mark. Where these two lines intersect is the required point (#2).

At the ½-second (180°) mark, the electromotive force has fallen to zero again and the point (#3) on the graph lies on the line of zero voltage. At the ¾-second (270°) mark, the electromotive force has risen once more to 10 volts, but this time it is in the opposite ($-$) direction. We find the corresponding point (#4) on the graph by dropping a vertical line down from the ¾-second mark and drawing a horizontal line to the right from the -10-volt mark. The point lies on the intersection of these two lines. At the 1-second (360°) mark, the electromotive force again has become zero. Accordingly, this point (#5) lies on the line of zero voltage.

Having thus established the points on the graph corresponding to the quadrants of the circle, how do we go about filling in the rest of the curve? Well, you will recall that as the loop rotated from its original position, as illustrated in Figure 8-4A, a quarter-revolution (90°), as illustrated in Figure 8-4B, the induced electromotive force rose gradually from zero to its maximum value ($+10$ volts). Accordingly, we indicate this rise by the curve appearing between points #1 and #2 on the graph illustrated in Figure 8-5.

During the next quarter-revolution (Figure 8-4B to Figure 8-4C) the electromotive force dropped gradually from its maximum value to zero. This is indicated by the curve connecting points #2 and #3 on the graph. During the next quarter-revolution (Figure 8-4C to Figure 8-4D), the electromotive force reversed its direction and again rose to its maximum value (-10 volts), as indicated by the curve connecting points #3 and #4. During the next quarter-revolution (Figure 8-4D back to Figure 8-4A), the electromotive force dropped gradually to zero again (as indicated by the curve connecting points #4 and #5).

Thus we obtain a graph that illustrates the electromotive force generated by the armature coil as it makes one complete revolution (and the flow of current in the external circuit connected to the brushes resulting from this electromotive force). Now you must not get the impression that the current is flowing in this roller-coaster type of path. Actually, the current is flowing back and forth through the external circuit. What this curve does show, however, is the strength of the induced electromotive force (and the resulting current flow) and its direction (+ or −) at any instant during one revolution.

So at the ¼-second mark the electromotive force is 10 volts and is acting in the direction indicated by plus (+). At the ¾-second mark the electromotive force again is 10 volts, but this time it is acting in the opposite direction, as indicated by minus (−). In the interval between the ¼-second mark and the ¾-second mark the electromotive force changes from +10 volts to −10 volts, dropping to zero at the ½-second mark, at which instant the electromotive force changes its direction.

We call an electromotive force whose strength and direction are indicated by the curve of Figure 8-5 an **alternating electromotive force.** The current that is set flowing by such an electromotive force will show a similar curve and is called an **alternating current** (abbreviated **ac**).

Note well the curve of Figure 8-5. This curve is typical for alternating currents and is known as the **sine curve.** This curve is said to be the **waveform** of the alternating current produced by the generator. In many applications we desire alternating current to be of this **sinusoidal** (that is, in the form of a sine curve) waveform.

Each **cycle** of alternating current represents one complete revolution of the armature coil of the generator. The number of cycles per second, which is known as the **frequency,** depends upon the number of revolutions per second. In Figure 8-5 the alternating current has a frequency of one cycle per second. Alternating current supplied to house mains in this country generally has a frequency of 60 cycles per second. In radio we encounter frequencies that run into millions and billions per second. Of course, no generator can be rotated at such a tremendous number of revolutions per second, but we have other means of producing high-frequency currents. The symbol for the cycle is the sine curve $\sim$. Thus 60-cycle alternating current may appear as 60 $\sim$ ac. It is from this symbol of a sine curve that we obtain the symbol for the alternating-current generator, which is $-\!\otimes\!-$.

To facilitate discussing currents and voltages with high frequencies, the terms **kilocycle** (abbreviated to **kc** and meaning 1000 cycles per second) and **megacycle** (abbreviated to **mc** and meaning 1,000,000 cycles per second) had been employed. Recently, by international agreement, the term **hertz** (abbreviated **Hz**) was selected as the unit for **cycle per second.** The unit indicating 1000 cycles per second thus becomes a **kilohertz (kHz)** and the unit for 1,000,000 cycles per second, the **megahertz (MHz).** Hence a current whose frequency is 60 cycles per second would appear as 60 Hz, a current of 5000 cycles per second as 5 kHz, and a current of 5,000,000 cycles per second as 5 MHz.

Although 60-Hz alternating current is most widely used, in certain localities currents with frequencies of 25 and 50 Hz are employed. For household use the mains usually are supplied at between 110 and 120 volts, depending upon the locality. For industrial installations the mains may be at 110–120 volts, 208 volts, 220 volts, 550 volts, or higher. Aircraft generally employ alternating current at 28 volts, 400 Hz.

Since the human ear generally can hear sounds whose frequencies are up to about 15,000 vibrations per second, alternating currents whose frequencies are up to about 15 kHz are called **audio-frequency (a-f)** currents. Above such frequencies the alternating currents are called **radio-frequency (r-f)** currents.

The Federal Communications Commission (FCC) has allocated frequencies for the regular **amplitude-modulated** (AM) broadcast band between 550 kHz and 1600 kHz, and frequencies for the regular **frequency-modulated** (FM) broadcast band between 88 MHz and 108 MHz. Regular **television** (TV) broadcast is confined to a band of frequencies extending from 54 MHz to 216 MHz, excluding the band reserved for FM broadcast.

Alternating electromotive force and current are changing constantly in magnitude — that is, the **instantaneous** values are changing. From the sine curve, you can see that there are two **maximum,** or **peak,** values for each cycle; a positive peak and a negative peak. We call the magnitude of the peak values — that is, the values represented by the distance of these peaks from the zero line in the graph shown in Figure 8-5 — the **amplitude.** Thus, in our illustration, the amplitude of the generated voltage is 10 volts. If we wish to indicate both peak amplitudes of the cycle, the **peak-to-peak** value is employed. In our illustration this would be twice the amplitude, or 20 volts.

If you observe the sine curve of alternating electromotive force or current, you will see that the true average value for a full cycle is zero, because there is just as much of the curve above the zero line (+) as there is below it (−). But when we use the **average values** in connection with alternating electromotive force or current, we do not refer to the average of a full cycle, but to the average of a half-cycle (or **alternation**, as it is also called).

It can be proven mathematically that the average value of a half-cycle of a sine curve is equal to 0.637 times the peak value. Thus:

Average emf (or current) = 0.637 × peak emf (or current).

If we wish to convert the average value to the peak value, we may multiply the average value by 1.57.

In practice, we generally use neither the peak nor average values of the electromotive force or current. To make alternating current compare as nearly as possible to direct current, it is necessary to use an **effective value.** In other words, we must find the value for the sine curve of alternating electromotive force or current that would have the same effect in producing **power** as a corresponding direct-current value. You will recall that the direct-current formulas for power are

$$P = I^2R \quad \text{and} \quad P = \frac{E^2}{R}.$$

From this relationship, you can see that the power is proportional to the square of the current (I^2) or to the square of the electromotive force (E^2). Thus, we must get the average (or **mean**) of the instantaneous values squared (instantaneous value × instantaneous value), and then calculate the square root of this average.

Because of the method used to determine the effective value, it is known as the **root-mean-square** (abbreviated **rms**) value. By means of mathematics it can be proven that the effective value is equal to 0.707 times the peak value and that the peak value is equal to 1.41 times the rms, or effective, value. Also to convert average values to rms values, we multiply the average value by 1.11.

QUESTIONS

1. Explain what is meant by an **induced electromotive force**; an **induced current.**

2. Explain the left-hand rule for determining the direction of induced current flowing in a conductor.

3. State and explain Lenz's law.

4. What are the three factors that determine the strength of an induced current?

5. Draw a diagram of a simple generator and explain its action.

6. Explain the reversal in direction of current flow as the armature coil of the generator rotates through its cycle.

7. What is meant by the **waveform** of an electric current?

8. Draw a graph showing the waveform of the alternating voltage produced as the armature coil rotates through one cycle. (Assume the maximum induced voltage is 20 volts.)

9. What is meant by a **cycle** of alternating current? What is meant by the **frequency** of alternating current?

10. What is meant by the **instantaneous values** of alternating current; by the **maximum values?** What is meant by the **amplitude?**

11. What is meant by the **average value** of alternating current? What is its relationship to the maximum value of the current?

12. What is meant by the **effective,** or **root-mean-square, value** of alternating current? What is its relationship to the maximum value of the current?

CHARACTERISTICS OF
ALTERNATING CURRENT

9

A. IMPEDANCE

If a steady direct current flows through a circuit, it is opposed by the resistance of that circuit. For example, if the steady direct current flows through a coil, only the resistance of the wires of that coil opposes the current flow. But if an alternating current flows through the coil, it is opposed by an additional factor. Since the alternating current is changing constantly, so is the magnetic field set up around the coil by that current. This changing magnetic field, constantly cutting across the turns of the coil, induces in it an electromotive force. From Lenz's law (see Chapter 8, subdivision A) we know that this electromotive force acts in a direction counter to the electromotive force that set the original current flowing through the coil. Thus the induced electromotive force is a **counter electromotive force** which opposes the flow of the original current.

Hence alternating current flowing through such a circuit encounters the opposition of both the resistance and the induced counter electromotive force. (As we shall learn later, factors other than the effect

135

of a coil can produce a counter electromotive force.) The total opposition to the flow of alternating current through a circuit is called the **impedance** of that circuit. Since it represents an opposition to the flow of current, it has the same unit of measurement as resistance — that is, the **ohm**. The symbol used to represent impedance is *Z*.

You will recall that when we were studying Ohm's law for direct-current circuits (see Chapter 4, subdivision D), we found the relationship between current (*I*), electromotive force (*E*), and resistance (*R*) expressed in the following equations:

$$I = \frac{E}{R}, \quad E = I \times R, \quad \text{and} \quad R = \frac{E}{I}.$$

If we now substitute impedance (*Z*) for resistance (*R*), our Ohm's-law equations apply equally well to alternating-current circuits. Thus:

$$I = \frac{E}{Z}, \quad E = I \times Z, \quad \text{and} \quad Z = \frac{E}{I},$$

where *I* is the current (in amperes), *E* is the electromotive force (in volts), and *Z* is the impedance (in ohms).

B. PHASE RELATIONSHIPS

Flow of current through a circuit is caused by the electrical pressure, or electromotive force. If this electromotive force is direct, the current, too, is direct. If the electromotive force is alternating, the resulting current is alternating as well.

If an alternating electromotive force, or voltage, of sinusoidal waveform is applied to a circuit, the natural tendency will be to cause to

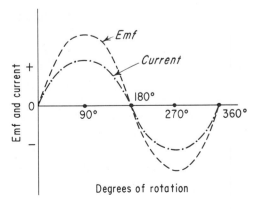

Fig. 9-1. Graph showing voltage and current in phase.

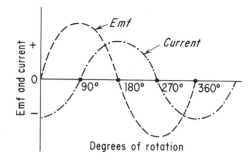

Fig. 9-2. Graph showing current lagging 90° behind voltage.

flow a current with a similar sinusoidal waveform. Current and voltage will reach zero together, rise and fall together, and attain peak values together. We say that the voltage and current are in step, or **in phase.** This relationship is illustrated in the graph shown in Figure 9-1. (Note that the relative sizes of the curves have no significance here. They are plotted for different units — the EMF curve in volts and the CURRENT curve in amperes. They are shown of different sizes primarily for the sake of clarity in presentation. What is important is the fact that they are in phase.)

In practical circuits, however, for reasons we will discuss later, the electromotive force and current may not be in step with each other. The current may either lag behind or lead the electromotive force. We then say that the electromotive force and the current are **out of phase** with each other.

In Figure 9-2 we see such a condition. Note that the EMF curve reaches its peak 90 degrees (90°) before the CURRENT curve and that it crosses the zero line 90° ahead of the current. We say the electromotive force is **leading** the current by 90°, or that the current is **lagging** 90° behind the electromotive force.

In Figure 9-3 the current is leading the electromotive force by 30°.

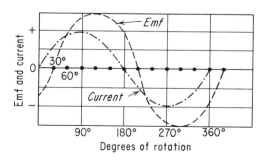

Fig. 9-3. Graph showing current leading voltage by 30°.

Another way of describing this difference in phase between electromotive force and current is to say that the **phase angle** is 30°. The electrical symbol for phase angle is the Greek letter θ **(theta).**

So far, we have been considering phase relationships between voltage and current coming from the same source. We may have phase differences between two or more currents (or voltages) coming from different sources.

In direct-current circuits (where there are no phase differences), if current is supplied from two or more sources, the resulting current (or voltage) is obtained merely by algebraic addition. Look at Figure 9-4.

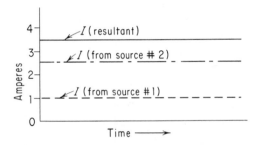

Fig. 9-4. Graph showing resultant of two direct currents.

Here, we have assumed that source #1 supplies a steady direct current of 1 ampere to the circuit. Source #2 supplies a similar current of 2.5 amperes. As a result, a steady direct current of 3.5 amperes flows through the circuit. (In Figure 9-4 we have assumed that the direction of current flow is the same from each source. Hence the two currents are added. If the directions of current flow are different for each source, we subtract the smaller current from the larger, and the direction of flow of the resultant is that of the larger.)

A similar effect is produced with alternating currents only when the currents from sources #1 and #2 are in step, or in phase, with each other. (See Figure 9-5.) The current from source #1 rises and falls in step with the current from source #2. The resultant current (which is found by the algebraic addition of the two) also is in phase.

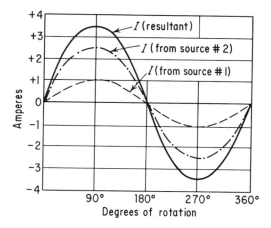

Fig. 9-5. Graph showing the resultant of adding two alternating currents in phase.

However, if the two currents are out of phase with each other, the result is different. Look at Figure 9-6. Here the two currents are out of phase, with the current from source #1 leading the current from source #2 by 90°. Not that the resultant current is out of phase with both of the others.

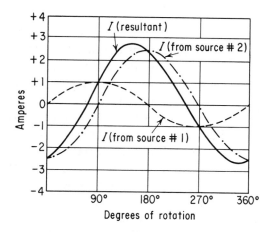

Fig. 9-6. Graph showing the resultant of adding two alternating currents 90° out of phase.

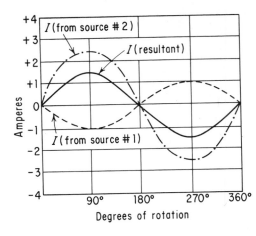

Fig. 9-7. Graph showing the resultant of adding two alternating currents 180° out of phase.

In Figure 9-7 you see the result when the currents are 180° out of phase with each other. The resultant current is obtained by subtracting one from the other (since one is always positive when the other is negative, and vice versa) and it is in phase with the larger current. But note that if the two currents were of equal value, the resultant would be zero since the positive and negative loops would cancel out. Of course, similar results are obtained when we add voltages from different sources.

C. POWER IN A-C CIRCUITS

In any circuit, the electrical power consumed at any instant equals the product of the voltage and current at that instant. The equation may be written as

$$p = e \times i,$$

where p is the instantaneous power (in watts), e is the instantaneous voltage (in volts), and i is the instantaneous current (in amperes). (It is common practice to use the small-letter equivalent of the capital letter to indicate an **instantaneous** value. Thus, whereas I stands for current, i stands for instantaneous current.)

The instantaneous power equation applies regardless of whether the current is direct or alternating. If the current is a steady direct current, the power, too, will be steady. But if the current is alternating, the power will vary from instant to instant with the changing current.

Let us examine the graph in Figure 9-8, which shows the relationship between voltage, current, and power in an alternating-current circuit. Note that the power curve is the result of the product of instantaneous values of voltage and current. Where both voltage and current are positive, the power, too, is positive since the product of two positive values is another positive value. Where both voltage and current are

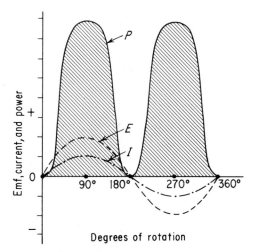

Fig. 9-8. Graph showing the relationship between voltage, current, and power when voltage and current are in phase. Shading is used to emphasize the power curve.

negative, the power, again, is positive since the product of two negative values is a positive value. Thus, except when it drops momentarily to zero, the power is always positive. That means that the source (which may be a generator) is constantly delivering electrical energy to the circuit.

Note, however, this situation exists if the current and voltage are in phase. Let us see what happens to the power in an out-of-phase circuit.

A graph depicting the relationship between current and voltage in an out-of-phase circuit was shown in Figure 9-2. By multiplying the instantaneous values of current and voltage we may obtain the power curve, as shown in Figure 9-9.

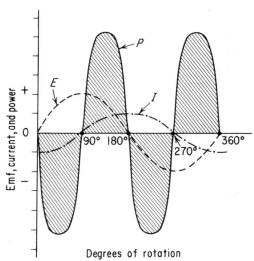

Fig. 9-9. Graph showing the relationship between voltage, current, and power when voltage and current are 90° out of phase.

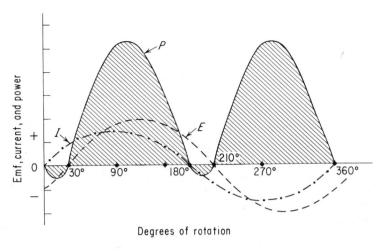

Fig. 9-10. Graph showing the relationship between voltage, current, and power when voltage and current are 30° out of phase.

For the first 90° of the cycle the voltage is positive and the current is negative. The product of a positive value and a negative value is a negative value. Accordingly, the resulting power is negative. That means that electrical energy is flowing from the circuit back to the source.

During the next 90° of the cycle both current and voltage are positive. Accordingly, the power, which is the product of the two, also is positive, and energy is flowing from the source into the circuit.

For the next 90° of the cycle the current is positive and the voltage is negative. The power is negative and energy again flows from the circuit back to the source. During the last 90° of the cycle, both current and voltage are negative. The power, thus, is positive and energy flows once more from the source into the circuit.

Examination of the graph shows that the positive power is equal and opposite to the negative power. As a result, they cancel out and the net result is zero. That is, the circuit consumes, or dissipates, no power. The electrical energy merely flows from the source into the circuit and back again. Note, however, that this condition holds only when the phase angle is 90°.

Now, what if the phase angle is some other value — such as, for example, 30°, as shown by the graph of Figure 9-3? Examine Figure 9-10.

Note that the power is negative only during the intervals between 0° and 30° and between 180° and 210°. During the rest of the cycle the power is positive. Thus, the average power is not zero, as in the case of a circuit where the phase angle is 90°, but some positive value. However, it is less than the average power dissipated by a comparable circuit where the current and voltage are in phase.

1. Power Factor

The instantaneous power consumed in any circuit, you will recall, is the product of the instantaneous voltage (e) and the instantaneous current (i). In alternating-current circuits, however, we generally deal with effective, or rms, values. As such, the $P = E \times I$ equation for power holds only when the current and voltage are in phase. (See Figure 9-8.) Under such conditions electrical energy flows only from the source to the circuit.

However, when the current and voltage are out of phase there are intervals when electrical energy is flowing back to the source from the circuit. (See Figures 9-9 and 9-10.) Accordingly, the **true power** consumed by the circuit is less than the **apparent power** calculated by multiplying the effective voltage (E) by the effective current (I).

The ratio of the true power to the apparent power is called the **power factor.** Thus:

$$\text{Power factor} = \frac{\text{true power}}{\text{apparent power}}.$$

This ratio may be expressed either as a fraction or as a percentage. We may say that the power factor is, for example, one-half, or 50 percent.

The power factor is a function of the phase angle. Where the current and voltage are in phase (phase angle = zero), the apparent power is equal to the true power. The power factor in this case is **unity** (that is, one) or 100 percent. Where the current and voltage are out of phase, the power factor is some lesser value, depending upon the phase angle.

The power factor should be stated as **leading** or **lagging.** This refers to the current with respect to the voltage. If, when discussing the characteristics of an electrical circuit, we say the power factor is 65 percent lagging, we mean that the current is lagging behind the voltage in the circuit.

The apparent power can be calculated by using a voltmeter to measure the rms voltage and an ammeter to measure the rms current. Multiplying the two values thus obtained gives us the apparent power.

The true power, on the other hand, can be obtained directly by measuring it with a wattmeter. To differentiate between the two, we generally measure the apparent power in units of **volt-amperes.** The true power we measure in units of **watts.**

What is the significance of the power factor? Well, consider an electric motor being run from an a-c line. Generally, we wish the motor to consume its full rated electrical power, because only in this way can we get the full rated mechanical power from the motor. If, during a portion of the cycle, electrical energy is being fed back to the line (power factor is less than 100 percent, or unity), the motor is not consuming its proper share of power and, hence, is not operating at full efficiency. Accordingly, we wish the power factor of the circuit to be as near 100 percent as possible.

Later in the book we shall consider methods for changing the phase angle between current and voltage in a circuit and, in this way, the power factor.

D. VECTORS

The magnitude and direction of factors such as force, pressure, and so on may be shown graphically by means of an arrow called a **vector.** The direction of the force is shown by the arrowhead. The magnitude of the force is indicated by the length of the arrow, choosing any convenient scale. Thus, a vector one inch long may be taken to represent, say, a force of 10 pounds. Then, using the same scale, a vector two inches long would represent 20 pounds.

When we wish to add the electromotive forces (or currents) from two direct-current sources, such as batteries, the process is a simple problem of vectorial addition or subtraction, as illustrated in Figure 9-11. In Figure 9-11A we have assumed that the two batteries have been connected so that their electromotive forces reinforce each other (that is, the direction of electromotive force is the same for each). The electromotive force of the first battery is represented by the vector E_1, using some suitable scale. The electromotive force of the second battery is represented by the vector E_2, using the same scale. The resultant vector ($E_{\text{Resultant}}$), then, is the sum of the two (drawn to the same scale).

In Figure 9-11B we have assumed that the two electromotive forces oppose each other, as indicated by the opposing directions of E_1 and E_2. The resultant, then, is the difference between the two.

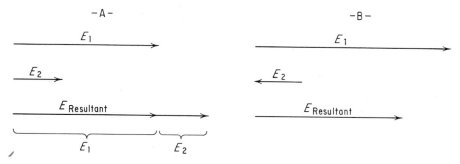

E_1

E_2

E Resultant

E_1 E_2

E_1

E_2

E Resultant

Fig. 9-11. A. Vectorial diagram showing two direct electromotive forces reinforcing each other.

      B. Vectorial diagram showing two direct electromotive forces opposing each other.

But when we wish to add together two alternating electromotive forces (or currents), we have a different problem. The electromotive forces (or currents) may not be in phase. For example, suppose we wish to add together two alternating electromotive forces with one (E_1) leading the other (E_2) by 30°.

The vectorial solution of this problem is illustrated in Figure 9-12. Using any convenient scale, draw a horizontal line (OB) to represent the vector for E_2. From point O and at an angular distance of 30° in a counterclockwise direction from OB, draw vector OA to represent E_1, using the same scale. Thus we have drawn the vectorial diagram of E_1 leading E_2 by 30°.

From point A draw a line parallel to OB and from point B draw a line parallel to OA. These lines intersect in point C. Line OC represents the vector for the resulting electromotive force (measured on the same scale as the other two electromotive forces). $E_{\text{Resultant}}$ leads E_2 by the angular distance between OC and OB, but lags behind E_1 by the angular distance between OC and OA. We follow the same procedure in adding together two alternating currents.

(Since vectors representing alternating voltages and currents also show phase relationships, they sometimes are called **phasors**. However,

Fig. 9-12. Vectorial diagram showing the addition of two alternating voltages 30° apart in phase.

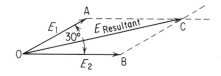

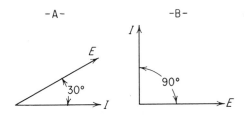

for the sake of simplicity, the term "vector" will be used throughout this book to indicate values of both direct and alternating voltages and currents.)

We can also show the phase relationship between the electromotive force and current from the same source by means of a vectorial diagram. Assume we wish to show the voltage leading the current by 30°. Draw a horizontal vector for the current, using any convenient scale. From the tail end of the current vector, and at the proper angular distance (30°), draw the voltage vector, as shown in Figure 9-13A. Since we are dealing here with two different factors, we need not use the same scale for the voltage vector.

Note that the diagram shows the voltage leading the current by 30°. Should we wish to show the voltage lagging, say, 90° behind the current, the vector diagram would appear as in Figure 9-13B.

E. WAVEFORMS

Basically, there are two types of electric current (and electromotive force, or voltage, which causes the current to flow). **Direct current** flows through its circuit only in one direction. **Alternating current** alternately flows first in one direction and then in the opposite direction. The **waveform** is a graphical representation showing the various characteristics, such as direction of electrical pressure or current flow, amplitude, frequency, and so on, of a voltage or current.

Various waveforms of direct current (or voltage) are illustrated in Figure 9-14. The waveform shown in Figure 9-14A is that of a **steady direct current** (or **voltage**). Since the current is direct, it flows through its circuit only in one direction. This is indicated by the fact that the line representing the current flow (called the **curve** of the graph) lies wholly in one half of the graph. The upper half of the graph is the positive (+) half. Therefore, the waveform shown here is of a **positive** direct current. That it is a **steady** direct current is indicated by the fact

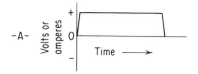

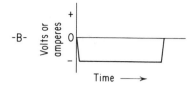

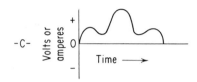

Fig. 9-14. Various d-c waveforms.
  A. Steady positive d-c waveform.
  B. Steady negative d-c waveform.
  C. Varying, or fluctuating, positive d-c waveform.

that its **amplitude,** as shown by the distance of the curve from the zero line, is constant (except for the instant of rise from zero to the steady state and the instant of fall from the steady state back to zero).

In Figure 9-14B the curve lies in the negative (−) portion of the graph. Hence the waveform of a steady, **negative** direct current is illustrated.

In Figure 9-14C the curve lies wholly in the upper half of the graph. Hence it is the waveform of a positive direct current. The amplitude, however, is not steady but varies continuously. Hence this is the waveform of a varying, or **fluctuating**, positive direct current (or voltage).

The waveform of a **symmetrical, sinusoidal alternating current** (or **voltage**) is illustrated in Figure 9-15A. That it is **alternating** is indicated by the fact that the curve lies in both halves of the graph. It is called **sinusoidal** because the variations of the curve resemble **sine curves.**

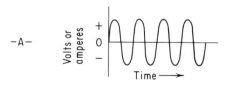

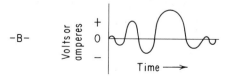

Fig. 9-15. Various a-c waveforms.
  A. Regular, sinusoidal a-c
   waveform.
  B. Irregular, nonsinusoidal
   a-c waveform.

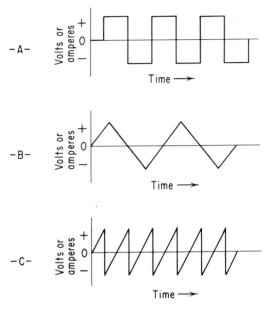

Fig. 9-16. Several types of nonsinusoidal waveforms.
A. Square or rectangular waveform.
B. Triangular waveform.
C. Sawtooth waveform.

It is **symmetrical** because the positive and negative alternations of each cycle are equal and opposite. Note that the **frequency** (that is, the number of cycles per second) and the **amplitude** (the distance of the peaks from the zero line) are constant.

If several symmetrical, sinusoidal alternating currents (or voltages) of different frequencies, amplitudes, or phases are added together, the result will be a current (or voltage) whose waveform is **irregular** and **nonsinusoidal**. An example of such a waveform is illustrated in Figure 9-15B.

The a-c waveform also may be nonsinusoidal and symmetrical. Three common types of such waveforms are illustrated in Figure 9-16. In Figure 9-16A is shown the **square**, or **rectangular**, **waveform**. The **triangular waveform** is shown in Figure 9-16B, and the **sawtooth waveform** is illustrated in Figure 9-16C.

Some interesting results may be obtained if a steady direct current (or voltage) is added to an alternating current (or voltage). In Figure 9-17A is shown the resultant of the graphical addition of a steady direct current and a symmetrical, sinusoidal alternating current, where the amplitude of the direct current is greater than that of the alternating current.

That the resultant current is direct is indicated by the fact that its waveform lies entirely in one half of the graph, But, since its amplitude varies continuously, it is a fluctuating direct current.

If the amplitude of the alternating current is greater than that of the direct current (see Figure 9-17B), the resultant is an alternating current. This is indicated by the fact that the waveform lies in both the positive and negative halves of the graph. But note that the two halves of the cycle are no longer equal in amplitude; hence the resultant is a non-symmetrical alternating current.

It is clear, then, that a fluctuating direct current or a nonsymmetrical alternating current contains both steady d-c and a-c components. (Where the waveform is symmetrical, the alternate half-cycles are equal in amplitude and opposite in direction. Hence they cancel out, leaving no d-c component.)

Note that the shape of the waveform of the resultant current (or voltage) is determined by the waveform of the a-c component. The steady d-c component merely affects the vertical positioning of the

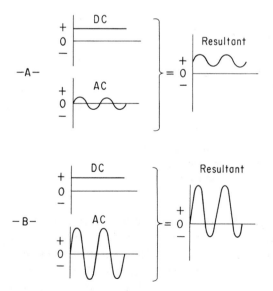

Fig. 9-17. A. How a steady direct current and a symmetrical alternating current are added to produce a fluctuating direct current.
 B. How a steady direct current and a symmetrical alternating current are added to produce a nonsymmetrical alternating current.

resultant curve of the graph. If the steady d-c component is more positive, the curve is moved higher up the graph. If the steady d-c component is more negative (or, what is the same thing, less positive), the curve is moved lower down the graph.

Similar results may be obtained from the addition of steady d-c and nonsinuosidal a-c components. Here, too, the shape of the waveform of the resultant current (or voltage) is determined by the waveform of the a-c component, and the vertical positioning of the curve of the graph is affected by the steady d-c component.

Under certain circumstances, which will be considered later in the book, composite currents (or voltages) may be separated into their various components.

The alternating currents (or voltages) we have discussed so far are all **continuous** — that is, a cycle starts where the previous one has left off and is repeated over and over. However, the currents (or voltages) need not be continuous. In Figure 9-18 are shown the waveforms of several types of bursts, or **pulses**, of current or voltage. These waveforms may be sinusoidal, spiked, triangular, or rectangular, as illustrated. Or they may be composites of two or more waveforms.

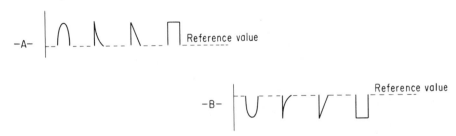

Fig. 9-18. Waveforms of various types of pulses.
    A. Positive-going pulses.
    B. Negative-going pulses.

Note in Figure 9-18A that the pulse starts at some reference value of current (or voltage), rises to a peak value, and falls back to its original reference value. Such a pulse is said to be **positive-going**. In Figure 9-18B the pulse starts at its reference value, falls to a negative (or less positive) peak value, and rises to its original reference value. Such a pulse is said to be **negative-going**.

The pulses may be **repetitive**, recurring at regular intervals of time, or they may be **transient**, appearing at irregular intervals.

QUESTIONS

1. What is meant when we say that an alternating current and voltage are **in phase?** Draw a graph illustrating this condition.

2. What is meant when we say that an alternating current and voltage are **out of phase?** Draw a graph showing the current **leading** the voltage by 90 degrees.

3. What is meant by **phase angle?**

4. Draw a graph showing the resultant of the addition of two steady direct currents, one 5 amperes and the other 6 amperes.

5. Draw a graph showing the resultant of the addition of two alternating currents, one whose maximum value is 2 amperes and the other with a maximum value of 3 amperes, that are in phase with each other.

6. Draw the graph showing the resultant of the same two currents as in question 5, if the phase angle between them is 90 degrees.

7. (*a*) What is a **vector?** (*b*) What two conditions does it describe?

8. Draw the vector diagram showing the resultant of the addition of two steady direct voltages, one +3 volts and the other −5 volts.

9. Draw the vector diagram showing the resultant of the addition of an alternating current whose maximum value is 3 amperes leading by 45° another alternating current whose maximum value is 5 amperes.

10. In any a-c circuit, what must be the phase angle between the voltage and current for the power to be (*a*) at its maximum; (*b*) at its minimum?
[*Ans*. (*a*) zero; (*b*) 90°.]

11. Draw a graph showing the power in an a-c circuit where a maximum voltage of 5 volts leads by 60 degrees a maximum current of 2 amperes.

12. What is meant by the **power factor** of a circuit?

13. What is the power factor of a circuit where the voltage and current are in phase? [*Ans.* 100 percent.]

14. In a certain circuit it was known that the current was lagging behind the voltage. When measured by means of a voltmeter and ammeter, the effective voltage was found to be 100 volts and the effective current to be 5 amperes. When measured with a wattmeter, the power was found to be 400 watts. What is the power factor of the circuit? [*Ans.* 80 percent.]

15. What is meant by **impedance?** State Ohm's law for a-c circuits.

16. Draw the waveforms of: (*a*) a steady direct current; a fluctuating direct current; (*b*) a regular, sinusoidal alternating current; an irregular, non-sinusoidal alternating current.

17. Draw the waveform of: (*a*) a continuous sawtooth a-c current; (*b*) a series of positive-going rectangular pulses.

FACTORS AFFECTING ALTERNATING CURRENT

10

A. RESISTANCE

Resistance, we have learned, opposes the flow of current through a circuit. This is true whether the current is direct or alternating. If the circuit contains nothing but resistance, Ohm's law for direct current ($I = E/R$) applies equally well for alternating current, and the impedance (Z) is equal to the resistance (R). However, when dealing with alternating-current circuits we encounter instantaneous, peak, average, and rms values. Accordingly, we must be careful to employ similar values for both current and voltage in each case.

Furthermore, resistance does not affect the phase relationship between current and voltage in an a-c circuit. That is, the current and voltage remain in phase. Since the voltage and current are in phase, the power factor of the circuit is unity, or 100 percent, and the apparent power is equal to the true power.

B. INDUCTANCE

In discussing induced currents (Chapter 8, subdivision A), we have seen that if a coil is cut by a magnetic field, an induced voltage is developed in the coil. It makes no difference how the magnetic field is produced. It may be produced by a permanent magnet, as illustrated in Figure 8-1. Or else, it may be produced by current set flowing through the turns of the coil by some voltage source. As the current starts flowing through it, a magnetic field is built up around the coil. As this field is built up, it cuts across the turns of the coil, inducing therein an electromotive force. When the current reaches a steady value, the magnetic field, too, becomes steady, and the induced electromotive force drops to zero. When the current drops, the magnetic field collapses, cutting across the coil again, and the induced electromotive force comes into being once more.

We have also learned, from Lenz's law, that the induced electromotive force and the induced current it sets flowing are always in such a direction as to oppose any change in the existing magnetic field. Thus, if the original (source) current is increasing and is causing the magnetic field around the coil to expand, the induced electromotive force sets the induced current flowing in such a direction as to build up a magnetic field in opposition to the one set up by the source current. If the source current is decreasing and causing the magnetic field to collapse, the induced electromotive force sets the induced current flowing in such a direction as to build up a magnetic field that aids the original magnetic field and thus tends to prevent its collapse. (See Figure 10-1.) Because it acts in opposition to any change in the source current and the source voltage that causes this current to flow, the induced electromotive force is called a **counter voltage,** or **counter electromotive force** (abbreviated **cemf.**)

The property of a circuit to oppose any change in the current flowing through it is called **inductance.** This property is due to the voltages induced in the circuit itself by the changing magnetic field. Components of the circuit that produce inductance, such as the coil we have been discussing, are called **inductors** or **reactors.** As we have learned, the induced voltage depends upon the strength of the magnetic field set up around the coil, the number of its turns, and the speed with which the changing magnetic field cuts across these turns. Hence the inductance of the inductor depends upon these factors.

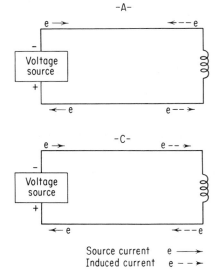

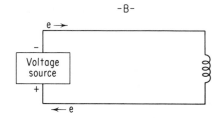

Fig. 10-1. Flow of current in a circuit containing an inductor.

A. Source current is increasing. Induced current opposes source current.

B. Source current is steady. There is no induced current.

C. Source current is decreasing. Induced current aids source current.

Anything that affects the magnetic field also affects the inductance of the inductor. For example, increasing the number of turns of the inductor increases its inductance. Similarly, substituting an iron core for an air core also increases its inductance.

The speed with which the magnetic field around an inductor is changing also affects its inductance. Thus, if a steady direct current flows through a circuit, there is no inductance except for those instants when the circuit is either completed or broken. However, when alternating currents are flowing in a circuit, the current is constantly changing and inductance becomes a factor.

The faster the current changes (that is, the higher its **frequency**), the larger the inductance will be. Since even a straight wire has a surrounding magnetic field when current is flowing through it, the wire has some inductance. However, unless the frequency of the current is very high or the wire extremely long, the inductance of a straight wire is small enough to be neglected.

The symbol for inductance, when used in electrical formulas, is L. Its unit is the **henry** (abbreviated **h**) which is the inductance of a circuit or component that will produce an induced electromotive force of one volt with a change of one ampere per second in the current flowing through it. (Since current variations seldom are at a uniform rate, a circuit or component has an inductance of one henry if it develops an **average** induced electromotive force of one volt as the current changes at an

average rate of one ampere per second.) Where the henry is too large a unit, we may use the **millihenry (mh)** which is one-thousandth (10^{-3}) of a henry, or the **microhenry (μh)** which is one-millionth (10^{-6}) of a henry.

Inductors employed for high-frequency applications generally have no cores (that is, are air-cored) or else have one made of powdered iron. Those employed for low-frequency operation have cores made of stacks of sheet iron. The symbol for the air-core inductor is ⌇⌇⌇. If the inductor has a core of powdered iron its symbol becomes ⌇⌇⌇. If its core is made of stacks of sheet iron its symbol becomes ⌇⌇⌇. (See Figure 10-2.)

Inductors, like resistors, may be connected in series, in parallel, or in series-parallel circuits. The total inductance of several inductors connected in series (provided the magnetic field of one inductor cannot act upon the turns of another) is equal to the sum of the inductances of the individual inductors. Thus:

$$L_{\text{total}} = L_1 + L_2 + L_3 +, \text{ and so forth.}$$

If two or more inductors are connected in parallel (again, providing

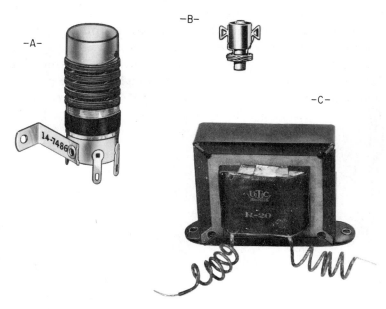

Fig. 10-2. A. Air-core inductor.
      B. Powdered iron-core inductor.
      C. Sheet iron-core inductor.

there is no interaction, or **coupling,** of their magnetic fields) we can find the total inductance from the following formula:

$$\frac{1}{L_{\text{total}}} = \frac{1}{L_1} + \frac{1}{L_2} + \frac{1}{L_3} +, \text{ and so forth.}$$

Note the similarity to the formulas for resistors connected in series and parallel.

As in the case of resistors, the total inductance of inductors connected in a series-parallel circuit (if there is no coupling of magnetic fields) may be obtained by first finding the joint inductance of the inductors in parallel and then adding this inductance to the inductances in series with it as though it were a straight series-inductor circuit.

1. Effect of Inductance on Phase Relationship

In an a-c circuit containing only resistance the voltage and current constantly remain in phase. This relationship is shown graphically in Figure 10-3A. (The relative amplitudes of the voltage and current curves have no significance here.) Now let us consider an a-c circuit that contains nothing but inductance. (Such a circuit is only theoretically possible, since all circuits contain some resistance.)

As you know, the induced electromotive force always opposes the change in the source current. When this current is rising, the induced electromotive force tends to keep the value of current less than the source voltage alone would do. Accordingly, the rise in the source current takes place later than does the rise in source voltage. When the current is decreasing, the induced electromotive force tends to oppose

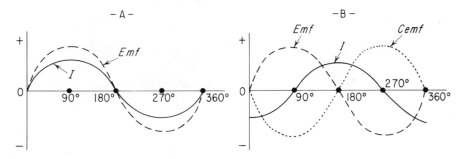

Fig. 10-3. A. Phase relationship between current and voltage in a circuit containing only resistance.

B. Phase relationship in a circuit containing only inductance.

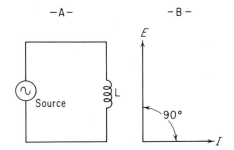

Fig. 10-4. A. A-C circuit containing only inductance.
B. Vector diagram showing the phase relationship between current (*I*) and voltage (*E*) in this circuit.

the decrease. Hence, the fall of source current, too, takes place later than does the fall in source voltage. Thus the source current lags behind the source voltage throughout the entire cycle.

The induced electromotive force (**cemf**) is greatest when the current is changing at its fastest rate. If you examine Figure 10-3B, you will see that this occurs as the current passes through zero at the 90° position. Since the source voltage in a circuit containing only inductance is equal and opposite to the induced electromotive force, the source voltage is at its positive peak at the same 90° position.

The induced electromotive force drops to zero as the current reaches its lowest rate of change. This occurs at the 180° position where the current reaches its positive peak. Since the source voltage is equal and opposite to the induced electromotive force, it, too, is zero at that point. Thus, you can see that, in an a-c circuit containing only inductance, the current lags behind the source voltage by 90° or, what is the same thing, the source voltage leads the current by 90°. In practical circuits, because of the presence of resistance, the current lag is some value less than 90° behind the voltage.

We have already learned that the use of vectors furnishes us with a convenient means for picturing the relationships between currents and voltages. Thus, if an alternating-current circuit has, theoretically, nothing but inductance in it, the vector diagram appears as in Figure 10-4B.

In this diagram the length of the voltage vector (*E*) is independent of the length of the current vector (*I*), and the length of each vector depends upon its own scale. Further, because of the inductance, the voltage leads the current (we always read vector diagrams in a counterclockwise direction). The phase angle is 90°.

If a wattmeter, which measures **true** power consumed by a circuit, is connected in a circuit containing only inductance, the meter will register zero. You can verify this by looking at Figure 9-9, which shows the true power in a circuit wherein the current lags behind the voltage by 90°. Note that the positive and negative loops are equal and, hence, cancel out. In such a circuit the electrical energy is merely transformed into magnetic energy in the inductor and back to electrical energy again. Thus no power is consumed. Since

$$\text{Power factor} = \frac{\text{true power}}{\text{apparent power}}$$

and the true power is zero, the power factor, too, must be zero.

2. Mutual Inductance

We have seen that when a changing current flows through a circuit that possesses the property of inductance, an induced voltage is generated in that circuit which opposes the changes in current. The ability of the circuit to act in this manner is called **self inductance.** Since the counter electromotive force developed across an inductor depends, in part, upon the speed with which the magnetic field around it changes (that is, the speed of changes in the current), if the rate of change is great, the counter electromotive force, too, may be large. In fact, if the rate of change is great enough, the counter electromotive force so developed may be many times as large as the voltage of the source driving the current through the circuit. This is the principle upon which the ignition coil of the automobile operates, producing the high-voltage spark used to ignite the gasoline mixture in the cylinders.

We have seen, too, that when two or more inductors are connected in series in such a way that their magnetic fields do not interact, the total inductance is the sum of the inductances of all the inductors. That is to say, the total inductance is the sum of the self inductances of all the inductors.

If, however, two inductors are placed close to each other, their magnetic fields will interact and the magnetic field of one may induce a voltage in the other as it cuts across the turns of wire. We say, then, that the inductors are **coupled.** Each inductor has its own self inductance, but, in addition, there is a further inductance due to the induced voltage produced by coupling between the inductors. We call this further inductance **mutual inductance.** We say the two coils are coupled together by mutual inductance. The terms **magnetic,** or **inductive, coupling** are sometimes used.

Mutual inductance, whose electrical symbol is M, is measured in the same unit as self inductance, the **henry.** When a change of one ampere

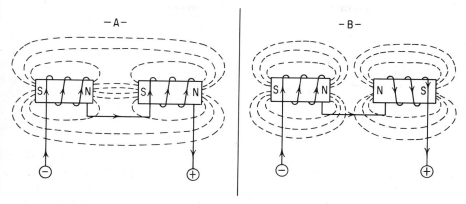

Fig. 10-5. Two inductors connected in series.
　　　　A. Magnetic fields aid each other.
　　　　B. Magnetic fields oppose each other.

per second in one inductor induces one volt in the other, the two inductors have a mutual inductance of one henry.

Inductors can be series-connected in two ways. Figure 10-5A shows the inductors connected so that the two magnetic fields aid each other. The effect of mutual inductance is to increase the total inductance, and our formula becomes

$$L_{total} = L_1 + L_2 + 2M.$$

The two inductors may also be connected in series in such a way that the magnetic fields oppose each other. Figure 10-5B shows this circuit. The effect of the mutual inductance is to decrease the total inductance, and our formula becomes

$$L_{total} = L_1 + L_2 - 2M.$$

A similar relationship holds true for two inductors connected in parallel. If the two magnetic fields aid each other, the formula for the total inductance becomes

$$\frac{1}{L_{total}} = \frac{1}{L_1 + M} + \frac{1}{L_2 + M}.$$

Where the magnetic fields oppose each other, the formula becomes

$$\frac{1}{L_{total}} = \frac{1}{L_1 - M} + \frac{1}{L_2 - M}.$$

3. Inductive Reactance

In Chapter 9, subdivision A, we learned that the impedance of an a-c circuit is the total opposition that circuit offers to the flow of current. Where only pure resistance is present in the circuit, the impedance is equal to the resistance. But we have seen that the presence of an inductor in the circuit causes a counter electromotive force to be built up which further opposes the flow of current. Under such conditions the impedance of the circuit is greater than the resistance.

The factor which, in an a-c circuit, causes the impedance (Z) to be larger than the resistance (R) is called the **reactance** (X). Since this reactance is due to the presence of inductance, we call it the **inductive reactance**. To show that it is inductive reactance, we add the subscript L to the symbol for reactance (X) and we now get X_L as the symbol for inductive reactance.

[This method of adding a subscript to identify an electrical value is commonly used. Thus, the current (I) flowing through the inductor is shown as I_L. The voltage (E) across the inductor becomes E_L. This type of notation is not restricted to inductors. Thus, for example, the voltage drop across a resistance may be designated as E_R and the current flowing through it as I_R.]

Since impedance represents an opposition to the flow of current, it has the ohm as its unit. The inductive reactance, which increases the impedance, also has the ohm for its unit.

The inductive reactance depends upon the magnitude of the induced voltage. This voltage, in turn, depends upon two factors: the inductance of the circuit (L) and the rate or frequency (f) at which the current (and, therefore, the magnetic field) is changing.

The formula for inductive reactance is

$$X_L = 2\pi fL,$$

where X_L is the inductive reactance in ohms, f is the frequency in cycles per second, and L is the inductance in henrys. The factor 2π is necessary to make the result come out in ohms. Since π is equal, approximately, to 3.14, 2π therefore equals 6.28.

EXAMPLE. What is the inductive reactance of a coil of 2 henrys as a 60-Hz alternating current flows through it?

$$X_L = 2\pi fL = 6.28 \times 60 \text{ Hz} \times 2 \text{ henrys}$$
ANSWER. $= 753.6$ ohms.

If we assume a theoretical circuit that has only inductance, we may substitute the inductive reactance (X_L) for the impedance (Z) in the formulas that state Ohm's law for a-c circuits. These formulas are expressed as follows:

$$I = \frac{E}{Z}, \quad E = I \times Z, \quad Z = \frac{E}{I}.$$

Substituting inductive reactance for impedance, we get for a theoretical circuit with inductance only,

$$I = \frac{E}{X_L}, \quad E = I \times X_L, \quad X_L = \frac{E}{I}.$$

C. CAPACITANCE

The outermost electrons of a conductor, we have learned, are loosely held and easily removed. Insulators, or **dielectrics,** on the other hand, have their electrons more firmly fixed. If a dielectric is placed between two conductors (see Figure 10-6A), a **capacitor** is formed. (Formerly, a capacitor was called a **condenser.** However, since the latter term is misleading, the term **capacitor** is preferred.)

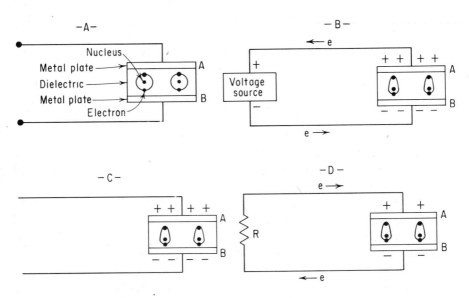

Fig. 10-6. A. Uncharged capacitor.
       B. Charging the capacitor.
       C. Voltage source removed, capacitor retains its charge.
       D. Discharging the capacitor.

Normally, the electrons of the atoms of the dielectric revolve around their nuclei in their regular orbits (indicated as circular), as shown in Figure 10-6A. If, however, the capacitor is connected to a voltage source, as in Figure 10-6B, electrons will flow onto plate B and away from plate A. Thus plate B will receive a negative charge and plate A a positive charge. This is called **charging the capacitor.**

In spite of these charges, electrons cannot flow through the dielectric. However, its electrons will be attracted toward the positive plate A and their orbits will be distorted, as indicated in Figure 10-6B. If the voltage source now is removed (Figure 10-6C), the charges will remain on the plates. Accordingly, the electrons of the dielectric will retain their distorted positions. The capacitor remains in its charged condition.

If, now, a resistor (R) is placed across the capacitor, it will start to **discharge** through this resistor (Figure 10-6D). Electrons will move from plate B around the circuit toward plate A until they are distributed equally over the entire circuit. As the charges are removed from the plates of the capacitor, the electrons of the dielectric will gradually resume their normal unstrained positions.

Note what happened. As the capacitor was charged, electrical energy from the voltage source was stored in the electrostatic field across the dielectric. As the capacitor discharged, the energy of this electrostatic field furnished the electromotive force that sent current flowing through the circuit. The property of a capacitor (or a circuit) to store electrical energy in this way is called **capacitance,** the symbol for which is *C*.

(Note that we have mentioned the capacitance of a circuit as well as that of a capacitor. Any two conductors separated by a dielectric may form a capacitor. Thus two of the connecting wires separated by air may constitute a capacitor. Even two adjacent turns of a wire coil, separated by their insulation, exhibit the property of capacitance. Such capacitance is called **stray,** or **distributed,** capacitance in contrast to the **concentrated** capacitance of a capacitor.)

The unit of capacitance is the **farad,** abbreviated **f.** The farad can be defined as being the capacitance present when one coulomb of electrical energy is stored in the electrostatic field of the capacitor or circuit as one volt is applied. The farad generally is too large a unit for ordinary purposes. Accordingly, we have the **microfarad** (abbreviated μ**f**) which is one-millionth (10^{-6}) of a farad. Where even the microfarad is too large a unit, we may use the **picofarad** (**pf**) which is one-millionth of a microfarad, or 10^{-12} farad. (Previously, the term **micromicrofarad,** abbreviated $\mu\mu$**f,** had been used instead of the picofarad.)

So far, we have been discussing capacitance in direct-current terms. When considering a-c circuits we must take a somewhat different point of view. In Figure 10-6B, as the capacitor was charged, electrons flowed from the voltage source toward plate B, placing a negative charge on that plate, and from plate A toward the voltage source, placing a positive charge on that plate. Because of these charges, a counter electromotive force was set up that was opposed to the voltage from the source. Electrons, then, would continue to flow until the counter electromotive force of the capacitor equaled the voltage of the source.

If, for any reason, the voltage of the source should now rise, more electrons would flow from the source to the capacitor. If the source voltage should fall below that of the counter electromotive force, electrons would flow back from the capacitor to the source. Thus, if the source voltage were changing, electrons would flow back or forth, depending upon which voltage were higher. (Remember that in a-c circuits the voltage is changing constantly.)

If the capacitor had a capacitance of one farad, a change of one volt in the source would cause the charge on the plates to increase or decrease by one coulomb. That is, it would cause one coulomb to flow through the circuit. Should the voltage change by one volt per second, it would cause one coulomb per second to flow through the circuit. Since one coulomb per second is equal to one ampere, the change of one volt per second would cause one ampere to flow through the circuit. We now can say that a capacitor (or circuit) has a capacitance of one farad if one ampere of current flows through the circuit when the applied voltage changes at the rate of one volt per second.

If a capacitor is placed in series with a source of steady direct current, current flows for an instant until the capacitor is sufficiently charged to develop a counter electromotive force equal to the voltage of the source. Then there would be no more current flow in the circuit.

If the d-c source is replaced by an a-c one, a different situation arises. As the voltage rises from zero to its maximum positive value, current flows from the source to the capacitor, building up a counter electromotive force across its plates. As the source voltage starts decreasing, this counter electromotive force sends current flowing back from the capacitor to the source.

The source voltage next reverses its direction and rises to its maximum negative value. Now the current set flowing by the counter electromotive force and the current from the source are flowing in the same direction. As a result, the currents flow through the circuit to the capacitor, building up a new counter electromotive force across its plates (though opposite in direction to the original counter electromotive force). As the source voltage decreases from its negative peak value, the counter electromotive force sends current from the capacitor back to the source. Then the entire cycle is repeated.

Note that during the entire cycle current is flowing through the circuit (excepting the dielectric of the capacitor). If a lamp, or any other device, were placed in series in this circuit it would indicate a continuous current flow through it. It is for this reason that we say that current can flow through an a-c series circuit containing a capacitor.

1. Effect of Capacitance on Phase Relationship

We have seen that the effect of inductance on an a-c circuit is to make the current lag 90° behind the voltage. Now let us see the effect of capacitance on such a circuit. Assume we have a circuit containing nothing but capacitance (again, this is only theoretical). We know that current will flow from the source to the capacitor only while the source voltage is rising, and that the greatest current will be flowing when the voltage is rising most rapidly. If we examine Figure 10-7, we will see that the source voltage is rising most rapidly at the 0° position (and again at the 180° position, though in a negative direction). Thus the current flow at the 0° position is at its maximum positive value.

From the 0° to the 90° position the source voltage continues to rise at an ever-decreasing rate. At 90° this rise has come to a halt. Hence, the current gradually falls from its maximum positive value to zero at the 90° point. Meanwhile the capacitor has become fully charged and the counter electromotive force (cemf) has reached its maximum value.

Now the source voltage begins to fall. Since the counter electromotive force finds itself greater than the source voltage, current starts flowing from the capacitor to the source, as indicated by the negative-current loop of Figure 10-7. At the 180° point the source voltage falls to zero

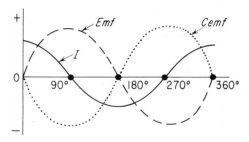

Fig. 10-7. Phase relationship between current and voltage in a circuit containing only capacitance.

again and the current reaches its maximum negative value. The capacitor, however, has become discharged and the counter electromotive force falls to zero.

The source voltage now changes its direction and continues to rise, at an ever-decreasing rate, and the current starts to decrease from its maximum negative value towards zero. Meanwhile, the capacitor becomes charged again (this time, in the opposite direction to the original charge) and the counter electromotive force rises once more.

At the 270° mark the rise of the source voltage ceases and the current reaches zero. As the source voltage starts to fall, the counter electromotive force starts the current flowing in the opposite (positive) direction until this current reaches its positive peak at the 360° mark. Then the entire cycle starts over again.

If you examine Figure 10-7, you will see, then, that the effect of capacitance is to make the current **lead** the source voltage by 90° or, what is the same thing, to make the source voltage lag behind the current by 90°. In practical circuits, because of the presence of resistance, the current lead is some value less than 90° ahead of the voltage.

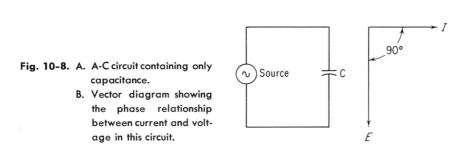

–A–   –B–

Fig. 10-8. A. A-C circuit containing only capacitance.
B. Vector diagram showing the phase relationship between current and voltage in this circuit.

The vector diagram for an a-c circuit that, theoretically, contains only capacitance is illustrated in Figure 10-8B. As in the case of the vector diagram shown in Figure 10-4B, the length of the voltage vector (*E*) is independent of the length of the current vector (*I*) and the length of each vector depends upon the scale selected for each. As in the case for inductance, the phase angle is 90°. But this time the current leads the voltage.

In a circuit where the phase angle between the voltage and current is 90°, the true power is zero. It makes no difference whether one leads or lags behind the other. Since, in a circuit containing only capacitance the phase angle is 90°, such a circuit has a power factor of zero and consumes no power. The electrical energy is merely converted to the energy of the electrostatic field across the capacitor and back to electrical energy again.

2. Capacitive Reactance

Just as inductance, because of its counter electromotive force, increases the opposition to current flow in a circuit by a factor known as inductive reactance (X_L), so capacitance, for the same reason, increases the opposition by a factor known as **capacitive reactance** (X_C). As is true for inductive reactance, the unit for capacitive reactance is the **ohm.**

As a capacitor is charged, it builds up a counter electromotive force across its plates. The larger the plates of the capacitor, the more thinly the charge (that is, the number of electrons) is spread — that is, the smaller the counter electromotive force becomes. But the smaller the counter electromotive force, the smaller is the capacitive reactance. Hence, the larger the plates of the capacitor (or, what is the same thing, the larger its capacitance), the smaller is its capacitive reactance.

If a capacitor is placed in series in a d-c circuit, current flows shortly until the capacitor is fully charged and then the counter electromotive force becomes equal and opposite to the applied electromotive force. Current ceases to flow in the circuit. We might say that a capacitor in a d-c circuit has infinite capacitive reactance.

But if we place the capacitor in a rapidly-alternating circuit, it has no time to charge up fully before the source voltage starts falling and the capacitor starts discharging. Thus, the more quickly the current is alternating — that is, the higher its frequency — the less the counter electromotive force and, hence, the less the capacitive reactance of the capacitor will be.

From the above, we may say that the capacitive reactance is inversely proportional to the capacitance of the capacitor and the frequency of the current. This relationship may be stated in the following formula:

$$X_C = \frac{1}{2\pi f C},$$

where X_C is the capacitive reactance in ohms, f is the frequency of the current in cycles per second, and C is the capacitance in farads. The constant 2π (6.28) is necessary to make the result come out in ohms.

EXAMPLE. What is the capacitive reactance of a 10-microfarad capacitor in a 60-Hz alternating-current circuit?

ANSWER. $X_C = \dfrac{1}{2\pi f C} = \dfrac{1}{6.28 \times 60 \times 0.00001} = 265.3$ ohms.

As is true of a purely inductive circuit, in a purely capacitive circuit we may substitute the capacitive reactance (X_C) for the impedance (Z) in the formulas that state Ohm's law for a-c circuits. Thus, instead of

$$I = \frac{E}{Z}, \quad E = I \times Z, \quad \text{and} \quad Z = \frac{E}{I},$$

we may state

$$I = \frac{E}{X_C}, \quad E = I \times X_C, \quad \text{and} \quad X_C = \frac{E}{I}.$$

3. Capacitors

A **capacitor** is a device that is purposely constructed and inserted in the circuit to introduce the desired capacitance. Any two conductors separated by a dielectric will have the property of capacitance. Thus, to make a capacitor, all that is necessary is to have two or more metallic plates separated by air or some other insulating material. The dielectrics in general use are air, mica, waxed paper, glass, ceramics, oil, and, in certain types, gas films.

Assume we have a capacitor made of two metal plates separated by a dielectric sheet. The ability of this capacitor to store electrical energy — that is, its capacitance — depends upon the electrostatic field between the plates and the degree of distortion of the orbits of the electrons of the dielectric. Thus, increasing the area of the plates increases the capacitance. Also, if the plates are brought closer together, thus intensifying the electrostatic field, the capacitance increases.

The degree of distortion of the orbits of the electrons of the dielectric depends upon the nature of the substance and is known as the **dielectric constant.** The dielectric constant of a substance is a measurement of its effectiveness when used as the dielectric of a capacitor. Air is taken to have a dielectric constant of unity, or 1. If, in a certain capacitor using air as a dielectric, as the air is replaced by mica and all other things remain equal, the capacitance becomes six times as great, the mica is said to have a dielectric constant of 6. Paper has a dielectric constant of from 2 to 3; paraffin about 2; titanium oxide from 90 to 170. The electrical symbol for dielectric constant is K.

The formula for calculating the capacitance of a capacitor may be stated as follows:

$$C = \frac{0.0885 \times K \times A}{T},$$

where C is the capacitance in picofarads (pf), K is the dielectric constant of the dielectric, A is the area (in square centimeters) of one side of one of the plates that is in actual contact with the dielectric, and T is the thickness of the dielectric (in centimeters).

EXAMPLE. Calculate the capacitance of a capacitor having two tin-foil plates each 2.5 centimeters wide and 250 centimeters long. The waxed paper that separates these plates has a thickness of 0.025 centimeter and has a dielectric constant of 2.

Substituting these values in the formula, we get

$$C = \frac{0.0885 \times 2 \times 625}{0.025} = 4,425 \text{ pf.}$$

ANSWER. Thus the capacitance of this capacitor is 4425 picofarads or 0.004,425 microfarad.

If the voltage across the plates of the capacitor becomes too large, the dielectric may be ruptured and the capacitor ruined. The ability of the dielectric to withstand such rupture is called its **dielectric strength** and is measured in the maximum number of volts a one-centimeter thickness of the dielectric can withstand. Air has a dielectric strength of about 30,000 volts and mica a dielectric strength of about 500,000 volts.

Capacitors may be fixed or variable. In its simplest form the fixed capacitor consists of two metal plates separated by a dielectric of mica or some ceramic material and enclosed in a plastic case for protection. Wire leads through the case make contact with the plates. Sometimes the metallic plates are deposited electrolytically directly upon the opposite sides of the dielectric sheet.

To increase the plate area, and, hence, the capacitance, a number of plates may be sandwiched between a number of sheets of dielectric. Alternate plates are connected together, thus forming two sets with a larger effective area. The capacitances of the capacitors we have described generally are quite small.

A fixed capacitor of greater capacitance can be made by placing a strip of waxed paper between two strips of tin foil about an inch wide and several feet long. The large area of the tin-foil plates will permit this capacitor to have a large capacitance. To save space, the whole is rolled up and encased in cardboard. This is called a **paper capacitor.**

Another type of fixed capacitor, commonly used where larger capacitances are required, is the **electrolytic capacitor.** In such a capacitor, a sheet of aluminum is kept immersed in a borax solution (called the **electrolyte).** An extremely thin coating of aluminum oxide and oxygen gas forms on the surface of the aluminum. If we consider the aluminum as one plate of the capacitor and the borax solution as the other, the coating of aluminum oxide and oxygen gas, which will not conduct electricity, becomes the dielectric. The aluminum need not be a straight sheet, but may be folded over many times or loosely rolled to save space. Because the "plates" are separated by an extremely thin dielectric, the capacitance of such a capacitor is very high.

A variation of this type of electrolytic capacitor (called a **wet** type because of the solution) is the **dry** type. Although this capacitor is not strictly dry, it is so called because, instead of the liquid electrolyte, a gauze saturated with borax solution is used. This "dry" electrolytic capacitor has a definite advantage in that the solution cannot spill.

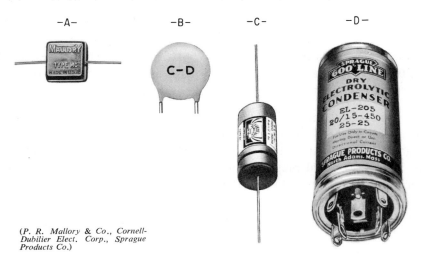

(*P. R. Mallory & Co., Cornell-Dubilier Elect. Corp., Sprague Products Co.*)

Fig. 10-9. Fixed capacitors.
- A. Mica capacitor.
- B. Ceramic disk capacitor.
- C. Paper capacitor.
- D. Electrolytic capacitor.

Care must be taken to always connect the aluminum plate of the electrolytic capacitor to the positive (+) side of the line; otherwise, the dielectric will be punctured and the capacitance destroyed. Fortunately, the "wet" type of capacitor is self-healing and a new coating of oxide and gas will form once the proper connections are made. In the "dry" type, however, puncturing the dielectric may permanently damage the capacitor.

If the electrolytic capacitor is connected in an a-c circuit, the dielectric will be punctured constantly since the voltage continuously reverses its direction. There is a variation of the electrolytic capacitor, however, that may be used safely in a-c circuits. This is the **a-c capacitor** formed by mounting two self-healing electrolytic capacitors back to back — that is, with the solution joining one to the other. Connections are made to the two outside plates. Now, whatever the direction of the voltage, one capacitor will always be connected properly (and the other punctured).

When rating a capacitor, we must take into consideration its capacitance and the dielectric strength of its dielectric. The **breakdown voltage** is the maximum voltage that the dielectric of a particular capacitor can stand without being punctured or "breaking down." The **working voltage** of the capacitor is the maximum safe d-c voltage that the manufacturer recommends be placed across its plates. Thus, for example, a capacitor may be rated as "0.1 μf, 600 volts, d-c working voltage. [Remember that, unless otherwise stated, a-c voltages generally are stated in rms values. Thus an a-c voltage of 600 volts (rms) has peak values of 846 volts. If this voltage is applied to a capacitor rated at 600 volts, d-c working voltage, the capacitor may be destroyed.]

The capacitance of a capacitor may be varied by changing the effective areas of the plates or the distance between them. One type of variable capacitor is the **trimmer** or **padder** capacitor illustrated in Figure 10-10A. Two metal plates are arranged like the pages of a book. The springiness of the metal keeps the "book" open. Between the "pages" is a sheet of mica or other dielectric material. The metal plates may be brought closer together or further apart by the adjustment of a screw. Thus the capacitance may be varied.

The **ceramic variable capacitor,** illustrated in Figure 10-10B, is another type of trimmer. A plate of silver, in the shape of a half-moon, is deposited upon the underside of a ceramic disk to form the fixed plate of the capacitor. The movable plate is a similarly shaped piece of metal which can be rotated on the upper surface of the ceramic disk which acts

−A−

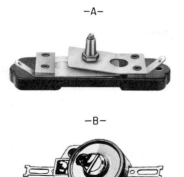

−B−

−C−

Fig. 10-10. Variable capacitors.
A. Compression-type trimmer.
B. Ceramic trimmer capacitor.
C. Variable air capacitor.

as the dielectric between the two plates. The capacitance is adjusted by rotating the movable plate so that the effective area between it and the fixed plate may be varied. The entire assembly is contained in a ceramic housing about ¾ inch in diameter and ½ inch thick.

Another type of variable capacitor that frequently is employed to "tune in" radio stations in the radio receiver is illustrated in Figure 10-10C. Here, two sets of meshing metal plates use air as their dielectric. The effective areas of the opposing sets of plates (and, hence, the capacitance of the capacitor) are varied by sliding one set of plates (called the **rotor**) between the other set (called the **stator**).

When used in electrical diagrams, the symbol for the fixed capacitor is ⊣⊢ . The curved element of this symbol represents the plate that is connected to the negative portion of the circuit. Sometimes, where electrolytic capacitors are involved, the polarity is further identified by plus and minus signs. The symbol for the variable capacitor is ⊣⊬ . The curved element generally indicates the movable portion of the capacitor.

Capacitors may be connected in series, in parallel, or in series-parallel circuits. Where capacitors are connected in series, they act as though we were adding to the thickness of the dielectric. Accordingly, the total capacitance decreases. Thus, for capacitors connected in series, the following formula applies:

$$\frac{1}{C_{\text{total}}} = \frac{1}{C_1} + \frac{1}{C_2} + \frac{1}{C_3} +, \text{ and so forth.}$$

If we connect capacitors in parallel, they act as though we were adding to the areas of their plates. Accordingly the total capacitance increases. Thus, for capacitors connected in parallel, the following formula applies:

$$C_{\text{total}} = C_1 + C_2 + C_3 +, \text{ and so forth.}$$

Note that this is the reverse action of resistors and inductors connected in series and parallel. Where capacitors are connected in series-parallel circuits, we first find the joint capacitance for the series-connected capacitors and add it in series to the joint capacitance of the parallel-connected capacitors.

QUESTIONS

1. What is the effect of resistance upon the phase relationship between current and voltage in an a-c circuit? What is the power factor of a circuit containing only resistance?

2. What is **inductance?** To what is it due?

3. What is the unit of inductance? What does it mean?

4. Assuming there is no interaction between their magnetic fields, what is the total inductance of an inductor of 4 henrys and one of 2 henrys if (*a*) they are connected in series; (*b*) they are connected in parallel?

[*Ans.* (*a*) 6 h; (*b*) 1⅓ h.]

5. What is the effect of inductance upon the phase relationship between the voltage and current in an a-c circuit?

6. What is the phase angle between the voltage and the current in an a-c circuit that contains only inductance? What is the **true** power consumed in such a circuit? [*Ans.* 90°, zero.]

7. What is **inductive reactance?** In what units is it measured?

8. What is the inductive reactance of a coil of 10 henrys as a 400-Hz alternating current flows through it? [*Ans.* 25,120 ohms.]

9. In an a-c circuit containing only inductance, what will be the current if the voltage is 100 volts and the inductive reactance is 25 ohms?

[*Ans.* 4 amps.]

10. Explain what is meant by **mutual inductance.**

11. What is **capacitance?** To what is it due?

12. What is the unit of capacitance? What does it mean?

13. What is a **capacitor?** What factors determine its capacitance?

14. A capacitor is made by placing two sheets of brass, each 60 centimeters long by 50 centimeters wide, at either side of a sheet of glass that is 3 millimeters thick. If the dielectric constant of the glass is 7, what is the capacitance of the capacitor thus formed?   [*Ans.* 6195 pf or 0.006195 μf.]

15. Two capacitors, one 6 microfarads and the other 3 microfarads, are connected in series. (*a*) What will be the resulting capacitance? (*b*) What will be the resulting capacitance if they are connected in parallel?

[*Ans.* (*a*) 2 μf; (*b*) 9 μf.]

16. What is the effect of capacitance upon the phase relationship between the voltage and the current in an a-c circuit?

17. What is the phase angle between the voltage and current in an a-c circuit that contains only capacitance? What is the **true** power consumed in such a circuit?   [*Ans.* 90°, zero.]

18. What is **capacitive reactance?** In what units is it measured?

19. What is the capacitive reactance of a 5-microfarad capacitor placed in a 400-Hz a-c circuit?   [*Ans.* 79.6 ohms.]

CIRCUITS CONTAINING RESISTANCE, INDUCTANCE, AND CAPACITANCE

11

When we consider practical electrical circuits, especially those operating on alternating currents, we are confronted with resistance, inductance, and capacitance. All three phenomena are present at the same time. Thus, in the case of an inductor we have, in addition to its inductance, the resistance of the wire constituting its turns and the capacitance between turns. Similarly, the plates and dielectric of the capacitor offer some resistance and, as current flows through them, are surrounded by a magnetic field and, therefore, exhibit some inductance. Even the connecting wires exhibit resistance, inductance, and capacitance.

Ordinarily, we may neglect these stray inductances and capacitances, and the incidental resistances may be minimized by using heavy conductors of short lengths. Under certain conditions, however, such stray inductances and capacitances play an important part. In addition, there are times when lumped resistance, inductance, and capacitance are deliberately inserted into the circuit.

A. CIRCUITS CONTAINING RESISTANCE AND INDUCTANCE

As we have learned, pure resistance has no effect on the phase relationship between current and voltage. Thus, if our circuit has nothing but resistance in it, the vector diagram shows us that the voltage and current are in phase (Figure 11-1B). The length of the voltage vector (E) is independent of the length of the current vector (I) and the length of each vector depends upon the scale selected for each. Here there is no phase angle between the two — that is, they are in phase.

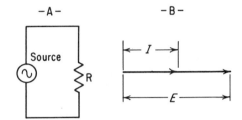

Fig. 11-1. A. A-C circuit containing only resistance.
B. Vector diagram showing the phase relationship between current and voltage in this circuit.

Another use of the vector diagram is to enable us to add voltages and currents. If we have two resistors in series, it is a simple arithmetical problem to add the voltage drop across each of the resistors in order to calculate the total voltage supplied by the source. See Figure 11-2B. (Although here the current and voltage vectors may employ different scales, the vectors of all the voltages must use the same scale.) Note that we merely have added the vector for the voltage drop across resistor

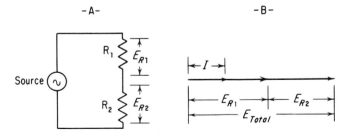

Fig. 11-2. A. A-C circuit containing two resistors connected in series.
B. Vector diagram showing the phase relationship between the current and the voltages in this circuit.

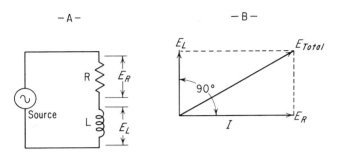

Fig. 11-3. A. A-C circuit containing a resistor and inductor connected in series.

B. Vector diagram showing phase relationship between current and voltages in this circuit.

R_2 (E_{R2}) to that for the voltage drop across R_1 (E_{R1}) to obtain the vector for the total voltage (E_{Total}).

If, however, we have an inductor and a resistor in series, we cannot simply add the voltage drop across each to give us the total voltage. We must take into consideration the fact that inductance affects the phase relationships. This situation appears in the vector diagram of Figure 11-3B. Note that I and E_R are in phase, but that I and E_L are 90° out of phase. To obtain the total voltage supplied by the source, we must make use of the parallelogram method described in Chapter 9, subdivision D.

Note, too, that the voltage of the source (E_{Total}) is out of phase with the current (I). But the phase difference between the two no longer is 90°. It is some lesser value depending upon the relative values of E_L and E_R. By varying these relative values, the phase difference between the current and the voltage of the source may be varied. Further, note that whereas E_L leads the source voltage, E_R lags behind it.

If you examine Figure 9-10, you will note that, if the phase angle between the voltage and current is some value less than 90° (it makes no difference which leads the other), the true power consumed by the circuit is some positive value, though less than if there were only resistance present. None of this power is consumed by the inductor (assuming a theoretically perfect inductor, that is, one that has no resistance). It all is consumed by the resistor. The power factor, too, is not zero (as it would be if there were only inductance present), but some value less than 1. (It would be unity if resistance alone were present.)

In a series circuit containing only resistance and inductance, both the resistance (R) and the inductive reactance (X_L), each expressed in ohms, impede the flow of current. However, since the effect of inductance is to

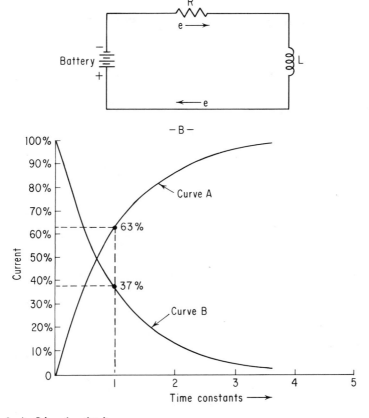

R

e →

Battery

L

– B –

← e

100%
90%
80%   Curve A
70%
60%   63%
50%
Current   40%   37%
30%   Curve B
20%
10%
0
        1   2   3   4   5
        Time constants →

Fig. 11-4. A. *R-L* series circuit.
B. Time-current curves for current flowing in the circuit.

cause the current to lag behind the voltage, we cannot use simple addition to find the impedance (Z). Instead, we apply the following formula:

$$Z = \sqrt{R^2 + (X_L)^2}.$$

EXAMPLE. What is the total impedance of a series circuit containing a resistor of 3 ohms and an inductor whose inductive reactance is 4 ohms?

ANSWER. $Z = \sqrt{R^2 + (X_L)^2} = \sqrt{9 + 16} = 5$ ohms.

1. The R-L Time Constant

If a source of steady direct current, such as a battery, is connected in series with a resistor and an inductor (see Figure 11-4A), the moment the circuit is completed, a counter electromotive force is built up in the inductor which opposes the flow of current. But since the voltage of the battery is constant, the counter electromotive force dies away, and more current flows until its maximum value is reached.

However, this process is not instantaneous. The time required to overcome the resistance of R and for the decay of the counter electromotive force of L is shown graphically by curve A of Figure 11-4B. The time interval required for the current to reach approximately 63 percent of its maximum value is known as the **R-L time constant.**

Thus the time constant is a function of the resistance and inductance of the circuit. Its value may be calculated from the following formula:

$$t = \frac{L}{R},$$

where t is the time in seconds, L is the inductance in henrys, and R is the resistance in ohms.

EXAMPLE.   What would be the *R-L* time constant of a series circuit containing a 4-ohm resistor and a 20-henry inductor?

ANSWER. $t \text{ (seconds)} = \dfrac{L \text{ (henrys)}}{R \text{ (ohms)}} = \dfrac{20}{4} = 5 \text{ seconds.}$

If, after the current has reached its maximum value (in the circuit illustrated in Figure 11-4A), the circuit is opened, the current will start to decrease. But the counter electromotive force so produced in the inductor will oppose this decrease in current. Thus the fall of current will be slowed up. After the counter electromotive force has decayed (in about five time constants), the current will fall to zero. An examination of the curve illustrating the decrease of current (curve B, Figure 11-4B) shows that the time required for the current to drop to approximately 37 percent of its maximum value is equal to one time constant.

The *R-L* time constant is important in the design of various electrical protective and control devices that employ circuits containing inductance. If quick-acting circuits (with small time constants) are required, the inductance should be kept small. If slower-acting circuits (with larger time constants) are required, the inductance should be larger.

B.   CIRCUITS CONTAINING RESISTANCE AND CAPACITANCE

As you have learned, in an a-c circuit containing only capacitance the current leads the voltage by 90°. (See Figure 10-8B.) If, however, we

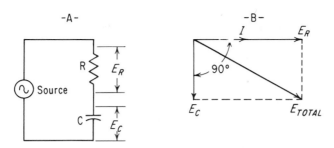

Fig. 11-5. A. A-C circuit containing a resistor and capacitor connected in series.
B. Vector diagram showing phase relationship between current and voltages in this circuit.

have a capacitor and resistor in series, the vector diagram appears as in Figure 11-5B. Note that I and E_R are in phase, but that I and E_C are 90° out of phase. Once again, we employ the parallelogram method to obtain the total source voltage (E_{Total}).

Note, too, that E_{Total} is out of phase with I and that, again, this phase difference is less than 90°, depending upon the relative values of E_C and E_R. Further note that, whereas E_R leads the source voltage, E_C lags behind it.

The effect of resistance in such a circuit is to reduce the voltage lag from 90° to some smaller value, depending upon the amount of resistance. Hence, the true power consumed by the circuit is not zero (as it would be if only capacitance were present), but some positive value, though less than if there were only resistance.

All this power is consumed by the resistor, none being consumed by the capacitor (assuming a theoretically perfect capacitor). The power factor, too, is not zero (as it would be if there were only capacitance present), but some value less than one.

As is true for the inductance-resistance circuit, the total impedance offered to alternating current by a series circuit containing resistance and capacitance may be found by means of the following formula:

$$Z = \sqrt{R^2 + (X_C)^2}.$$

1. The R-C Time Constant

Just as there is a time constant for the series resistance-inductance (*R-L*) circuit, there is one for the series resistance-capacitance (*R-C*) circuit. If a battery is connected in series with a resistor and capacitor

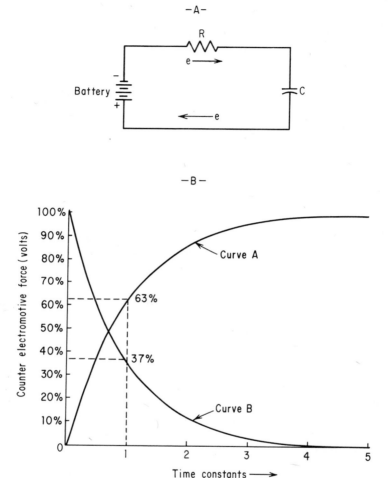

Fig. 11-6. A. *R-C series circuit.*
B. *Time-voltage curves for the counter electromotive forces in this circuit as the capacitor is charged and discharged.*

(see Figure 11-6A), the moment the circuit is completed, current starts to flow at its maximum rate, charging up the capacitor. As a result of this charge a counter electromotive force is produced, which opposes the current flow. When this counter electromotive force is equal to the electromotive force of the source, the current ceases to flow.

In many applications we are more interested in the counter electromotive force than in the current. The curves of Figure 11-6B illustrate the values of this counter electromotive force as the capacitor is charged (curve A) and as the capacitor is discharged (curve B). On curve A, the time it takes the counter electromotive force to reach approximately 63 percent of its maximum value is equal to one time constant. On curve B, the time it takes the counter electromotive force to fall to approximately 37 percent of its maximum value is also equal to one time constant.

The value of the **R-C time constant** may be calculated from the following formula:

$$t = C \times R,$$

where t is the time in seconds, C is the capacitance in farads, and R is the resistance in ohms.

EXAMPLE. What would be the *R-C* time constant of a series circuit containing a 0.001-microfarad capacitor and a resistor of 50,000 ohms?

t (seconds) = 0.000,000,001 farad $\times$ 50,000 ohms,

ANSWER. t = 0.000,05 second, or 50 microseconds.

The *R-C* time constant is employed in many electronic applications where a very small, accurate time interval is desired. The proper interval is obtained by using appropriate resistors and capacitors.

C. CIRCUITS CONTAINING RESISTANCE, INDUCTANCE, AND CAPACITANCE

1. Circuits Containing Inductance and Capacitance

If an a-c circuit contains only inductance and capacitance, the only opposition to current flow will be the inductive reactance (X_L) and capacitive reactance (X_C). (We assume here the theoretical position that no resistance is present.) Note, however, that whereas the inductive reactance tends to cause the current to lag 90° behind the voltage, the capacitive reactance tends to cause the current to lead by a similar amount. Hence the two reactances tend to cancel each other out. Thus, the total reactance (X) is equal to the difference between the inductive and capacitive reactances. This may be expressed in the following formula:

$$X = X_L - X_C.$$

Subtraction is performed algebraically. If the answer is a positive value, the resulting reactance (X) has the characteristics of an inductive reactance. That is, the current tends to lag behind the voltage. If the answer is a negative value, the resulting reactance has the characteristics of a capacitive reactance.

EXAMPLE. What is the reactance of a 60-Hz, a-c circuit containing an inductor of 2.5 henrys and a capacitor of 10 microfarads connected in series?

$$X_L = 2\pi fL = 6.28 \times 60 \times 2.5 = 942 \text{ ohms,}$$

$$X_C = \frac{1}{2\pi fC} = \frac{1}{6.28 \times 60 \times 0.00001} = 265.3 \text{ ohms,}$$

$$X = X_L - X_C = 942 - 265.3$$

ANSWER. $= 676.7$ ohms (inductive reactance).

We may illustrate the above vectorially as shown in Figure 11-7. Since the voltage across the inductor (E_L) leads the current by 90° and the voltage across the capacitor (E_C) lags 90° behind the current, the two voltages are 180° out of phase and their vectors are shown on the same straight line. Since these vectors are in opposite directions, we may subtract one from the other to get the vector for the resulting voltage (E_{total}).

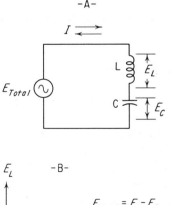

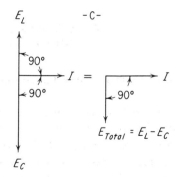

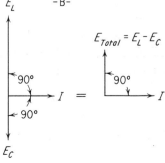

Fig. 11-7. A. A-C circuit containing inductance and capacitance.
B. Vectorial diagram when E_L is greater than E_C.
C. Vectorial diagram when E_C is greater than E_L.

Assume (as illustrated in Figure 11-7B) X_L is larger than X_C and hence the voltage drop (E_L) across the inductor is larger than the voltage drop (E_C) across the capacitor. The result, then, is as if we had nothing in the circuit but an inductor whose voltage-drop vector is the difference between vector E_L and vector E_C. If, as in Figure 11-7C, E_C is larger than E_L, the result is as if we had nothing in the circuit but a capacitor whose voltage-drop vector is the difference between vectors E_C and E_L.

Since inductive and capacitive reactances tend to cancel out, if a circuit contains inductive reactance, we may reduce the total reactance by adding some capacitive reactance. This may be done by placing a capacitor in series in the circuit. On the other hand, if the circuit contains capacitive reactance, we may reduce the total reactance by adding some inductive reactance (by placing an inductor in series).

Practical circuits always contain some resistance. To find the total impedance (Z) of a circuit containing resistance as well as inductance and capacitance, we use the following formula:

$$Z = \sqrt{R^2 + (X_L - X_C)^2}.$$

EXAMPLE.   What is the total impedance of an a-c circuit containing a resistor, inductor, and capacitor in series if the resistance is 30 ohms, the inductive reactance is 210 ohms, and the capacitive reactance is 250 ohms?

$$Z = \sqrt{R^2 + (X_L - X_C)^2}$$

ANSWER. $= \sqrt{(30)^2 + (210 - 250)^2} = 50 \text{ ohms.}$

2. Resonance

Assume we connect an inductor (L) and a capacitor (C) in series with a generator, as in Figure 11-8. Note that some resistance (R) is always

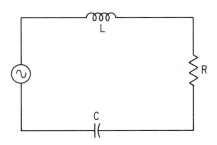

Fig. 11-8. Inductor L, resistor R, and capacitor C connected in series across a source of alternating current.

present. This is mainly due to the wire of the inductor as well as the connecting leads.

The opposition to the flow of current in this circuit comes from the inductive reactance (X_L), the capacitive reactance (X_C), and the resistance (R). Since the inductive reactance causes the current to **lag** behind the voltage and the capacitive reactance causes the current to **lead** the voltage, these two reactances tend to cancel out each other.

Since $X_L = 2\pi fL$ and $X_C = 1/(2\pi fC)$, the higher the frequency, the greater the inductance reactance and the smaller the capacitive reactance. The lower the frequency, the smaller the inductive reactance and the greater the capacitive reactance. Thus, we can see that at a certain frequency (depending upon the values of L and C), the two reactances will cancel out each other, leaving only the resistance (R) to oppose the flow of current. Since the resistance may be very small, the flow of current may become quite large. This condition is known as **resonance**, and the critical frequency as the **resonant frequency.**

Note that at resonance the inductive and capacitive reactances are not each equal to zero. Rather, X_L and X_C are equal and opposite to each other and thus cancel out. As a matter of fact, these reactances may be quite high. Since the current flowing in the circuit is quite large, the voltage drops across the inductor and capacitor may be very large, too. Under certain conditions, these voltage drops at resonance may even be greater than the applied voltage of the generator.

The resonant frequency (f_r) of the circuit depends upon the values of inductance (L) and capacitance (C). Since, at resonance, $X_L = X_C$, then substituting, we get $2\pi f_r L = 1/(2\pi f_r C)$. Solving this equation for the resonant frequency, we get

$$f_r = \frac{1}{2\pi\sqrt{L \times C}},$$

where the frequency is in cycles per second, inductance is in henrys, and capacitance is in farads.

EXAMPLE. What will be the resonant frequency of a series circuit containing an inductor of 200 microhenrys (0.0002 henry) and a capacitor of 200 pf (0.000,000,000,2 farad)?

$$f_r = \frac{1}{2\pi\sqrt{L \times C}} = \frac{1}{6.28 \times 0.000,000,2} = \frac{1}{0.000,001,256},$$

ANSWER. $f_r = 796,178$ Hz, or 796.178 kHz.

As you know, the inductive and capacitive reactances cancel out each other at the resonant frequency, leaving only the resistance to oppose the flow of current. Since this resistance may be very small, the flow of current may be very large. If the frequency is reduced below resonance, the capacitive reactance becomes larger than the inductive reactance, and the net reactance no longer is zero. If the frequency is increased above resonance, the inductive reactance becomes larger than the capacitive reactance. Again the net reactance no longer is zero. In both cases, the current flowing in the circuit decreases.

A graph showing the relationship between the current and the frequency may be drawn as in Figure 11-9. Note that the current reaches its maximum amplitude at the resonant frequency (f_r), falling off rapidly on either side of that point.

Note also the effect of resistance in the circuit. The smaller the resistance, the greater the current at resonance and also the sharper the curve — that is, the greater the discrimination against nonresonant frequencies. On the other hand, the greater the resistance, the lower the resonant peak and the flatter the curve.

We call a circuit such as illustrated in Figure 11-8, a **tuned circuit,** tuned to a resonant frequency determined by the values of inductance and capacitance. Thus, if alternating currents of various frequencies were being supplied by the generator, only those close to the resonant frequency would get through the tuned circuit without much opposition. If the frequencies were higher than the resonant frequency, the currents

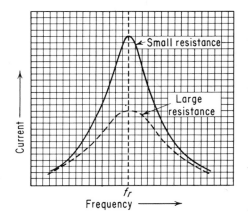

Fig. 11-9. Curves showing the current-frequency relationship around the resonant frequency f_r of a circuit containing inductance and capacitance in series. Note the effect caused by resistance.

would be opposed by the inductive reactance of L. If the frequencies were lower, they would be opposed by the capacitive reactance of C. The resistance (R) is the same for currents of all frequencies.

The ability of the tuned circuit to discriminate against currents of nonresonant frequency (as indicated by the sharpness of the resonant curves illustrated in Figure 11-9) is called **selectivity.** The sharper the curve, the greater is the selectivity of the circuit, and the greater is the discrimination against currents of the nonresonant frequency.

We employ this principle when we tune the antenna circuit of a radio receiver to receive signals of one frequency and to reject all others. Usually, the antenna circuit consists of a tuned circuit, such as described, with the voltage of the generator being replaced by an electromotive force induced in the antenna circuit as the radio wave crosses the antenna. The values of C and L are adjusted to produce resonance at the frequency of the desired signal.

The circuit described above is a series circuit; hence, at the resonant frequency, it is called a **series-resonant circuit.** At this frequency there is very little opposition to the flow of current from the source, and so very large currents may flow.

Another type of resonant circuit, illustrated in Figure 11-10, is called a **parallel-resonant circuit.** Current flowing from the generator divides into two paths: one through the capacitor and the other through the inductor and resistor. (This resistor represents the resistance of the wire of the inductor and will be disregarded for the time being.) The greater current will flow through the path of lower impedance.

At low frequencies the capacitive reactance will be higher, and thus more current will flow through the inductor. At high frequencies the inductive reactance will be higher, and now more current will flow through the capacitor. At a certain frequency (the resonant frequency) both reactances will be the same, and thus equal current will flow through both paths.

But the inductive reactance would tend to cause the current to lag

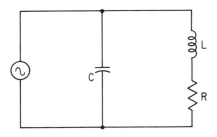

Fig. 11-10. The parallel-resonant circuit.

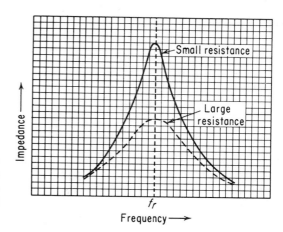

Fig. 11-11. Curves showing the imped- ance-frequency relationship around the resonant frequency f_r of a parallel- resonant circuit. Note the effect caused by resistance.

behind the voltage by 90°, and the capacitive reactance would tend to cause the current to lead the voltage by 90°. Thus, the currents flowing in each path would be 180° out of phase — that is, would be flowing in opposite directions. The net result would be that they would cancel out each other, and no current would flow from the generator. This is the same as saying that a **parallel-resonant circuit offers an infinite impedance to the source of current.**

Keep in mind, however, that there may be large currents flowing around the loop formed by the capacitor and the inductor. It is not that there are no currents present, but rather that the currents are equal and opposite, to explain why the net result is zero.

We may obtain a better picture, perhaps, if we consider what is happening in the closed loop at resonant frequency. As the capacitor becomes charged, a counter electromotive force is generated which causes current to flow through the inductor. This, in turn, produces a counter electromotive force across the inductor that charges the capaci- tor. Thus, electric energy is stored up in the electrostatic field of the capacitor and then in the magnetic field of the inductor. This energy continually changes from one field to the other. As a result, current flows back and forth. We call this back-and-forth flow of current, **oscillation** and, theoretically, it should continue indefinitely, once started.

The presence of resistance upsets this theoretical picture. As the current flows in the loop, some electric power is dissipated by this re- sistance, and this loss must be replaced from the outside source. Hence, some current flows from the generator, and the parallel-resonant circuit acts as a very high, rather than an infinite, impedance.

We may plot an impedance-frequency graph for this type of circuit just as we plotted a current-frequency graph for the series-resonant circuit. This is shown in Figure 11-11. As in the case of the series-

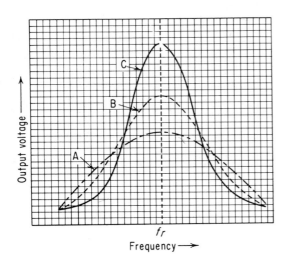

Fig. 11-12. Curves showing how selectivity is increased as tuned circuits are added. Curve A — one circuit. Curve B — two circuits. Curve C — three circuits.

Output voltage

f r

Frequency ⟶

resonant graph, the resistance affects the amplitude, sharpness, and the selectivity of the curve.

In radio receivers the signal generally is made to pass through a number of circuits in succession, all tuned to the same resonant frequency. The effect is to increase the discrimination of the receiver against unwanted signals whose frequencies are other than the frequency of the desired signal to which the circuits are tuned — that is, to increase the selectivity of the receiver. In Figure 11-12, curve A shows the resonant curve for a single tuned circuit; curve B, the resonant curve of two tuned circuits; and curve C, that of three tuned circuits, all tuned to the same resonant frequency. Note that the curves become sharper (more selective) as the number of tuned circuits is increased. The increased height of the curves is caused by amplification resulting from the amplifiers generally associated with the tuned circuits.

D. FILTERS

Electrically, **filters** are used to separate currents of certain frequencies from those of other frequencies. (For our discussion here, we may consider direct current as having a zero frequency.) Assume that you have as an electrical source a battery (whose symbol is ⊣⊢) supplying direct

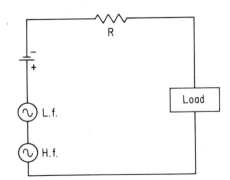

Fig. 11-13. Circuit showing direct, low-frequency, and high-frequency currents feeding a load through resistor R in series.

current, a generator of low-frequency alternating current ($-\sim_{LF}$), and a generator of high-frequency alternating current ($-\sim_{HF}$). Assume that they are all connected in series and supply current to a load through a resistor (R) in series, as shown in Figure 11-13. The resistor will not have any filtering action, since it impedes equally all currents that pass through it, regardless of frequency.

Now assume that you replace the resistor with a capacitor (Figure 11-14). The direct current will be completely filtered out, since the capacitor offers infinite impedance to its passage. Since $X_C = 1/(2\pi fC)$, the higher the frequency, the less will be the impedance. Hence, the low-frequency current will be more strongly impeded (or **attenuated**) than the high-frequency current.

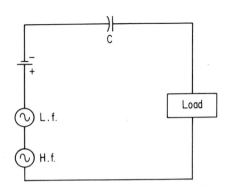

Fig. 11-14. Circuit showing direct, low-frequency, and high-frequency currents feeding a load through capacitor C in series.

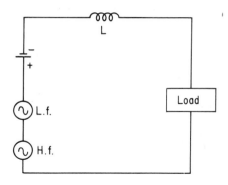

Fig. 11-15. Circuit showing direct, low-frequency, and high-frequency currents feeding a load through inductor L in series.

Now assume that the capacitor is replaced by an inductor (Figure 11-15). The direct current will be only slightly impeded, owing to the resistance of the inductor. Since $X_L = 2\pi fL$, the higher the frequency, the greater the impedance will be. Hence, the high-frequency current will be more strongly attenuated than the low-frequency current.

Suppose we connect the capacitor across the load (Figure 11-16). None of the direct current will flow through the capacitor, all of it going through the load. Since the capacitor offers a fairly high impedance to low-frequency current, most of this, too, will flow through the load. But since the capacitor offers a low impedance to high-frequency current, most of this current will flow through the capacitor rather than through the load. This type of circuit is called a **low-pass filter,** since it passes the low-frequency currents on to the load, bypassing the high-frequency currents (that is, high-frequency currents bypass the load).

In Figure 11-17 the capacitor is replaced by an inductor. This inductor offers a low-impedance path to the direct and low-frequency

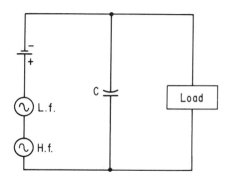

Fig. 11-16. Circuit showing direct, low-frequency, and high-frequency currents feeding a load with capacitor C in parallel.

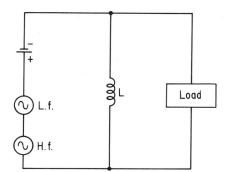

Fig. 11-17. Circuit showing direct, low-frequency, and high-frequency currents feeding a load with inductor L in parallel.

currents. Hence, most of these currents are bypassed and do not reach the load. But the high-frequency current finds the inductor a high-impedance path, and thus most of this current flows through the load. This type of circuit is called a **high-pass filter**, since it passes the high-frequency currents on to the load.

1. Resonant Circuits as Filters

Where the filtering action is to be limited to a single frequency or, at most, to a narrow band of frequencies, resonant circuits are used. Filters of this type are used extensively in electronic circuits.

Figure 11-18 shows such a circuit, which is an example of a **band-pass filter** used to pass only current of a single frequency (or a narrow band of frequencies) to a load. The values of the inductors (L) and capacitors (C) are such as to form resonant circuits at the frequency we wish to pass through the filter. The series-resonant circuit offers a very low im-

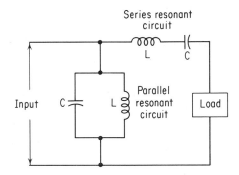

Fig. 11-18. Band-pass filter.

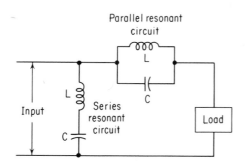

Fig. 11-19. Band-elimination filter.

pedance to currents of the resonant frequency and a relatively high impedance to currents of other frequencies.

The parallel-resonant circuit, on the other hand, offers a very high impedance to currents of the resonant frequency and a relatively low impedance to currents of other frequencies. Thus, currents of the resonant frequency pass easily through the filter while currents of all other frequencies are bypassed through the parallel-resonant circuit.

In Figure 11-19 is shown the circuit used to filter out current of a particular frequency (or a narrow band of frequencies). Here, too, L and C are shown to form resonant circuits at the desired frequency. Currents of this frequency will find the parallel-resonant circuit an extremely high-impedance path, whereas the series-resonant circuit forms a very low-impedance path.

Currents of other frequencies, however, will find the parallel-resonant circuit a fairly low-impedance path, but the series-resonant circuit furnishes a fairly high-impedance path. Thus, currents of the resonant frequency will be stopped and bypassed, while currents of all other frequencies will pass through. This type of filter is called a **band-stop,** or **band-elimination, filter.**

E. COUPLED CIRCUITS

In electrical circuits we generally pass power from a source to a load. The source may be a generator or another circuit through which an electric current is flowing. It can be shown that the greatest transfer of power takes place when the impedance of the load is equal to that of the source.

Examine Figure 11-20. The source is a generator producing a voltage which we may call E_{gen} and causing a current (I) to flow. For the sake of

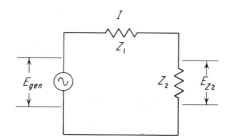

Fig. 11-20. Diagram to illustrate the transfer of electric power.

simplicity we show all impedances in the diagram as resistors (⎌). The impedance of the generator is represented by Z_1, which is in series with the impedance of the load (Z_2).

The voltage drop across the load $(I \times Z_2)$ is represented by E_{Z2}. Let us assume E_{gen} to be equal to 100 volts, Z_1 to be equal to 10 ohms, and Z_2 to be one ohm. The total impedance of the circuit $(Z_1 + Z_2)$ is 11 ohms. From Ohm's law we can determine the current (I) flowing through the circuit.

$$I = \frac{E_{gen}}{Z_{total}} = \frac{100}{11} = 9.09 \text{ amperes.}$$

The voltage drop (E_{Z2}) across the load can be determined as follows:

$$E_{Z2} = I \times Z_2 = 9.09 \times 1 = 9.09 \text{ volts.}$$

If we assume different values for the load impedance (Z_2), we can draw up the following table:

E_{gen} (volts)	Z_1 (ohms)	Z_2 (ohms)	I (amperes)	E_{Z2} (volts)	P_{Z2} (watts)
100	10	1	9.09	9.09	82.62
100	10	5	6.66	33.30	221.77
100	10	10	5.00	50.00	250.00
100	10	20	3.33	66.60	221.77
100	10	100	0.90	90.00	81.00

Note that the voltage of the generator was kept at 100 volts under all conditions of load. From this table we can see that when the impedance of the load (Z_2) is equal to the impedance of the source (Z_1), the maximum transfer of power from the source to the load occurs. This rule applies to any circuit where electric energy is transferred from one circuit to another.

1. The Transformer

At times, a maximum transfer of power is desired even though the impedance of the load does not match that of the source. Under such circumstances we may employ an impedance-matching device, called a **transformer.**

The transformer consists of two coils of wire that are magnetically coupled to each other. If an alternating current is set flowing in one of these coils (called the **primary winding**), the varying magnetic field set up around that coil will cut across the turns of the second coil (called the **secondary winding**). As a result, an induced voltage is produced across this secondary winding. In this way electric energy is transferred from the primary to the secondary circuit.

If the impedance of the primary winding matches that of the source and the impedance of the secondary winding matches that of the load, an impedance match is effected between the source and load (Figure 11-21).

The transfer of energy from the primary to the secondary winding of the transformer depends upon the magnetic lines of force around the primary winding cutting across, or **linking** with, the turns of the secondary winding. In an ideal transformer, all these lines of force would link up with all the turns of the secondary winding, achieving 100 percent (or unity) coupling. Since this is impossible to achieve in practice, a certain amount of the magnetic lines of force will leak off into the air. We call this **leakage flux,** and its effect is termed **leakage reactance.**

To reduce this leakage flux, the primary and secondary windings may be wound on a core of iron or other magnetic substance. This tends

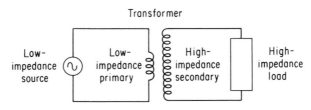

Fig. 11-21. How a transformer is used to match a high-impedance load to a low-impedance source.

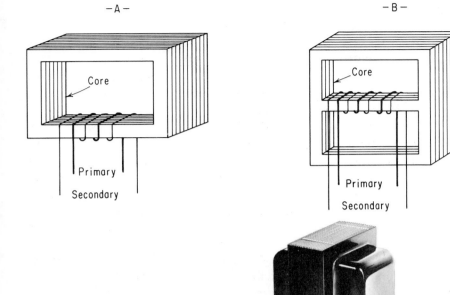

Fig. 11-22. Closed-core transformers.
A. Simple type.
B. Shell-core type.
C. Commercial shell-core transformer.

Chicago Std. Transformer Co.

to concentrate the lines of force and to keep them from leaking off. If this core is in the form of a straight bar, we have what is called an **open-core** transformer.

More frequently this core is in the form of a closed ring or square. Thus, a closed magnetic circuit is furnished, and the leakage flux is reduced further. We call this type a **closed-core** transformer (Figure 11-22). A variation that is greatly used is the **shell-core** type illustrated in Figure 11-22B. The primary and secondary windings are placed one over the other on the center arm of the core. The shell-core transformer can be designed to produce a coupling that closely approaches unity, or 100 percent.

In addition to the losses due to imperfect coupling we may have **copper** and **iron** losses in a transformer. The copper loss is due to the resistance of the wire making up the turns of the windings. The iron loss may be divided into two parts.

Since the core is in the magnetic field, it is magnetized. But the alternating current causes the iron core to change the polarity of its poles in step with the frequency of the current. A certain amount of energy is required to make this change. This energy comes from the electrical source and, therefore, is a loss. We call this loss the **hysteresis loss.** This loss may be partially reduced by using cores of **silicon steel** or of certain other alloys that are much more permeable than iron — that is, are easier to magnetize and demagnetize.

The other iron loss is due to the electric current that is induced in the iron core by the changing magnetic fields of the coils wound upon it. This induced current is called the **eddy current.** Since the eddy current must come from the electrical source, it, too, is a loss. To reduce eddy-current losses these cores are not made of solid metal, but are built up of very thin strips, called **laminations.** Each lamination is coated with an insulation of oxide or varnish so that the eddy current cannot circulate through the core.

If the current flowing through the windings of the transformer becomes great enough, a condition may be reached where the core reaches its utmost limit of magnetization. This condition is called **saturation.** Where this condition occurs, any increase in current in the windings cannot magnetize the core any more. As a result, the inductance falls off. To prevent saturation the cross-sectional area of the core must be made large enough to handle the currents present in the windings. Sometimes a very narrow air gap is left in the core to increase the reluctance of the magnetic path, and thus more current may flow before saturation takes place.

As the magnetic lines of force around the primary winding cut across the turns of the secondary winding, they induce a voltage. As a result, if there is a closed path, an induced current flows in the secondary winding. This, in turn, sets up magnetic lines of force around the secondary winding which cut across the turns of the primary winding, inducing a voltage. This voltage impedes the flow of current in the primary just as if an impedance were present. We call this effect the **reflected impedance** from the secondary to the primary circuit.

Suppose, in the circuit illustrated in Figure 11-21, the secondary circuit is open and, therefore, drawing no current. In the primary circuit, current flows from the generator, through the primary coil, and back to the generator.

Only a small current flows through the primary circuit. This is because the magnetic field, constantly cutting across the turns of the primary coil, induces in it a counter electromotive force that is nearly equal to the applied voltage. What power is consumed is used mainly to magnetize the core.

Suppose, now, that we complete the secondary circuit, permitting current to flow through the secondary winding and load. As current flows through this winding, a magnetic field is set up around it which, according to Lenz's law, tends to neutralize a certain amount of the magnetic field around the primary winding. This reduces the counter electromotive force induced in the primary and, as a result, more current flows through that coil from the source.

If the load is increased (that is, its impedance is reduced), more current flows through the secondary circuit. As a result, the counter electromotive force of the primary is further reduced and more current is drawn from the source through the primary circuit. If the load is reduced and less current flows through the secondary circuit, less current flows through the primary circuit. Thus the transformer automatically adjusts itself to changes in load. However, if the load is made too great, enough current may flow through the primary circuit to burn out its winding.

In addition to its energy-transfer and impedance-matching functions the transformer may perform another function. Assume that we have an ideal transformer with no losses and with coupling of 100 percent. Further assume that the primary winding contains 100 turns and that it is connected to an a-c generator producing 100 volts.

If we assume that the secondary circuit is open, no power will be consumed by this secondary circuit. Since, in the ideal transformer, the power consumed in both the primary and secondary circuits are equal, no power is consumed in the primary circuit either. Electrical energy is changed to magnetic energy and back again.

The counter electromotive force produced by the primary winding is equal to the voltage of the source, 100 volts. Hence, as the magnetic field cuts across this 100-turn winding, each turn has one volt induced into it. But the same magnetic field is cutting across the turns of the secondary winding as well. Hence each turn of this winding, too, has one volt induced in it. If the secondary winding also has 100 turns, the induced voltage across the secondary is 100 volts.

Suppose, however, we construct our transformer with only 10 turns in the secondary. Since one volt is induced in each turn, the secondary voltage will be 10 volts. This is called a **step-down** transformer. If we construct the transformer with 1000 turns in the secondary, the secondary voltage will be 1000 volts. This is a **step-up** transformer. Thus the transformer can be used to step up or step down alternating voltage.

From the above, we can see that the ratio between the voltage across the primary winding and that across the secondary winding is equal to the ratio between the number of turns of the primary winding and the number of turns of the secondary winding. This relationship may be expressed in a formula as follows:

$$\frac{E_p}{E_s} = \frac{N_p}{N_s},$$

where E_p is the voltage across the primary winding, E_s is the voltage across the secondary winding, N_p is the number of turns in the primary winding, and N_s is the number of turns of the secondary winding.

EXAMPLE. A transformer is required to deliver a 330-volt alternating current from the secondary winding. Assume a primary winding of 1,000 turns connected across the 110-volt a-c line. How many turns must we have in the secondary winding?

$$\frac{E_p}{E_s} = \frac{N_p}{N_s} \quad \text{or} \quad \frac{110}{330} = \frac{1000}{N_s},$$

ANSWER. $\quad N_s = \dfrac{1000 \times 330}{110} = 3000$ turns.

In our ideal transformer we assume no losses. Thus, the power $(E \times I)$ of the secondary circuit is equal to the power of the primary circuit. If, as in the above example, the voltage across the secondary winding has been stepped up three times, the current set flowing in the secondary winding will be reduced to one-third that of the primary. From this we may obtain the following formula:

$$\frac{I_p}{I_s} = \frac{N_s}{N_p},$$

where I_p is the current flowing in the primary winding, I_s is the current flowing in the secondary winding, N_s is the number of turns in the secondary winding, and N_p is the number of turns of the primary winding. Thus the transformer can be used as well to step up or step down alternating current. Of course, if the voltage is stepped up, the current is stepped down, and vice versa.

Transformers may be constructed with two or more secondary windings to achieve both step-up and step-down relationships. Thus, for example, the power transformer of a radio or television receiver may contain a single primary winding and two separate secondary windings, all wound upon the same core. One of the secondary windings may be of the step-down type, supplying low-voltage, high-current electricity for the filaments of the electron tubes. The other is a step-up, high-voltage,

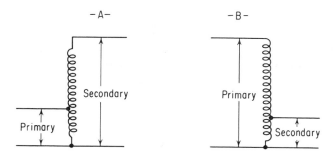

Fig. 11-23. The autotransformer.
  A. Voltage step-up connection.
  B. Voltage step-down connection.

low-current secondary winding whose output, when rectified, supplies
the plates and screen grids of these tubes.

A variation of the two-winding transformer is the **autotransformer**
illustrated in Figure 11-23. This transformer consists of a single tapped
coil. The turns between the tap and one end constitute one winding of
the transformer, and the entire coil constitutes the other winding. In
Figure 11-23A a step-up transformer is illustrated; in Figure 11-23B we
have the step-down version. The turn ratios hold for the autotransformer
just as they do for the two-winding type.

Since the iron losses increase with the frequency of the current flowing through the windings, the laminated, iron-core transformer can be
used only with currents of relatively low frequencies (15 kHz and lower).
At higher frequencies air-core or powdered iron-core transformers must
be used.

In air-core transformers only a small portion of the magnetic field
links the primary and secondary windings. Thus the voltage and current

Fig. 11-24. A. Air-core transformer.
  B. Powdered iron-core transformer.

Meissner Mfg. Co.

ratios described for the ideal transformer do not hold true. The powdered iron-core transformer is another high-frequency type employing a special core of powdered iron. Although this type does not achieve the coupling that is possible with laminated cores of silicon steel, it provides much better coupling than does the air-core type.

The symbol for the air-core transformer is ⊰⊱ ; that for the powdered iron-core transformer is ⊰⫴⊱ ; and the symbol for the laminated iron-core transformer is ⊰‖⊱ .

2. Coupling for High-Frequency Circuits

In high-frequency circuits (such as used in radio receivers) it often becomes necessary to transfer energy from one resonant circuit to another. A number of methods exist by which this may be accomplished.

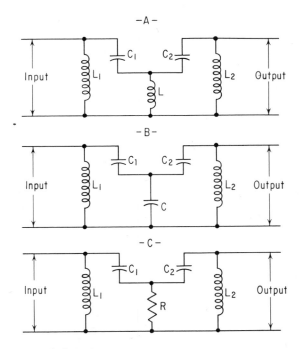

Fig. 11-25. Direct-coupled circuits.
   A. Inductor L is the common component.
   B. Capacitor C is the common component.
   C. Resistor R is the common component.

In the **direct-coupled** method a component, such as an inductor, capacitor, or resistor, is made common to both resonant circuits (Figure 11-25). The input resonant circuit (composed of L_1 and C_1) is called the **primary** circuit. The output resonant circuit (composed of L_2 and C_2) is called the **secondary** circuit. The inductors are so placed that there is no inductive coupling between them.

As current flows in the primary circuit, a voltage drop occurs across the common component (L, C, or R). This voltage drop, in turn, causes current to flow in the secondary circuit.

Another method of coupling is the **capacitive-coupling** method illustrated in Figure 11-26. Energy from the primary circuit is passed on to secondary circuit through the coupling capacitor C. As before, L_1 and L_2 are not inductively coupled.

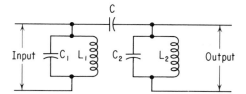

Fig. 11-26. Capacitive-coupled circuits.

Still another method of coupling is the **inductive-coupling** method illustrated in Figure 11-27. In this case coils L_1 and L_2 are inductively coupled, forming the primary and secondary windings, respectively, of a transformer.

Since such a transformer operates at high frequencies, it is of the air-core or powdered iron-core type. As such, the coupling between the primary and secondary windings is considerably less than unity, and the step-up or step-down effects arising out of turn ratios do not hold. Coupling between the inductors can be varied by placing the windings closer or farther apart.

The reflected impedances or resistances from one circuit to the other become quite important. Assume that both the primary and secondary circuits are tuned to the same resonant frequency and that the coils are

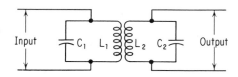

Fig. 11-27. Inductive-coupled circuits.

placed a considerable distance apart. As current flows in the primary circuit, a certain amount of voltage will be induced in the secondary circuit. Because the coupling is very small, the energy transferred will be small, too, and so will be the voltage output of the secondary circuit.

Since, at resonance, the inductive reactance cancels out the capacitive reactance, both the primary and secondary circuits are resistive in nature. But at all frequencies above and below the resonant frequency (f_r), X_L does not cancel out X_C, and these circuits, therefore, become reactive in nature.

At the resonant frequency the reflected resistance from each circuit reduces the **selectivity** of the other. At frequencies other than the resonant, the reflected impedance from one circuit to the other upsets the **tuning** and makes each circuit resonant for some other frequency than the original resonant frequency. Thus, the voltage output becomes maximum for these new frequencies, whereas it drops off at the original resonant frequency because of the effect of the reflected resistance.

This is illustrated in the series of curves shown in Figure 11-28. Curve A shows the effect of having the coils far apart. Both the reflected resistance and reflected impedance are quite small. The selectivity, therefore, is fairly high, as witnessed by the sharp peak of the curve. But since the energy transfer is low, the output voltage is not very great.

As the coils are brought closer and closer together, the energy transfer becomes greater and the output voltage increases (curves B and C). Note that the curves tend to flatten out, indicating the loss of selectivity resulting from the increased reflected resistance.

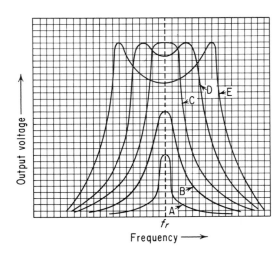

Fig. 11-28. Resonance curves showing the effects of changes in coupling.

Curve C represents the maximum coupling possible before the effect of reflected impedance becomes apparent. This point is called the **critical coupling.** Note that the output voltage has reached its maximum.

As the coupling is increased, the reflected impedance becomes greater. The selectivity is reduced, flattening the top of the curve, and a double peak appears, one on each side of the original resonant frequency. This indicates the new resonant frequencies. (See curve D.)

As the coupling is increased still further, the effect becomes more pronounced. (See curve E.) The curve becomes flatter and broader, and the two new resonant peaks are further apart. Because of the greater reflected resistance from the closer coupling, the dip at the original resonant frequency becomes greater.

The flow of current in the secondary circuit depends, not only on the output voltage, but on the resistance of the load as well. If the resistance of the load is low, a relatively large current will flow. This larger flow of current will create a more powerful magnetic field around the secondary inductor which, in turn, will result in a larger reflected resistance into the primary circuit. As a result, the selectivity of the circuit will be reduced.

QUESTIONS

1. What is the effect of resistance upon the phase angle between the voltage and current in an a-c circuit containing inductance?

2. In an a-c circuit containing a resistor of 12 ohms in series with an inductor whose inductive reactance is 16 ohms, how much current will flow if the voltage is 100 volts?   [*Ans.* 5 amperes.]

3. Explain what is meant by the **R-L time constant.**

4. What is the *R-L* time constant of a series circuit containing a 10-henry inductor and a 100-ohm resistor?   [*Ans.* 0.1 second.]

5. What is the effect of resistance upon the phase angle between the voltage and current in an a-c circuit containing capacitance?

6. A capacitor and a 4000-ohm resistor are connected in series in an a-c circuit. If, with a voltage of 100 volts, 20 milliamperes of current flows through this circuit, what is the reactance of the capacitor?

[*Ans.* 3000 ohms.]

7. Explain what is meant by the **R-C time constant.**

8. What is the *R-C* time constant of a series circuit containing a 0.005-μf capacitor and a 2-megohm resistor?   [*Ans.* 0.01 second.]

9. What is the total reactance in an a-c circuit containing an inductor whose inductive reactance is 2500 ohms and a capacitor whose capacitive reactance is 3000 ohms, connected in series? What will be the nature of this reactance? [*Ans.* 500 ohms capacitive reactance.]

10. What is the total impedance of an a-c circuit containing a resistor, inductor, and capacitor in series, if the resistance is 150 ohms, the inductive reactance is 300 ohms, and the capacitive reactance is 500 ohms?
[*Ans.* 250 ohms.]

11. What is meant by the **resonant frequency** of an a-c circuit?

12. At the resonant frequency, what is the theoretical impedance of (*a*) a series-resonant circuit; (*b*) a parallel-resonant circuit?

13. What is the effect of resistance on (*a*) the current flowing in a series-resonant circuit; (*b*) the impedance of a parallel-resonant circuit?

14. Find the resonant frequency of an a-c circuit containing a 5-henry inductor and a 5-microfarad capacitor connected in series. [*Ans.* 31.8 Hz.]

15. What is the function of an electrical filter? Explain.

16. Draw the circuit of resonant circuits used as a **band-pass filter.** Explain its action.

17. Draw the circuit of resonant circuits used as a **band-stop filter.** Explain its action.

18. Describe the construction of an iron-core transformer. Explain the function of each of its parts.

19. What is meant by the **leakage flux** of a transformer? How may it be reduced? Explain.

20. What are the **copper** and **iron** losses of a transformer due to? What can be done to reduce these losses? Explain.

21. Explain what is meant by (*a*) a **step-up transformer**; (*b*) a **step-down transformer.**

22. A transformer has a primary winding of 500 turns and a secondary winding of 5000 turns. If the primary winding is connected across a 120-volt a-c line, what will be the voltage across the secondary winding?
[*Ans.* 1200 volts.]

23. State three functions of the transformer. Explain.

24. Explain the operation of a **step-up autotransformer,** using a simple diagram. How may this transformer be used as a **step-down** type?

25. Explain why the iron-core transformer is not suitable for coupling high-frequency circuits.

26. Describe three methods for coupling high-frequency circuits.

27. What is meant by **reflected impedance** in coupled high-frequency circuits? What is its effect?

A-C
MEASURING
INSTRUMENTS

12

A. METER MOVEMENTS THAT MAY BE USED FOR A-C OR D-C MEASUREMENT

Most of the d-c movements described in Chapter 7 may be used for a-c measurement as well. In the case of the hot-wire movement (see Figure 7-1) the wire is heated as current flows through it, regardless of whether the current is direct or alternating. In the inclined-coil movement (see Figure 7-2) the iron vane tends to line itself up with the magnetic lines of force around the coil, regardless of the direction of these lines of force. Similarly, the vanes of the repulsion-vane movement (see Figure 7-3) tend to repel each other, regardless of whether this repulsion is due to two north poles or two south poles.

In the solenoid-type movement (see Figure 7-4) the solenoid is energized and the plunger attracted, regardless of the direction of current flow through the coil. Hence, this type of movement can be used on either direct or alternating current. In the dynamometer movement (see Figure 7-8) the movable and fixed coils are connected in series. Hence, the relative attraction or repulsion between coils remains the same, regardless of the direction of the current that flows through all of them. For this reason this movement, too, can be used for a-c as well as d-c measurement.

When they are connected in an a-c circuit, all of the above movements, except the hot-wire movement, suffer from a common fault. As the current reverses its direction of flow, the movable member of the movement tends to vibrate, causing the pointer to flutter and making it difficult to obtain an exact reading. To overcome this flutter, a process called **damping** is employed.

One method, called **air damping,** is to attach a small, light vane to the same shaft that carries the pointer. This vane is permitted to swing in a closed box that is just large enough to accommodate the vane. As the shaft rotates, the vane swings in the box, compressing the air in front of it. The pressure of the compressed air on the vane tends to slow up its swing and thus reduce the flutter of the pointer.

The other method, called **magnetic** damping, employs a device similar to that used in the watthour meter discussed in Chapter 7 (see Figure 7-16). A small, light vane of aluminum or copper is attached to the shaft and passes between the poles of a permanent magnet. As the movement vibrates, the vane cuts across the field of the magnet. The resulting induced current in the vane (the **eddy current**) sets up its own magnetic field which, interacting with the field of the magnet, slows up the swing of the vane, thus reducing the pointer flutter.

Air damping generally is used for movements such as the inclined-coil, repulsion-vane, and solenoid types, since they are most sensitive to the disturbing effect of the magnetic field of a nearby magnet. Care must be taken with all the movements, except the hot-wire type, to keep them away from large metal objects. Otherwise, the varying magnetic fields that surround their coils as alternating current flows through them would induce currents in these metals. These induced currents represent losses that produce false readings by the meters.

All of the above movements, it has been stated, will operate on alternating current. However, if the frequency is too high, before the pointer has a chance to move from its zero position, the current reverses itself and is flowing in the opposite direction. The pointer, under such conditions, merely vibrates around zero. Hence the movements can be used only with alternating currents of relatively low frequencies, such as the 60-Hz current used in most house circuits.

The hot-wire movement is an exception. Since the heating effect of the current does not depend upon its frequency, currents of any frequency can be measured.

As is true for d-c operation, all of these movements may be used as a-c ammeters or voltmeters, provided that they are properly connected into the circuits. Ammeter ranges may be extended by means of shunts, and voltmeter ranges through the use of multipliers. There is still another method for extending the ranges of a-c ammeters and voltmeters. This is through the use of a **transformer.**

As you know, the transformer is a device that may be used to step up (increase) or step down (decrease) an alternating current or voltage. Suppose we wish to measure a very large alternating current. By means of a step-down transformer (known as a **current transformer**) we step this current down, say, 200 times. Assume that, as the output of the transformer is measured by an a-c ammeter, a reading of 5 amperes is indicated. The original current, then, is 5×200, or 1000 amperes.

Current transformers may have different step-down ratios and usually are designed to produce a maximum output of 5 amperes. Accordingly, they generally are used with a-c ammeters having a 0–5 ampere scale.

Other types of step-down transformers, known as **potential transformers,** may be used similarly to measure large alternating voltages. These transformers, too, may have different step-down ratios and generally are designed to produce a maximum output of about 100 volts.

Except for some specially designed types, all a-c voltmeters and ammeters indicate the **effective,** or **rms,** values of voltage and current. (See Chapter 8, subdivision C.) Theoretically, the a-c wattmeter measures the product of the **instantaneous** values of voltage and current (Chapter 9, subdivision C, 1). However, since the meter cannot follow the quick variations of these instantaneous values, it indicates, instead, the **average** value of true power.

B. ADAPTING THE D'ARSONVAL-TYPE MOVING-COIL MOVEMENT TO A-C MEASUREMENT

The d'Arsonval-type moving-coil movement (see Figure 7-6) has a number of advantages over the other types. It is efficient, may be very accurate, and may have great sensitivity. Besides, it has the desired uniform scale. It has one main disadvantage: it cannot be used to measure alternating current. With the current flowing through it in one direction, the pointer moves from its left-hand zero position to the right across the scale. However, as the current reverses and flows through the movement in the other direction, the pointer attempts to move left from the zero position. Thus it may be bent or the instrument otherwise damaged.

Accordingly, two methods have been devised for adapting this movement to a-c measurement. These operate by converting the alternating current to direct current and then using the instrument to measure this direct current.

1. Rectifier Method

The first method changes the alternating current directly into direct current by means of a process called **rectification.** It has been found that if a copper disk is coated on one side with a thin layer of copper oxide, current will flow quite easily from the copper to the copper oxide, but will encounter an extremely high resistance if it attempts to flow in the opposite direction. Thus the copper-copper-oxide disk furnishes a one-way passage for electric current. We call this combination a **copper-oxide rectifier.** The symbol for the rectifier is ➤⊦ . The flow of current is taken to be from the heavy vertical bar to the arrow tip.

Now let us see what happens if the rectifier is placed in series in an a-c circuit (see Figure 12-1A). The waveform of the output current from

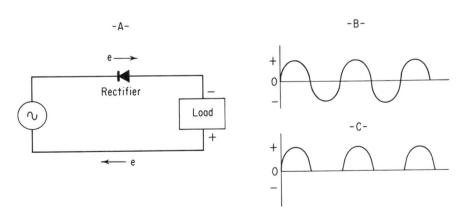

Fig. 12-1. A. Rectifier connected in series in an a-c circuit.
B. Waveform of input to rectifier.
C. Waveform of output from rectifier.

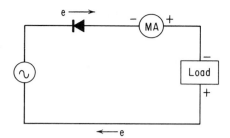

Fig. 12-2. How a moving-coil milliammeter is connected in a half-wave rectifier circuit.

the a-c generator is shown graphically in Figure 12-1B. For one half of each cycle, current flows through the rectifier and load. During the other half-cycle, current cannot flow through the rectifier and, hence, there is no current flow through the circuit. This is shown graphically in Figure 12-1C.

Note that the graph of Figure 12-1B indicates an alternating current — that is, a flow first in one direction and then in the other. The graph of Figure 12-1C, on the other hand, shows a flow in only one direction. Hence it is a direct current. It is true that it is not a **steady** direct current. Rather, the effect of a rectifier is to convert an alternating current to a **pulsating direct current.** Because only half the cycle is utilized (the other half being blocked out), the circuit shown in Figure 12-1A is called a **half-wave rectifier circuit.**

Now let us connect a moving-coil milliammeter in series in this circuit, as shown in Figure 12-2. Because a direct current flows in this circuit, the meter will indicate this current. However, the inertia of the movement is such that it cannot follow the rapid series of pulses (there are 60 pulses per second if the current furnished by the generator has a frequency of 60 Hz). Instead, it registers the **average** value of these pulses. Since most a-c meters are calibrated to indicate **effective,** or **rms,** values, we must multiply the average values registered on this meter by 1.11 to obtain the rms values. This is done on the instrument scale, which is marked directly in rms values.

The ratio between average and rms values holds true only if the waveform of the current is sinusoidal. If it be some other shape, the 1.11-relationship does not hold. Accordingly, care must be taken to use the moving-coil meter, which is marked in rms values, only with currents that have the sinusoidal waveform.

If you examine Figure 12-2, you will see that current flows through the circuit only during half of each cycle. A more efficient circuit, called

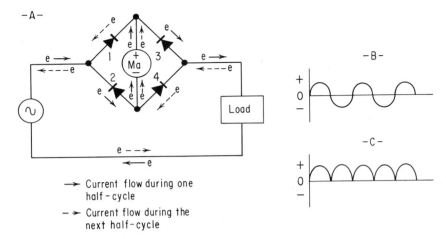

−A−

Load

Ma

1 2 3 4

−B−

−C−

⟶ Current flow during one half-cycle

─ ➤ Current flow during the next half-cycle

Fig. 12-3. A. Bridge-rectifier circuit.
   B. Waveform of input to rectifier.
   C. Waveform of output from rectifier.

the **full-wave,** or **bridge, rectifier circuit,** utilizes both halves of each cycle, as shown in Figure 12-3A. It employs four rectifiers.

During one half-cycle, current flows (as indicated by e ⟶) from the generator, through rectifier 2, through the meter, through rectifier 3, through the load, and back to the generator. Current is prevented from flowing through the other portions of the bridge circuit by the high resistances of the alternative paths.

During the next half-cycle, current flows (as indicated by e ---➤) from the generator, through the load, through rectifier 4, through the meter, through rectifier 1, and back to the generator. Again, current is prevented from following alternative paths through the bridge circuit by the high resistances of these paths.

Note that, although the current flows in both directions through the load, current flow through the meter during the entire cycle is always in the same direction. Hence the current flowing through the meter is a direct current, which it is able to register.

The output of the full-wave rectifier circuit is shown graphically in Figure 12-3C. Note that there are now two pulses produced during each cycle. Accordingly, the average d-c values registered by the meter are higher than for an equivalent half-wave rectifier circuit. As before, the

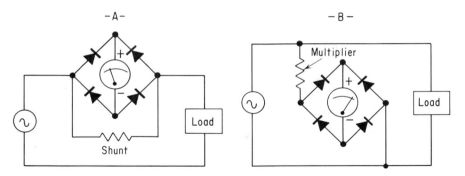

Fig. 12-4. A. Circuit of bridge-rectifier meter used as ammeter.
B. Circuit of bridge-rectifier meter used as voltmeter.

scale is marked in rms values. Also, the calibrations of the scale hold true only if the current is sinusoidal in waveform.

The rectifier-type meter can be used as an ammeter or as a voltmeter, as illustrated in Figure 12-4. Where the meter is used as an ammeter and a shunt is employed, the shunt is placed across the entire rectifier-movement circuit (Figure 12-4A) so that the bulk of the line current flows through the shunt. If the shunt were placed merely across the meter movement, the entire line current would flow through the rectifiers, thus

Fig. 12-5. Multirange a-c ammeter.

Weston Instruments Div.,
Daystrom, Inc.

damaging them. Of course, when calculating the value of the shunt, we now must take into consideration the total resistance offered by the rectifiers and meter movement.

Similarly, when calculating the resistance of the voltage multiplier for the voltmeter (see Figure 12-4B), we must consider the total resistance of the meter movement and rectifiers, too. As is true of other types of meters, the range of the rectifier-type meter may be varied by using different values for the shunts and multipliers.

Note that when the rectifier is in its current-opposing position, it acts as a sort of capacitor. As a result, it offers a certain amount of capacitive reactance to the current. If the frequency of this current is low, the capacitive reactance is very high and very little current can flow through the rectifier. If, however, the frequency is high, the capacitive reactance is low and considerable current can flow through the rectifier, upsetting its rectifying action. Accordingly, the rectifier-type meter is not suitable for currents of very high frequencies. It generally is not used where the frequency of the current is higher than about 20 kHz.

The copper-copper-oxide disks used as meter rectifiers usually are about a half inch in diameter. Because the copper-oxide layer is very thin, it cannot stand high voltage without breaking down. Accordingly, several such disks may be sandwiched together (connected in series with the copper side of one disk in contact with the copper-oxide side of another) to form a larger rectifier unit that can withstand a higher voltage. The entire unit is bolted together, care being taken to insulate the disks from the bolt on which they are mounted. Suitable terminals and metallic fins for radiating away some of the heat produced are provided. (See Figure 12-6.) Where a bridge-type rectifier is employed, all four rectifiers may be mounted in one such stack with suitable insulation between units and terminals for each.

Fig. 12-6. Meter rectifier.

Another type meter rectifier frequently used is made of selenium coated on one side of an iron disk. Its action is the same as that of the copper-copper-oxide type and it is constructed in the same way.

Since an ammeter must be placed in series with the circuit whose current is to be measured, this circuit must be opened every time we wish to insert the measuring instrument. Sometimes this is difficult to do. Accordingly, a clever device, called a **clamp-type ammeter,** may be used to measure the current in an a-c circuit without the necessity for opening the circuit.

Essentially, it consists of a transformer that has a primary winding coupled to a secondary winding by means of an iron core. As a result of the magnetic field set up by the alternating current flowing through the primary winding, a current is induced in the secondary winding. The strength of the magnetic field around the primary winding depends upon the number of turns in that winding and the current flowing through it. Thus, if all other factors remain constant, the greater the current flowing through the primary winding, the greater will be the current induced in the secondary winding.

The primary winding is the conductor in the circuit whose current we wish to measure and may be considered to consist of a single turn. This is coupled to the secondary winding (contained within the clamp-type ammeter illustrated in Figure 12-7A) by means of the iron core. To avoid the necessity for opening the circuit, this core is split and hinged. Thus the core may be opened and placed around the conductor. Then it is closed and the primary is effectively coupled to the secondary winding.

Hence a current, proportional to the amount of current flowing through the conductor, is induced in the secondary winding. This induced current is then measured by means of a rectifier-type, moving-coil ammeter. The scale of the ammeter is calibrated to read the current flowing in the conductor. As in any other type of ammeter, shunts may be used to vary the range of the instrument.

Frequently, the same moving-coil instrument may be used as a voltmeter as well, by employing suitable multipliers. As in ordinary voltmeters, the instrument then must be connected in shunt with the circuit to be measured. The circuit of a typical clamp-type voltmeter-ammeter is shown in Figure 12-7B. Current is measured by clamping the core around the current-carrying conductor. Voltage is measured by a pair of leads connecting the terminals in the side of the instrument across the circuit.

–A–

Weston Instruments Div.,
Daystrom, Inc.

–B–

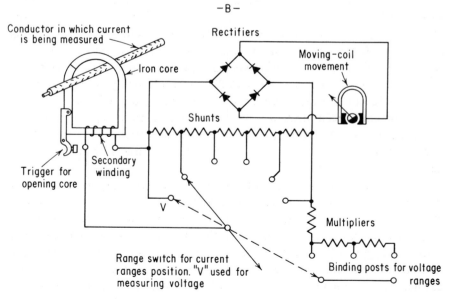

Fig. 12-7. A. Clamp-type a-c volt-ammeter.
B. Circuit of clamp-type a-c volt-ammeter.

The ammeter range is chosen by means of the switch on the front of the instrument. When voltage is to be measured, the switch is turned to VOLTS. The voltmeter range is determined by selecting the proper set of terminals. The trigger is used to open or close the core.

2. Thermocouple Method

The other method for changing alternating current into direct current is an indirect process that first converts the alternating current into heat and then uses this heat to generate a direct current. It has been found that if two dissimilar metal wires or strips are joined at one end and this junction heated, a small direct voltage will appear between the cool, open ends. Further, this voltage is directly proportional to the difference in temperature between the hot and cold ends. This phenomenon is known as the **thermoelectric effect,** and the combination of the metal wires or strips is called a **thermocouple.**

If we connect a resistor (called the **heater**) in series with the a-c line, the current heats the resistor as it flows through it. The amount of heat is proportional to the square of the current (heating effect $= I^2R$). This heat is applied to the junction of a thermocouple. As a result, a direct voltage, directly proportional to the heat, appears across the cool, open ends. We can measure this voltage (or the current set flowing by it) by means of a moving-coil millivoltmeter (or milliammeter). The reading on the meter, then, will be an indication of the alternating current flowing in the line. (See Figure 12-8.)

Any two dissimilar metals may be used for the thermocouple. How-ever, two different alloys, **constantan** and **manganin,** are frequently em-

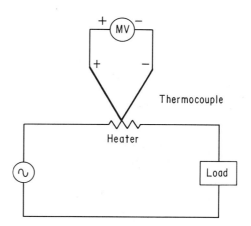

Fig. 12-8. Thermocouple meter used as an ammeter.

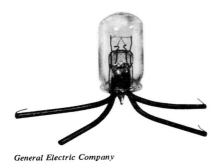

General Electric Company

Fig. 12-9. Thermocouple enclosed in glass bulb.

ployed. The thermocouple and heater generally are formed into one unit. For measuring currents up to about 100 milliamperes, the entire unit usually is sealed into an evacuated glass tube to reduce heat loss to the surrounding air. Wire leads sealed into the glass make contact with the various elements. For higher currents, where the heat loss is not serious, the entire unit may be exposed to the air.

The combination of thermocouple, heater, and millivoltmeter is called a **thermocouple meter.** Such meters can be used to measure current or voltage. Their ranges may be extended through the use of shunts or voltage multipliers. Because the heating effect of the current is not affected by its frequency, thermocouple meters can be used in high-frequency circuits.

Note that the scale of the meter is of the square-law type, since the heating effect is proportional to the square of the current. The scale can be modified to a uniform type by changing the shape of the pole pieces to that illustrated in Figure 12-10. Note that when the moving coil is in

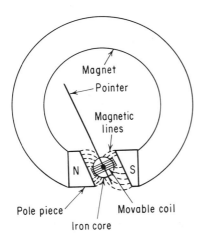

Magnet

Pointer

Magnetic lines

N   S

Pole piece   Movable coil

Iron core

Fig. 12-10. How the pole pieces of a thermocouple meter are modified to make the scale linear.

its low-scale position (pointer to the left), it is cutting across the stronger portion of the magnetic field (as indicated by the concentration of the magnetic lines). When the coil is in its high-scale position (pointer to the right), it is in the weaker portion of the field. This weakening of the magnetic field as the coil moves to its high-scale position reduces the sensitivity of the meter for that portion of the scale and tends to change the square-law relation to a linear one.

An interesting variation of the thermocouple meter is the **pyrometer,** an instrument used to measure temperature. Here, only the thermocouple and millivoltmeter are used. The junction of the thermocouple becomes the probe that is applied to the object whose temperature is to be measured; the meter scale is marked in degrees of temperature.

C. THE A-C WATTHOUR METER

The a-c watthour meter operates on the same general principle as the d-c type described in Chapter 7, subdivision D, 5. An a-c motor (which will be discussed later in the book) revolves at a speed that is determined by the amount of power being consumed by the house circuits. The length of time this power is being consumed determines the length of time this motor rotates. The shaft of the motor is attached to a set of dials that record the electrical energy (power × time) in kilowatthours consumed by the circuits.

A simplified drawing of the a-c watthour meter is shown in Figure 12-11A. The motor consists of the current coils, the potential (voltage) coil, and the aluminum disk between them. The current coils are wound with few turns of heavy wire on a set of soft-iron cores and are connected in series with the power line. The potential coil is wound with many turns of fine wire on an iron core and is connected across the power line. Note that these coils are fixed. The aluminum disk is the rotor. As current flows through the coils, the aluminum disk rotates between them. The damping magnets perform the same function here as they do in the d-c watthour meter.

D. THE FREQUENCY METER

In a-c work it sometimes is necessary to keep the frequency of the current constant. Accordingly, a **frequency meter** may be connected to the circuit to indicate the frequency at any given time. There are a num-

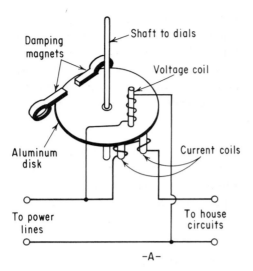

Damping magnets

Shaft to dials

Voltage coil

Aluminum disk

Current coils

To power lines

To house circuits

–A–

–B–

Westinghouse Electric Corp.

Fig. 12-11. A. Simplified diagram of a-c watthour meter.
B. Commercial a-c watthour meter.

ber of different types of frequency meters, but the simplest, perhaps, is the type illustrated in Figure 12-12.

A coil of many turns of fine wire wound on a soft-iron core is connected across the line. Near it is a soft-iron plate held in place by a flexible support. As current flows through the coil, the iron plate is attracted twice each cycle as the current reaches its positive and negative peaks. Between these peaks the springiness of the support restores the plate to its original position. Thus the plate makes two vibrations for every cycle of the current.

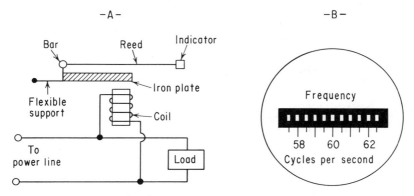

Fig. 12-12. A. Frequency meter.
B. Dial of frequency meter.

Attached to the plate, and moving with it, is an iron bar. A set of thin reeds are attached to the bar in such a way that they are able to vibrate freely. These reeds have natural frequencies that differ in sequence by two vibrations per second. (If a reed, or any other object, is free to vibrate, it normally will do so at a frequency that is determined by its physical characteristics, such as kind of material, length, thickness, etc. This frequency is known as the **natural frequency** of the object.)

As the iron plate vibrates, the reeds vibrate also. But the reed whose natural frequency matches the rate of vibration of the plate will vibrate most strongly. Thus, by noting which reed vibrates most vigorously, we may determine the rate of vibration of the iron plate and, hence, the frequency of the current.

Light-weight indicators are attached to the free ends of the reeds so that their vibrations may be seen more easily. The vibrating-reed frequency meter does not cover a broad band of frequencies but, rather, a narrow range of frequencies from a few cycles per second below the desired frequency to a few cycles per second above. Thus, if the frequency of the current falls or rises a few cycles per second, the deviation may be noted and steps taken to restore the current to the desired frequency.

E. BRIDGES

Although the ohmmeter method for measuring resistance is the simplest, more accurate measurement may be obtained by means of the

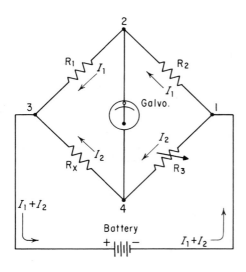

Fig. 12-13. Basic circuit of the Wheat-stone bridge.

Wheatstone bridge (see Figure 12-13). Two fixed resistors R_1 and R_2, a potentiometer R_3, and the unknown resistance whose value is to be measured (R_x) are connected to form a diamond-shaped bridge circuit. The values of R_1 and R_2 are known, and the dial of R_3 is calibrated in ohms so that the value of R_3, too, is known. A battery is connected to points 1 and 3 and a center-zero galvanometer is connected between points 2 and 4.

Current flowing from the battery divides at point 1. Part of this current (I_1) flows through resistors R_2 and R_1 to point 3, and back to the battery. The other part of the current (I_2) flows through resistors R_3 and R_x to point 3 and joins I_1 in flowing back to the battery.

If points 2 and 4 are at the same potential (that is, the bridge is **balanced**), no current will flow through the center-zero galvanometer connected across these two points. If there is a difference of potential between these two points (the bridge is **unbalanced**), current will flow through the meter in the direction determined by the unbalance, and in an amount determined by the degree of unbalance.

To obtain the condition of balance, the voltage drop across R_2 must be equal to that across R_3, and the voltage drop across R_1 must be equal to that across R_x. Since the voltage drops across the various resistors are equal to the product of each resistance and the current flowing through it, then

$$I_1 \times R_2 = I_2 \times R_3$$

and

$$I_1 \times R_1 = I_2 \times R_x.$$

Solving for I_1 in the first equation, we get

$$I_1 = \frac{I_2 \times R_3}{R_2}.$$

Substituting this value of I_1 in the second equation, we get

$$\frac{I_2 \times R_3}{R_2} \times R_1 = I_2 \times R_x.$$

Dividing both sides of this equation by I_2, we get

$$R_x = \frac{R_1}{R_2} \times R_3.$$

From the above we see that $R_x = K \times R_3$, where K represents the fixed ratio of R_1/R_2. Should R_1 equal R_2, K = 1 and $R_x = R_3$. Should R_1 be ten times as great as R_2, K = 10. Then $R_x = 10 \times R_3$, and so forth.

The bridge is balanced by adjusting R_3 until the galvanometer reads zero. Then, knowing the value of K and noting the value of R_3 (from its calibrated dial), the value of the unknown resistance (R_x) may be ascertained. The point at which the bridge is balanced is called the **null point,** and the galvanometer that shows when this point is reached is called a **null indicator.**

A practical variation of the Wheatstone bridge is illustrated in Figure 12-14. Instead of fixed resistors R_1 and R_2, a potentiometer is connected between points 1 and 3 of the bridge. Instead of R_3, R_m (whose value

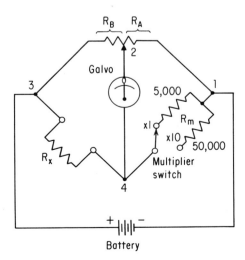

Fig. 12-14. Practical Wheatstone bridge circuit for measuring resistance.

may be either 5000 or 50,000 ohms, depending upon the position of the **multiplier switch)** is employed.

Let us call the resistance between the slider and the end of the potentiometer connected to point 1, R_A. The resistance between the slider and the end connected to point 3 may be called R_B. Our formula for the balanced bridge now becomes

$$R_x = R_m \times \frac{R_B}{R_A}.$$

The bridge may be balanced by throwing the multiplier switch to its 5000-ohm position and varying the ratio between R_A and R_B. This may be done by manipulating the slider arm of the potentiometer until the galvanometer reads zero. The dial that operates the slider arm may be calibrated in terms of the resistance of R_x. If the switch is thrown to the 50,000-ohm position, the dial reading then must be multiplied by 10.

As shown above, the Wheatstone bridge is d-c operated. It would work equally well if a source of alternating current were substituted for the battery. However, alternating current would flow through the galvanometer, hence a rectifier must be placed in series with it to convert the alternating current to a direct current which the instrument can measure. Also, care must be taken that the resistors in all the arms are noninductive, otherwise the additional inductive reactance would interfere with the operation of the bridge. Such resistors may be of the composition type or, if they are wire wound, special noninductive windings may be employed. Half the winding is in one direction and half in the other, hence the inductances balance out.

Where alternating current is employed, a pair of headphones may be substituted for the galvanometer as a null indicator. When the bridge is unbalanced, alternating current flows through the headphones, producing an audible hum. When the bridge is balanced, this hum is reduced to a minimum or disappears altogether. When headphones are used it is customary to generate the alternating current by means of an oscillator, hummer, or buzzer that produces an output at a frequency of about 1000 Hz, because the human ear can discern this sound most easily.

In the a-c Wheatstone bridge it is the balancing of the impedances of the arms that produces the null condition. Hence it makes no difference

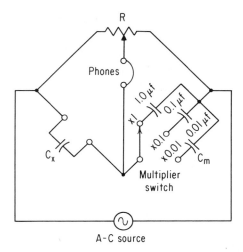

Fig. 12-15. Practical a-c bridge circuit for measuring capacitance.

whether we employ resistance, capacitive reactance, inductive reactance, or combinations of the three. Accordingly, we may use resistors, capacitors, inductors, or combinations of these in the arms of the bridge.

This latitude provides us with an opportunity to employ the bridge circuit to measure capacitance or inductance. A practical capacitance-measuring circuit is shown in Figure 12-15. Note that this circuit resembles the resistance-measuring circuit of Figure 12-14, except that fixed capacitors are substituted for the fixed resistors in the multiplier section and the bridge is a-c operated, using an a-c source instead of a battery and headphones as a null indicator. The dial of potentiometer R is calibrated to indicate the value of the unknown capacitance (C_x) when the multiplier switch is thrown to its 1.0-μf position. When the switch is in the 0.1-μf position, the reading on the dial is divided by 10. When the switch is in the 0.01-μf position, the reading on the dial is divided by 100.

In an ideal capacitor the electrical energy used to charge up the capacitor is returned without loss as the capacitor discharges. There is no true power in the circuit and the power factor is zero (see Chapter 9, subdivision C). Actually, however, there always are some losses, owing to resistance and leakage. Hence there is some true power in the circuit, and the power factor rises above zero. The greater the losses, the higher is the power factor.

To measure the power factor of a capacitor, another type of bridge circuit may be employed. (See Figure 12-16.) Capacitor C_s is a standard capacitor having a known, low power factor. Capacitor C_x is the capacitor under test. Potentiometer R_{pf} is the power-factor potentiometer, whose dial is calibrated to indicate the power factor of that capacitor.

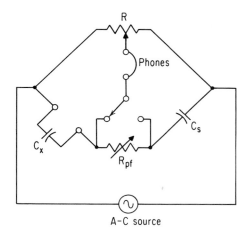

Fig. 12-16. Bridge circuit for measuring the power factor of a capacitor.

To operate this bridge, it is balanced as far as is possible by means of potentiometer R, switching the power-factor potentiometer (R_{pf}) from one arm to the other. Then R_{pf} is adjusted for a better balance. It may be necessary to readjust R to complete the balance. When the bridge is balanced, the power factor of C_x may be read on the dial of R_{pf}.

In a method similar to that used for measuring capacitance, the a-c Wheatstone bridge may be modified to measure inductance. In this case, fixed multiplier inductors are used instead of fixed capacitors, and the dial of the potentiometer is calibrated to read in units of inductance.

Where the unknown frequency of a current is within a range up to about 20,000 Hz, the frequency can be measured by a variation of the Wheatstone bridge circuit known as the **Wien bridge.** (See Figure 12-17.) It can be proved mathematically that if the components of the bridge are such that $C_1 = C_2$, $R_3 = R_4$, and $R_1/R_2 = 2$, then the bridge is balanced at a frequency determined by the following formula:

$$f = \frac{1}{2\pi R_3 C_1}.$$

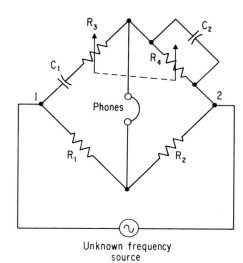

Fig. 12-17. Circuit of the Wien bridge for the measurement of frequency.

Unknown frequency source

Accordingly, the bridge is constructed with C_1 and C_2 equal in value and with the resistance of R_1 twice that of R_2. R_3 and R_4 are identical slide-wire potentiometers, both mounted and operated by the same shaft. The dial that controls this shaft may be calibrated in units of frequency. The unknown frequency source is connected to points 1 and 2 of the bridge and the dial controlling R_3 and R_4 rotated until the null point, as indicated by the headphones, is reached. Then the frequency of the unknown source can be read from the dial.

QUESTIONS

1. Explain why the d'Arsonval-type moving-coil movement cannot be used directly for a-c measurement. Why can the dynamometer movement be so used?

2. Describe two methods commonly used to convert the d'Arsonval-type moving-coil movement to a-c measurement.

3. Draw and explain a half-wave rectifier circuit using a copper-oxide rectifier.

4. Draw and explain a full-wave bridge-rectifier circuit using copper-oxide rectifiers.

5. Explain the operation of a thermocouple as applied to a meter movement.

6. Explain the operation of an a-c watthour meter.

7. Explain the operation of a vibrating-reed frequency meter.

8. Draw and explain the operation of the basic Wheatstone bridge circuit.

9. Draw and explain the operation of a bridge circuit for the measurement of capacitance.

10. Draw and explain the operation of the Wien bridge for measurement of the frequency of an unknown current.

SOURCES OF ELECTRICAL ENERGY

IV

MECHANICAL GENERATORS

13

A. ALTERNATING-CURRENT GENERATORS

Michael Faraday discovered, you will recall, that if a conductor cuts across a magnetic field, an electromotive force is generated between the ends of the conductor. This is the principle of the simple generator illustrated in Figure 8-3. A single loop of wire revolves in a magnetic field, cutting across the magnetic field as it rotates. As a result, an electromotive force is induced in the loop.

We also now know (see Chapter 8, subdivision A) that the induced voltage (and the induced current resulting from it) will be greater if:

1. The magnetic field is made stronger.

2. The number of conductors cutting across the magnetic field is increased.

3. The speed of relative motion between the magnetic field and conductors is increased.

229

1. Generators Using Permanent Magnets for Field

A practical application of the above principles is the **magneto** illustrated in Figure 13-1A. The stronger magnetic field is obtained through the use of several horseshoe magnets so mounted that all similar poles are together, producing the effect of a large north pole and a large south pole. The number of conductors cutting across the magnetic field is increased by the use of a coil of many turns of wire instead of a single loop. Each turn of the coil adds its share of induced voltage to that of the others, resulting in a larger total induced voltage. The speed of relative motion between the magnetic field and conductors is increased by means of a system of gear wheels which multiplies the speed at which the crank is turned.

Figure 13-1B shows the **armature** of the magneto. Many turns of insulated copper wire (the **armature coil**) are wound on the iron armature

–A–

Fig. 13-1. A. Magneto.
   B. Armature and other ro-
      tating portions.
   C. Cross-sectional view.

Western Electric Co., Inc.

–B–

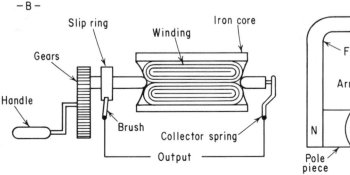

–C–

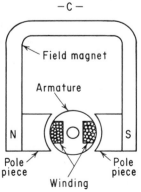

core. This core not only supports the coil, but also furnishes an easy path for the magnetic flux from the north field pole to the south. One end of the coil is connected to a metal **collector,** or **slip, ring** mounted on the armature shaft, but insulated from it. A metallic **brush** makes a wiping contact with this ring. The other end of the coil is attached to the armature shaft which, in turn, makes a wiping contact with the **collector spring.** The induced voltage appears between the brush and the collector spring.

A cross-sectional view of the magneto appears in Figure 13-1C. The armature coil is wound in slots in the armature core. Note the **pole pieces,** which are made of soft iron. Since magnetic lines of force will travel much more readily through soft iron than through air, these pole pieces concentrate the magnetic flux near the armature. Thus the magnetic field being cut by the armature coil is increased and, as a result, the induced voltage is greater.

A generator such as the one illustrated in Figure 13-1A is used in portable telephone systems to ring a bell at the far end of the line; it is called a **bell-ringing magneto.** Somewhat similar magnetos are used by certain types of gasoline engines (for airplanes, motorcycles, motorboats, and the like) to generate the high voltage required for ignition purposes. In some of these the field magnets are stationary and the armature revolves. In others the armature is stationary and the field magnets revolve. The results are the same — a voltage is induced as a magnetic field is cut by a conductor.

2. Generators Using Electromagnets for Field

For simple, low-current purposes the magneto described above may be suitable. However, where large amounts of current are required — as, for example, for lighting and for operating machinery — electromagnets are used instead of permanent magnets to produce the necessary magnetic field. Not only can we obtain stronger magnetic fields by means of electromagnets, but we can control the field of the electromagnet much more easily, simply by varying the strength of the current flowing through it. In this way the output voltage of the generator may be varied.

A steady magnetic field is required. Accordingly, a steady direct current must be sent through the electromagnet. In some installations this direct current is obtained from storage batteries. In others it may be obtained from a direct-current generator (which will be described later in this chapter).

The direct current for the field is called the **exciting current.** Where a d-c generator is used to furnish this current, it is called an **exciter.** Frequently the mechanical energy that rotates the exciter comes from the same source that drives the a-c generator. Often both generators are mounted on the same shaft.

The alternating-current generator is also known as an **alternator.** When used in circuit diagrams its symbol is –Ⓐ– . An armature, consisting of an armature coil wound upon an iron core, rotates in a magnetic field set up by the field electromagnets. The induced current is led to the external circuit through a set of brushes that make a wiping contact with a set of slip rings connected to the ends of the armature coil. (See Figure 13-2.)

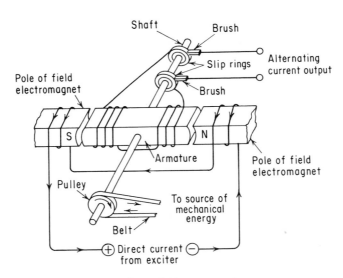

Fig. 13-2. Simplified illustration of practical a-c generator.

Here we have an improved version of the simple generator illustrated in Figure 8-3. The armature coil consists, not of a single loop, but of many turns of wire wound upon an iron core. The armature is rotated by mechanical energy from some source such as a steam or water turbine. The magnetic field is set up by direct current flowing through the field electromagnets. As in the simple generator, a single cycle is generated as the armature coil makes one complete revolution, passing by a single set of poles (that is — a north pole and a south pole).

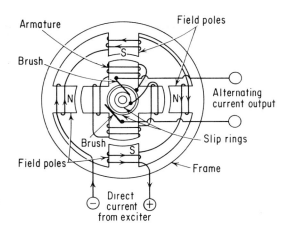

Fig. 13-3. Alternator with two sets of field poles.

Current in the United States usually is generated at a frequency of 60 Hz. Accordingly, the generator illustrated in Figure 13-2 would have to make 60 revolutions per second (60 rps) or 3600 revolutions per minute (3600 rpm). The design of a generator rotating at this speed presents a number of difficulties, particularly if the machine is a large one. Accordingly, we generally seek to reduce the speed of rotation.

You will recall that a cycle is generated as the armature passes a set of north and south poles. If there is only one such set of poles, there can be only one cycle per revolution. But if there are several sets of poles, a cycle will be generated as the armature passes by each set of north and south poles. Accordingly, there will be more than one cycle generated per revolution. Thus, to obtain the 60-Hz current, we would have a generator rotating at less than 3600 rpm.

In Figure 13-3 you see a generator that has two sets of poles. Note that the field winding is such that north and south field poles are set up alternately. Thus, in half a revolution the armature rotates past a north and south pole, and a cycle is generated. Accordingly, there are two cycles per revolution. To generate a 60-Hz current, the armature need rotate only 30 rps, or 1800 rpm.

Practical alternators may have even more sets of poles, thus permitting slower rotation to produce the desired 60-Hz current. There is a general equation which may be applied here:

$$f = \frac{\text{number of poles}}{2} \times \frac{\text{speed in rpm}}{60},$$

where f is the frequency in cycles per second. If we wish to determine the speed of rotation of an alternator to generate 60-Hz current, we may use the following equation:

$$\text{Speed (rpm)} = \frac{3600}{S},$$

where S stands for the number of **sets** of alternate north and south poles.

EXAMPLE. How many revolutions per minute must a 6-pole alternator make to generate a 60-Hz current?

Since 6 poles means 3 sets, then

ANSWER. $\qquad \text{Speed} = \dfrac{3600}{3} = 1200$ rpm.

Our large generating plants generally are of two types. Where falling water is used to drive a relatively slowly rotating water turbine which, in turn, furnishes the mechanical energy to turn the generator, the alternator, too, is of the slow-speed type, employing four or more poles to produce the 60-Hz current. Where the heat of an atomic reactor or burning fuel (such as coal, oil, or gas) is used to produce steam to drive a rapidly rotating steam turbine, which, in turn, rotates the generator, the alternator usually is of the high-speed type, employing only two or four poles.

Where the currents and voltages generated by the alternator are moderate, we may employ machines similar to those we have just discussed. However, some of our larger alternators are called upon to generate thousands of volts and to set flowing thousands of amperes of current. The wires of the armature coil must be very heavy, as must be their insulation. Thus it becomes quite unwieldy to rotate the armature. Also, the wiping contact between the slip rings and brushes produces serious losses at these high voltages and currents.

On the other hand, the currents and voltages for the field electromagnets are much smaller. Accordingly, it becomes feasible to have the armature remain stationary and to rotate the field coils around it. (You will remember that it makes no difference whether the magnetic field is stationary and the conductor cuts through it, or the conductor is stationary and the magnetic field moves across it.)

The armature (now called the **stator**) is made stationary and the output is taken from the ends of the armature coil by means of heavy, fixed connectors. The mechanical energy is applied to rotate the field coils (now called the **rotor**) inside the armature. The exciting current is applied to the field coils by means of brushes and slip rings.

3. Polyphase Generators

Suppose we wind two separate armature coils upon the same core of the generator. Assume that these coils are wound one over the other but are electrically separate and that each has its own set of slip rings and brushes. As the armature rotates, an alternating voltage will be induced in each coil. Because the coils are wound over each other and rotate together, the voltages induced in them will be in step, or **in phase.**

Now suppose that, instead of winding both coils over each other, we wind them at right angles to each other, as illustrated in Figure 13-4. At the instant armature coil A is cutting the maximum lines of force (and, hence, its induced voltage is at maximum), armature coil B is not cutting any lines and its induced voltage is at zero. The two induced voltages are **out of phase** with each other. Because the two coils are at right angles to each other (90° apart), the phase difference between the two induced voltages, too, is 90°.

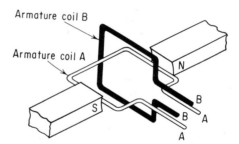

Fig. 13-4. Two armature coils at right angles to each other.

Assume we wind three armature coils upon the same core, each coil being wound 120° from its neighbor. The phase difference between the three induced voltages then is 120°. These relationships are shown graphically in Figure 13-5. In Figure 13-5A you see the graph of the alternating voltage produced by the generator with a single armature coil. Such a generator is called, appropriately, a **single-phase** generator. In Figure 13-5B you see the graph of the voltages produced by a generator with two armature windings 90° apart (the **two-phase generator**). The voltage of phase B starts 90° behind that of phase A and all the variations of the former are 90° behind those of the latter.

In Figure 13-5C you see the graph of the voltages produced by a generator with three armature windings, each 120° apart (the **three-phase**

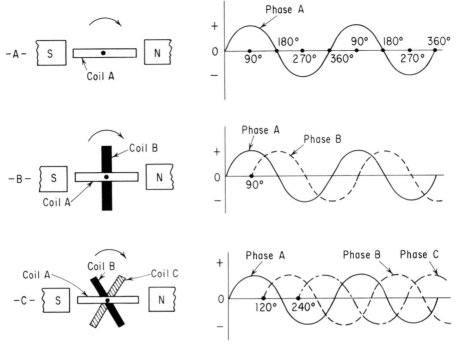

Fig. 13-5. A. Single-phase generator.
  B. Two-phase generator.
  C. Three-phase generator.

generator). Voltage of phase B is 120° behind that of phase A, and the voltage of phase C is 120° behind that of phase B.

Generators that have more than one set of armature coils are called **polyphase** generators; the voltages and currents they produce are called **polyphase** voltages and currents. Polyphase generators offer certain advantages, especially when they are called upon to produced power for certain types of a-c motors (which will be discussed later in the book). Most modern polyphase generators are of the three-phase type.

Many industrial machines employ three-phase alternating current. Since each armature coil of a three-phase generator has two ends, we might expect that six lines, two for each coil, would be needed to transmit the three-phase current from the generator to the machine. However, it is possible to join one end of each coil at the generator and then to transmit the current over three lines, one for each phase.

−A−   −B−

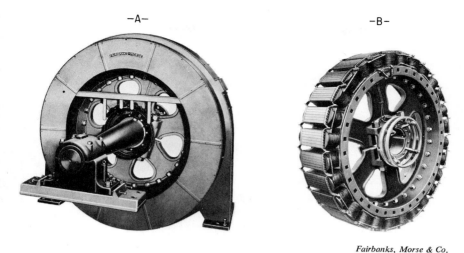

Fairbanks, Morse & Co.

Fig. 13-6. A. Three-phase alternator designed to operate at 60 Hz.
B. Revolving field of this alternator.

There are two variations of this connection. (See Figure 13-7.) The circuit shown in Figure 13-7A is called a **star,** or **Y (wye),** connection. The one shown in Figure 13-7B is called a **delta** connection. The advan-

−A−

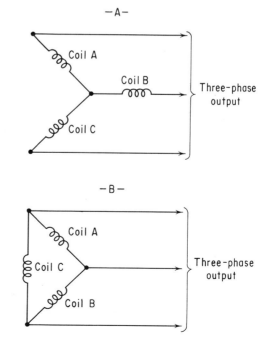

Fig. 13-7. Three-phase connections.
A. Star or Y connection.
B. Delta connection.

tages of polyphase alternating current and the characteristics of the different connections will be discussed in a later chapter.

B. DIRECT-CURRENT GENERATORS

Although alternating current is used practically everywhere in our country, direct-current generators are required to furnish current for a number of industrial processes such as electroplating and battery charging. Essentially, the direct-current generator resembles the alternator previously described except that it employs a device that mechanically changes the alternating current generated in the armature coil to a direct current, which is delivered to the brushes. This device is called a **commutator.**

Look back to the a-c generator illustrated in Figure 13-2. As the armature rotates through one cycle, the current flows first out of one brush and then, a half-cycle later, reverses and flows out of the other brush. If, at the end of each half-cycle, we could transpose the connections between the ends of the armature coil and the brushes, current would always flow out of the same brush. This changeover is performed by the commutator.

Look at Figure 13-8. Note the absence of the slip rings. In their place we have the commutator. This consists of a split metal ring mounted

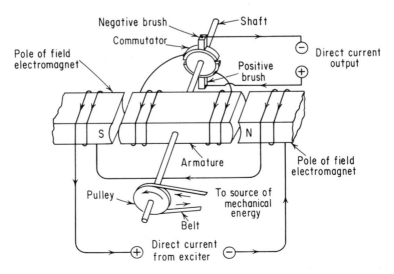

Fig. 13-8. Simplified illustration of d-c generator.

around the rim of an insulator disk. The two halves of the ring are separated from each other by a small gap that usually is filled with mica, which insulates one half from the other. Each end of the armature coil is attached to one of these halves and the entire commutator rotates with the armature. The brushes make contact with opposite points on the commutator.

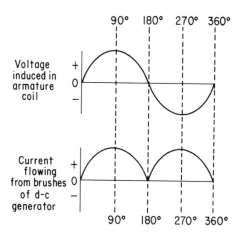

Fig. 13-9. Graph showing relationship between voltage induced in armature coil and current flowing from brushes of the d-c generator.

For one half-cycle the direction of the induced voltage in the armature is such as to force electrons to stream out of the negative brush to the external circuit. Then the direction of the induced voltage in the armature is reversed. However, at the same instant, the rotating commutator transposes the connections between the half-rings and the brushes. Accordingly, electrons again stream out of the negative brush. Such a reversal takes place at the end of each half-cycle, and so the electrons always stream from the negative brush to the external circuit. This is direct current. (When used in circuit diagrams, the symbol for the d-c generator is —Ⓖ—. The symbol for the armature, commutator, and brushes is shown as —◯—.)

The relationship between the induced voltage in the armature and the current flowing from the brushes of the d-c generator can be shown graphically. (Look at Figure 13-9.) The voltage induced in the armature coil is alternating, just as in the a-c generator. If we consider the current flowing from the brushes of the d-c generator, we see that for the first half-cycle we get a positive loop, just as in the alternator. During the

– A –

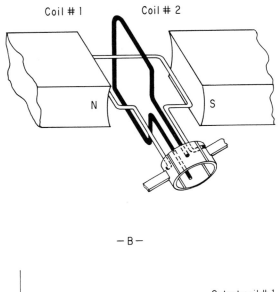

Coil # 1   Coil # 2

N   S

– B –

Output

0   90°   180° 270° 360°

– – – Output coil # 1

– · – Output coil # 2

———— Resultant output

Fig. 13-10. A. D-C generator with two armature coils.

B. Graph showing the resultant output.

next half-cycle, instead of a negative loop (as in the alternator), the commutator action produces another positive loop.

Because the current flow is always in the same direction, it is a direct current. However, as you can see from the graph, it is not a steady direct current but one that rises and falls. We call such a current a **pulsating direct current.**

The fluctuations indicated in Figure 13-9 are called **ripples.** Generally, a steadier and less-fluctuating d-c output is required from the generator. The ripple effect may be reduced by increasing the number of windings, or **coils,** of the armature of the generator. As shown in Figure 13-10A, the armature contains two coils at right angles to each other. The commutator is divided into four **segments,** or bars. Each coil terminates in two opposite commutator bars.

The two brushes make contact with the two bars connected to the ends of the coil that, at that moment, is cutting the magnetic lines of force and, accordingly, is producing an induced voltage (in our illustration, coil #1). Coil #2 is not cutting any lines of force and, hence, is producing no induced voltage.

After 90° of rotation the coils change places. Now the voltage of coil #1 has dropped to zero and that of coil #2 has reached a maximum. At the same time the commutator, too, has revolved 90° and the brushes now make contact with the commutator bars connected to the ends of coil #2.

The combined effect of the two armature coils is shown graphically in Figure 13-10B. Note that the ripple frequency of the resultant is twice that of each coil and that the output variations from maximum to minimum are smaller than for each individual coil. The more coils (and commutator bars) we employ, the smaller will be the ripple.

Note, in Figure 13-10, that the induced voltage of only the armature coil that connects to the commutator bars in contact with the brushes is producing a current. The induced voltages (if any) of the other coils are wasted. Hence, such a system is not practical.

If we were able to connect all the coils in series in such a way that the induced voltage of each would add to all the others, we would have a more efficient system. Such an arrangement is shown in Figure 13-11. The armature core is in the shape of a ring and the wire of the armature is wound continuously in the same direction around this ring, the ends being connected together. Hence this is called a **ring-wound armature.** The commutator bars are connected to the winding in such a way that an equal number of turns lie between adjacent bars.

As the armature rotates, current flows through the windings as indicated by the arrows on the wires. If you examine the current flow in the armature winding you will notice it follows two paths. One path is

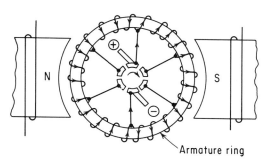

Fig. 13-11. Ring-wound armature.

Armature ring

from the top of the ring, through the left-hand winding, to the bottom of the ring. The other is from the top of the ring, through the right-hand winding, to the bottom of the ring. The top and bottom junctions of these paths are the **neutral points** on the armature winding.

The neutral points lie midway between the field poles and correspond to the points on the armature winding where no lines of force are being cut and, hence, where no voltage is being induced. Because of this, the brushes are placed in contact with the commutator bars that correspond to the neutral points on the windings. As the armature rotates, each armature turn (and its associated commutator bar) occupies, in turn, the neutral position.

Note that the current in both paths of the armature winding flows toward one of these neutral points (in our illustration, the bottom one), which we call the **negative** neutral point. The current flows away from the other neutral point, which is called the **positive** neutral point. The negative brush (marked $\ominus$) and the positive brush (marked $\oplus$) make contact with their respective neutral points through the commutator bars. You can see that all the turns of the winding are contributing to the output of the generator and none of the induced voltage is wasted.

As the armature rotates, different portions of the winding will pass the two field poles. However, as the armature revolves, so does the commutator. Accordingly, current will always flow to the negative brush and away from the positive brush.

The ring-wound armature is seldom used in modern generators. For one, the inside portion of the winding, shielded as it is from the magnetic lines of force by the iron of the armature core, produces no induced voltage. Also, it is quite difficult to wind such an armature since it means threading the wire in and out of the core. Most generators, today, employ the **drum-type** armature illustrated in Figure 13-12.

As the name implies, the drum-type armature is in the form of a cylinder, or drum. The armature winding consists of a series of coils lying in slots on the surface of the core and connected in series. Note that each slot contains the conductors of two adjacent coils.

Connections to the commutator bars are similar to those for the ring-wound armature and the operation, too, is similar. However, there are no shielded (and, hence, wasted) portions of the winding, and the drum-type armature is easier to wind. The present-day generator employs a drum-type armature with slots for many coils and a commutator with many segments or bars.

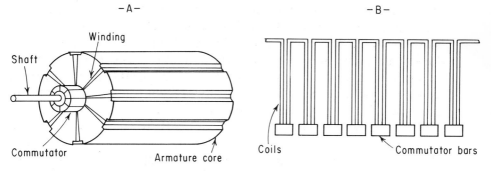

Fig. 13-12. A. Drum-type armature.

B. Commutator bars and armature coils drawn in linear form to show how they are connected.

A two-pole d-c generator employing a drum-type armature is shown in simplified form in Figure 13-13. Note that all the circles representing the cross-sectional view of the armature wires at the left-hand side of the

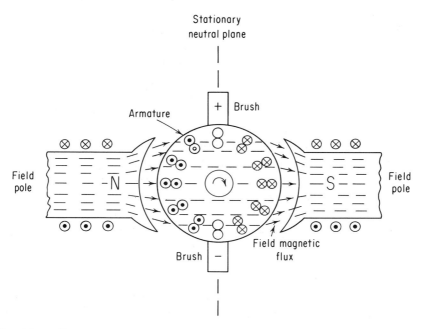

Fig. 13-13. Two-pole, drum-type, d-c generator showing the stationary neutral plane.

drum have a dot (·) in their centers, whereas all the wires at the right-hand side have a cross (✗) in their centers. This is a convention for showing the direction of current flow through a conductor when we look at a cross-sectional view of it. The current is considered as an arrow. If it is flowing out of the page towards you, you see the tip of the arrow, which resembles a dot. If it is flowing away from you, you see the feathered end of the arrow, which resembles a cross.

As the armature rotates, its coils cut across the main magnetic field produced by the field coils and an electromotive force is induced in them, as indicated in the illustration. Since the top and bottom armature wires move parallel to the magnetic flux, no voltage is induced in the coils they form (as represented by the absence of both dot and cross). Thus the armature coils that at any instant are at the top and bottom (and the commutator bars connected to them) are at the positive and negative neutral points of the generator, respectively. Hence the brushes may short-circuit these bars without producing sparks, which would wear both the brushes and the commutator.

Note the uniform appearance of the main magnetic field. The plane of the neutral points, and the brushes connected to them, is a vertical plane midway between the two field poles. This is called the **stationary neutral plane.**

As long as the generator is not connected to a load — that is, as long as no current is being drawn from it — the situation is as described. But if current is drawn from the generator, this current must also flow through the armature coils. The flow of this current, in turn, sets up its own magnetic field around the armature. This field, reacting with the main magnetic field, causes the latter to become distorted. This action is called **armature reaction.**

The effect of armature reaction is illustrated in Figure 13-14. Note that the magnetic flux of the field is crowded (stronger) at one of the pole tips of the field poles and thinned out (weaker) at the other pole tip of the same pole. The effect of armature reaction on the generator, then, is to tip the neutral plane in the direction of rotation to a new plane, called the **running neutral plane.** To minimize sparking, then, the brushes, too, must be shifted to this running neutral plane.

There is another factor that may cause sparking between the brushes and the commutator. Note that, as an armature coil approaches the neutral point of the generator, a voltage is induced in it in one direction. At the neutral point the induced voltage drops to zero. Then the voltage

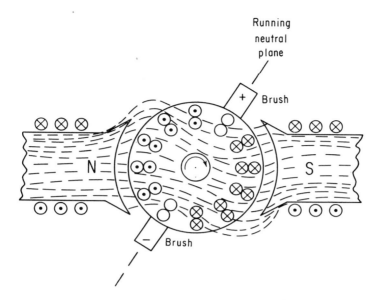

Fig. 13-14. Two-pole, drum-type, d-c generator showing the running neutral plane.

is induced in the opposite direction as the coil passes beyond the neutral point.

The induced current in the coil tends to follow the induced voltage. But when, at the neutral point, the current drops to zero, the self-induction of the coil will produce an induced voltage that can cause considerable sparking at the brushes.

To reduce sparking between the brushes and the commutator, most modern generators employ **commutating poles.** (See Figure 13-15.) These poles are located between the main field poles. Their windings are of relatively few turns of heavy wire and are connected in series with the armature coils and the brushes. Thus the armature current flows through these windings.

The commutating poles are so wound that the polarity of each pole is opposite from that of the field pole it follows in the direction of rotation of the armature. The effect of the magnetic flux produced by these poles is to oppose and balance out the armature flux at the normal neutral plane. Hence the effects of armature reaction and the self-induction of the armature coils at the neutral plane are minimized. As a result, there is very little sparking at the brushes.

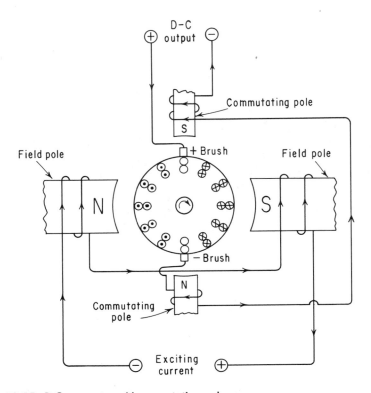

Fig. 13-15. D-C generator with commutating poles.

Since the armature winding of the generator is in series with the external load, the current drawn by that load also flows through the armature winding. Hence the maximum current that can be drawn from the generator is limited by the amount of current that can flow through the wire of the armature winding without causing it to overheat.

We can, of course, increase the current-carrying capacity of the armature wires by increasing their size. However, there is a method whereby we can increase the maximum current that can be drawn safely from a generator without increasing the size of its wires.

Consider the two-pole generator shown in simplified form in Figure 13-11. The armature current, you will recall, flows along two parallel paths through the armature winding. Thus the individual wires of the winding are called upon to carry only one-half of the total armature current.

If we were able to increase the number of parallel paths for the current to follow through the armature winding, we would be able to have a greater total armature current without increasing the size of the wire. This is accomplished by the use of generators having more than two field poles.

In Figure 13-16 you see the simplified drawing of a four-pole generator. Note that the field poles are arranged so that north and south poles alternate. A conductor on the armature undergoes a complete cycle (360 electrical degrees) as it rotates past two field poles of opposite polarities. In the case of the two-pole generator, this means a complete revolution. But in the case of the four-pole generator, this means a half-revolution.

Midway between each set of poles are the neutral points. Hence, we have four neutral points — two positive and two negative — and, accordingly, two positive brushes and two negative brushes. If we connect the two positive brushes together and the two negative brushes together, we have **four** parallel paths for the armature current to flow through the armature winding. Thus the individual wires of the winding are called upon to carry only one-fourth of the total armature current.

By increasing the number of field poles, we may have even more parallel paths for the armature current and, hence, a greater total armature current that the wires can safely handle. Generators that are called upon to supply very large amounts of current generally have six or more field poles. Medium-size generators usually have four. Small generators, such as those used in automobiles to charge the storage battery, generally have two.

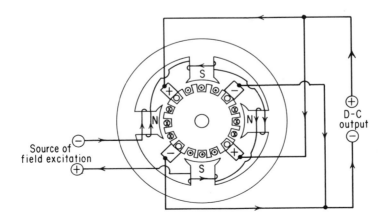

Fig. 13-16. Four-pole generator.

Fig. 13-17. Open-type d-c generator.

There is another advantage for the multipolar generator. As previously indicated, a steady, nonfluctuating output generally is desired from the d-c generator. You will recall that one method for reducing the ripple of the d-c output is to increase the number of armature coils of the generator. Another method is to increase the number of field poles.

When we discussed the alternator we found that increasing the number of field poles increased the frequency of the voltage induced in the armature, the speed of rotation being kept constant. Accordingly, when the alternating current in the armature is changed to direct current by commutator action, the ripple frequency of the output of a multipolar d-c generator is greater than that of the two-pole type. Thus the d-c pulses are closer together and, hence, the output is a steadier direct current.

Even with many coils of armature winding and many field poles, the output of the d-c generator may not be steady enough for certain applications. Accordingly, **filters** may be employed to smooth out the ripple. These filters consist of capacitors and inductors connected as shown in Figure 13-18.

The pulsating d-c output of the generator contains both steady d-c and a-c components. (See Chapter 9, subdivision E.) The inductor (L) offers a high impedance to the a-c component but a relatively low resistance to the d-c component. The capacitor (C), on the other hand, offers a low impedance to the a-c component but an infinitely high resistance to the d-c component, Accordingly, the steady d-c component of the output is passed on to the load through the inductor and the a-c component is bypassed through the capacitors.

The armature of the generator must be symmetrical, both mechanically and electrically. The windings must be so placed that no unwanted vibrations are set up upon rotation. Also, since all the electrical paths are in parallel, the resistances, induced voltages, and currents of each path must be the same. Otherwise, spurious current will circulate through the windings. Not only will such currents be wasted as far as the output of the generator is concerned, but they also will produce undesirable heating effects.

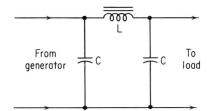

Fig. 13-18. Ripple filter.

Since the paths are all in parallel, the total resistance of the armature is equal to the resistance of one path divided by the number of paths. The output voltage is equal to the voltage across one path from brush to brush. The output current, on the other hand, is equal to the current of one path multiplied by the number of paths.

1. Types of Direct-Current Generators

a. Self-excited d-c generators

Inasmuch as the d-c generator delivers a direct current, it becomes obvious that we may use part of the delivered current to excite the field coils. Such generators are called **self-excited** types. There are three general classes of self-excited generators, depending upon the connections between the field and armature windings. These are the **series-wound generator,** the **shunt-wound generator,** and the **compound-wound generator.** These generators will be discussed below.

A question should be raised immediately. Since the generator does not generate any current until the armature starts rotating and cutting the magnetic lines of force, and if the field coils do not receive any current until the generator delivers it, from where, then, comes the magnetic flux at the start?

The answer lies in the **residual magnetism** of the field poles. These poles are made of iron and become magnetized as current flows through the field coils. When the generator stops rotating and this current ceases to flow, the field poles lose most of their magnetism. But a certain small amount (the residual magnetism) remains. When the armature starts rotating again, the cutting of the magnetic flux owing to this residual magnetism produces a weak induced current. All this current (or a portion, depending upon the type of generator) is fed back to the field coils, increasing the strength of the magnetic field. As a result, a greater current is generated, more current is fed back to the field coils, and the process continues to build up until the field reaches its normal strength. The entire procedure usually takes about 20 or 30 seconds. Where the generator has been standing idle for a long time, or where the residual magnetism has been lost because of some other effect, it may be necessary to use batteries to furnish field current at the start. Once the generator has started delivering current, the batteries may be removed.

(1) *Series-wound generator.* In the **series-wound** generator the field coils are connected in series with the armature windings and the load, as shown in Figure 13-19. Here the entire current furnished by the generator flows through the field coils. Accordingly, these coils must be wound with heavy wire. Since the magnetic field depends upon the ampere-turns of these coils, and since the current is large, these coils need have only a few turns to obtain the desired magnetic flux.

With the load disconnected there is no current flowing through the armature and field windings of the generator. Nevertheless, because of the residual magnetism of the pole pieces, lines of force are being cut by the armature and a relatively small voltage appears at the output terminals of the generator.

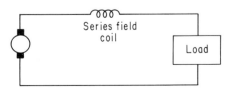

Fig. 13-19. Circuit of the series-wound generator.

As the load is connected, this voltage causes a flow of current through the armature winding, the load, and the field winding. This current flow, in turn, increases the magnetic flux of the field, causing an increase in the terminal voltage. As a result, there is an increase in the output current. This process is cumulative, causing the terminal voltage and output current to build up.

As the building-up process is continuing, however, a reverse process is taking place. The greater the output current of the generator, the greater become the *IR* losses in the armature and field windings. And the greater become the losses due to armature reaction. All this tends to reduce the terminal voltage.

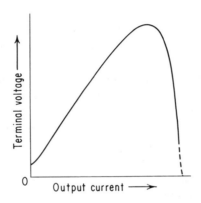

Fig. 13-20. Output characteristic curve of the series-wound generator.

At a certain point the two processes balance and the terminal voltage will not rise beyond this point. (See Figure 13-20.) A small increase in the load, then, will cause a sharp decrease in the terminal voltage.

An examination of the characteristic curve of the series-wound generator indicates that the terminal voltage will vary greatly with variations in the load current. For this reason, such generators are not generally employed. However, note that after the build-up reaches its peak, the output current remains fairly constant for large variations in the voltage. This makes this generator suitable for certain special applications such as arc-welding, where such a characteristic is desirable.

(2) *Shunt-wound generator.* In the **shunt-wound** generator the field coils are connected in parallel, or shunt, with the armature windings

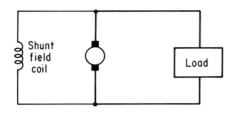

Fig. 13-21. Circuit of the shunt-wound generator.

and the load, as shown in Figure 13-21. The armature current now flows through two parallel paths, one through the load and the other through the field coils. Since the current flowing through the field coils is "lost," so far as the generator's output is concerned, it is necessary to keep this loss as small as possible. Less than 5 percent of the armature current generally is fed to the field coils.

To keep the current flowing through the field coils low, they must have a high resistance. Hence these coils are wound with fine wire. However, to supply an adequate magnetic flux, the shunt field coils must have many turns.

With the load disconnected, the rotating armature, cutting through the magnetic flux resulting from residual magnetism, produces an induced voltage. This voltage causes a current to flow through the armature windings and the windings of the shunt field coils. As a result, the magnetic flux of the field is increased, and so are the induced voltage and current. This process is cumulative and the magnetic field builds up rapidly until the normal voltage of the generator for a given speed and field resistance is attained.

As the load is connected, the IR drop in the armature and the armature reaction act to reduce the induced voltage applied to the field coils. As a result, the field current is reduced, producing a weaker field flux and a smaller terminal voltage. As the load is increased, the increased armature current causes a further reduction in the terminal voltage. (See Figure 13-22.)

Note that at light loads the characteristic curve is nearly level — that is, the terminal voltage remains fairly constant for variations in the load current. But as the load is increased further, the terminal voltage drops sharply. Beyond a certain point, increasing the load actually causes a decrease in the armature current. If the load becomes extremely large (as, for example, if a short-circuit occurs), the armature current will drop to practically zero.

Note that if it is not loaded too heavily, the shunt-wound generator produces a more constant terminal voltage with variations in load than

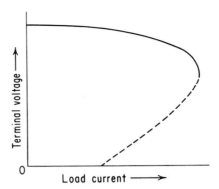

Fig. 13-22. Output characteristic curve of the shunt-wound generator.

does its series-wound counterpart. However, it has largely been replaced by the compound-wound generator (which will be discussed next), which has even better constant-voltage characteristics.

(3) *Compound-wound generators.* In the **compound-wound** generator both shunt and series field windings are employed in the same machine. The shunt field winding consists of many turns of fine wire wound over the field poles and is connected in shunt with the armature winding. The series field winding consists of a few turns of heavy wire wound over the shunt field winding and is connected in series with the armature winding. (See Figure 13-23.)

If we disregard, for the moment, the series field winding, we have, in effect, a shunt-wound generator. In such a generator, we have seen, the terminal voltage tends to drop as the load is increased. But, as load current flows through the series field winding, a magnetic flux is produced. If the series field winding is in the right direction, this flux will aid the magnetic flux set up by the shunt field winding. As a result, the terminal voltage will rise.

If the load is light, the terminal voltage tends to drop slightly from its no-load value, owing to the small armature *IR* drop and reaction. The relatively small increase in flux produced by the series field winding thus restores the terminal voltage to its no-load value. If the load is heavier,

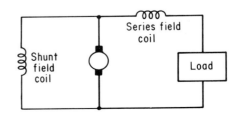

Fig. 13-23. Circuit of the compound-wound generator.

the armature IR drop and reaction, too, are greater. But so is the increase in flux produced by the greater current flowing through the series field winding. Hence the terminal voltage tends to remain constant for different values of load. (Note, however, that this holds true only within the normal operating limits of the generator. Large overloads will cause the terminal voltage to drop drastically.)

In order for the terminal voltage to remain constant for varying loads, the series field winding must contain the proper number of turns so that the induced voltage resulting from its magnetic flux may counterbalance the voltage reduction due to armature IR drop and reaction. A generator operating in this manner is said to be **flat-compounded.** (See Figure 13-24.)

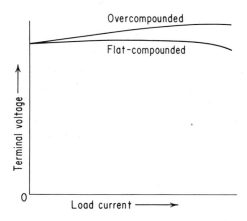

Fig. 13-24. Voltage-current characteristic curves for the compound-wound generator.

As a matter of fact, the series field winding usually has a few more turns than is necessary to produce a flat-compounded generator. Such a generator is said to be **overcompounded.** Thus, at greater load, the voltage of the overcompounded generator tends to rise slightly. This is to offset the increased IR drop in the wires connecting the output of the generator to its load owing to the increased current. Hence the voltage actually applied to the load tends to remain constant.

There are two ways in which we may wind the shunt and series field coils upon the pole pieces of the compound generator. One method is to wind these coils over each other in such a way that their polarities coincide. (See Figure 13-25A.) Such a generator is called a **cumulatively wound** compound generator and its behavior has been described above. This type of generator is most widely employed, chiefly because of its

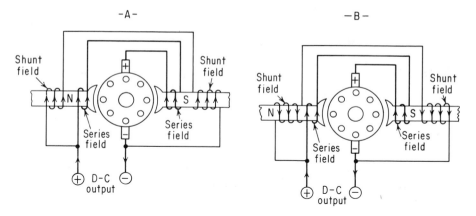

Fig. 13-25. A. Cumulatively-wound compound generator.
B. Differentially-wound compound generator.

tendency to maintain a fairly constant voltage under varying load conditions.

It is possible, however, to wind the shunt and series field coils in such a way that their polarities oppose each other. (See Figure 13-25B.) Such a generator is called a **differentially wound** compound generator. Generators of this type are used for special purposes, such as arc-welding, where sudden heavy loads may be applied. When such a heavy load occurs, the large magnetic field created by the series winding neutralizes the field created by the shunt winding. As a result, the overall field is reduced. The induced voltage drops, reducing the armature current and thus protecting the armature winding from overheating. In effect, this acts as an automatic overload control.

b. *Separately excited d-c generators*

Just as in the alternator, the direct current used to excite the field coils of the d-c generator may be obtained from a separate d-c generator or from storage batteries. (See Figure 13-26.) It is usual, in practice, to keep this exciting current at about 5 percent of the rated current of the main generator. Thus, if the rated output current of the main generator is, say, 100 amperes, its field current would be about 5 amperes.

Since the field current of the separately excited generator is independent of the armature current, the magnetic flux of the field is less affected by changes in the load than in self-excited generators. Hence the terminal

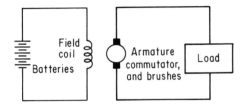

Fig. 13-26. Schematic diagram of separately-excited d-c generator.

voltage of the separately excited generator tends to be more constant than that of an equivalent self-excited, shunt-wound generator.

Generally, separately excited d-c generators are not used since the cumulatively wound compound generator is able to produce an even more constant terminal voltage. Also, self-excited generators are more economical both in cost and space. However, in certain applications where voltage control is important, the separately excited generator offers certain advantages.

A common method for controlling the terminal voltage of the generator is to control the strength of the field by controlling the current flowing through the field coils. Where the generator is called upon to deliver large output currents, the field current, although generally about 5 percent of the output current, nevertheless may be quite large. In the self-excited generator (as we shall see later) the device, usually a rheostat, that controls the field current must be heavy enough to handle this fairly large current.

But where the field of the main generator is separately excited by another generator (the **exciter**), this latter generator need deliver only enough current to excite the field of the main generator. Accordingly, the field current of the exciter generator then need be only about 5 per-

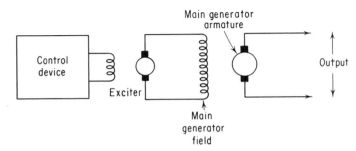

Fig. 13-27. How the output of a separately-excited generator may be controlled.

cent of the field current required for the main generator. Hence the device that controls the field current of the exciter generator may be proportionally lighter. This is particularly advantageous in applications where automatic control is desired since the automatic-control devices then may be smaller and lighter. (See Figure 13-27.)

C. EFFICIENCY AND REGULATION OF GENERATORS

The power output of the d-c generator normally is rated in watts or kilowatts. In alternating-current circuits, however, the inductance and capacitance of the load affect the phase relationship between the current and voltage and, hence, the power (see Chapter 9, subdivision C). Since the inductance and capacitance of the load are variable factors that cannot be determined in advance, the power output of alternators usually is rated in volt-amperes (v-a) or kilovolt-amperes (kv-a).

As is true of all machines, the efficiency of the generator is the ratio between the power output and the power input. But note that the input to the generator is the **total** input, which includes the power of the machine that drives the generator as well as the power used to excite it (if it is separately excited). If we wish to find the efficiency (in percent) of a generator, we apply the following formula:

$$\text{Percent efficiency} = \frac{\text{power output}}{\text{power input}} \times 100.$$

EXAMPLE. A separately excited d-c generator whose power output is rated at 1000 kilowatts is turned by a 1500-horsepower diesel engine and uses 50 kilowatts to excite its field. What is its efficiency?

Power input = power of diesel engine + power for field excitation. To convert horsepower to kilowatts, multiply by 0.746. Thus:

$$\text{Power of the diesel engine} = 1500 \times 0.746$$
$$= 1119 \text{ kilowatts.}$$

Therefore:

$$\text{Power input} = 1119 + 50 = 1169 \text{ kilowatts.}$$

$$\text{Percent efficiency} = \frac{\text{power output}}{\text{power input}} \times 100$$

ANSWER.
$$= \frac{1000}{1169} \times 100 = 85.5\%.$$

The efficiency of a commercial generator generally ranges between 80 and 95 percent.

Aside from losses within the source that supplies field excitation (if the generator is separately excited) the losses within a generator fall into three general categories — **mechanical** losses, **copper** losses, and **iron** losses.

MECHANICAL LOSSES. The largest mechanical losses are caused by friction at the bearings that support the rotating parts and by friction between the brushes and slip rings or commutators. In addition, there is the loss due to wind resistance (called **windage**) encountered by the rotating members. These losses are kept low by proper design of the bearings and rotating parts.

The brushes generally are made of powdered carbon and graphite, held together by some suitable binder. The graphite acts as a lubricant, cutting down friction. In some low-voltage generators (such as the automobile generator that charges the storage battery) powdered copper is added to the carbon and graphite to lower the resistance of the brush.

COPPER LOSSES. These losses are due to the power consumed by the heating effect (I^2R) of the current that flows through the wires of the armature and field windings. The wire must be heavy enough to keep this loss at a minimum and to prevent overheating.

IRON LOSSES. These losses are due, essentially, to the fact that the iron armature core and field poles are located within a rapidly changing magnetic field. Thus they are alternately magnetized and demagnetized. Whenever a substance is alternately magnetized and demagnetized, the magnetizing force encounters in the substance a sort of "resistance" which causes the magnetizing effect to lag behind the magnetizing force. This lagging is called **hysteresis.**

The energy loss due to this "resistance" shows up as heat produced within the substance. To minimize the hysteresis loss the armature core and field poles generally are made of soft iron, annealed steel, or certain other alloys that have a high permeability.

In addition to the hysteresis loss there is another caused by the changing magnetic field. The armature core and field poles are conductors and, like other conductors, the changing magnetic field will induce a current within them. This is called the **eddy current** and, since it comes from the generated current and is not available as output, it represents a loss. Eddy-current losses are reduced by building up the core and poles of thin sheets, called **laminations,** instead of making them of solid metal.

Each lamination is coated with an insulating varnish and so the flow of eddy current is broken up.

Incidentally, great care must be taken to preserve the magnetic flux of the field. The frame of the generator, which supports the field poles, is made of soft iron to furnish an easy path for the circulation of the flux between these poles. The armature and field poles are so mounted that only a very small air gap exists between them. The ends of the poles are curved and flared out to provide a more even distribution of the magnetic flux over the armature.

The mechanical and iron losses vary, essentially, with variations in the speed at which the generator is rotated. Since generators generally are driven at constant speed, these losses, too, tend to remain constant. The copper losses, on the other hand, rise rapidly with increases in the load and fall when the load is decreased.

We have seen how the terminal voltage of the generator tends to drop with an increase in the current drawn by the load, owing to the increased *IR* drop in the armature winding and to armature reaction. The inherent change in the voltage of a generator with changes in load is known as its **voltage regulation.** For rating purposes, we generally take the voltage change between no load and full load. The voltage regulation usually is expressed in percent as determined by the following formula:

Percent voltage regulation =
$$\frac{\text{voltage at no load } - \text{ voltage at full load}}{\text{voltage at full load}} \times 100.$$

EXAMPLE. What is the voltage regulation of a generator whose no-load voltage is 110 volts and whose full-load voltage is 100 volts?

ANSWER. Percent regulation $= \dfrac{110 - 100}{100} \times 100 = 10\%.$

The lower the percentage, the better is the regulation of the generator.

D. CONTROL OF GENERATORS

We can vary the voltage of a generator in three ways. First, we may change the number of conductors of the armature winding. Increasing the number of conductors cutting a magnetic field, you will recall, increases the induced voltage. Obviously, this method creates mechanical difficulties; hence, it generally is not used.

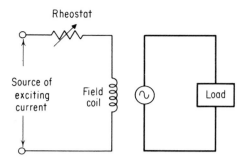

Fig. 13-28. A rheostat in the field circuit is used to vary the voltage of the a-c generator.

Second, we may vary the speed of rotation of the armature. If we increase the rate at which the magnetic lines of force are cut, we increase the induced voltage. Because generators usually are designed to be run at a constant speed, this method, too, is not in general use.

The method most frequently used is one that varies the strength of the magnetic field. The greater the number of magnetic lines of force being cut, the greater is the induced voltage. Since the field of the generator is created by current flowing through the field coils, it is quite easy to vary this current and, hence, the magnetic field.

In the alternator, the magnetic field may be varied by means of a variable resistor, called a **rheostat,** placed in series with the field coil and the source of exciting current. (See Figure 13-28.) When the resistance of the rheostat is increased, the field-coil current is lowered. This results in a smaller magnetic field and a decreased generator voltage. Decreasing the resistance of the rheostat increases the field current and, hence, the generator voltage. (The electrical symbol used to represent the rheostat is either ⌇ or ⌇ .)

Generally, the field current for the alternator is supplied by the output from a d-c exciter generator. The output current of the exciter then is

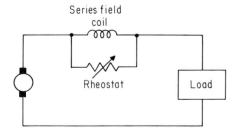

Fig. 13-29. Voltage control for series field generator.

controlled by a rheostat in its field circuit. Thus, by varying the field current of the exciter, the field current of the alternator and its output is controlled in like degree. The advantage of this arrangement is that a lighter rheostat may be used, since the field current of the exciter is much smaller than the field current of the alternator.

The rheostat may be operated manually. In some installations automatic relays that cut fixed resistors into or out of the circuit are used. These relays are operated by the generator voltage. Should the voltage rise, it operates a relay that inserts a resistor into the circuit, thus reducing the voltage. Should the voltage fall, another relay cuts a resistor out of the circuit. In this way the voltage is kept constant.

For direct-current generators there are several methods for varying the voltage, depending upon the type of machine. If the generator is separately excited, we may use the same system of control as for the alternator.

In the self-excited, series field type we may place a rheostat in parallel, or in shunt, with the field coil, as shown in Figure 13-29. The current flowing from the generator to the load has two paths to follow — one through the series field coil and the other through the rheostat. The greater the resistance of the rheostat, the less current will flow through it and the more current will flow through the field coil. Hence, the greater the magnetic field and the larger will be the generator voltage. Reducing the resistance of the rheostat causes less current to flow through the field coil; hence, the voltage is reduced. (This rheostat sometimes is called a **diverter rheostat** since it shunts, or diverts, current from the series field.)

In the shunt field generator we may place a rheostat in series with the field coil, as shown in Figure 13-30. If we increase the resistance of the rheostat, the current through the field coil is reduced. This causes a reduction in the magnetic field and, accordingly, the generator voltage

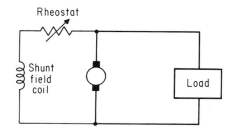

Fig. 13-30. Voltage control for shunt field generator.

is lowered. Reducing the resistance of the rheostat results in a greater field current, a stronger magnetic field, and a larger voltage.

In the compound generator we may use either or both of the methods of voltage control just described. In all instances the rheostat may be adjusted manually or else automatic control may be achieved by relays that cut resistors in or out of the proper circuits as the voltage rises or falls.

The automobile generator, which supplies direct current to charge up the storage battery, presents a special problem. Because it is driven by the engine, it does not rotate at a constant speed, as do most other types of generators. Rather, its rate of rotation depends upon the speed of the car. The problem is one of keeping its voltage constant, else it may damage the lights and other appliances. Further, its output current, too, must be kept constant if the storage battery is not to be overcharged or undercharged.

A two-pole, shunt field generator of special design is frequently used. It has three, instead of two, brushes, as shown in simplified form in Figure 13-31. The shunt field coil is connected across the negative and third brushes. Because the third brush is not at the neutral point, the field coil does not receive its full share of field current, as it would were

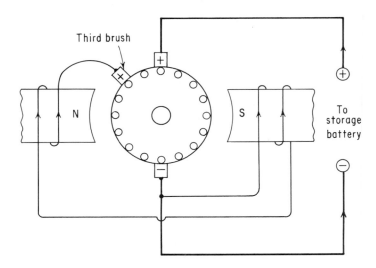

Fig. 13-31. Automobile generator with a third brush.

it connected to the positive brush. Hence the magnetic field is weaker and the generator output is lowered.

The third brush is movable. The nearer it approaches the positive brush, the nearer it gets to the neutral point, and the greater is the field current. Hence the output of the generator rises. If the brush is moved farther away from the neutral point, the output drops. Thus, by adjusting the position of the third brush, we may control the rate at which the generator will charge the battery.

The generator has no commutating poles. Hence, the faster it rotates and the greater the armature current, the greater becomes the armature reaction and the more the output current tends to drop. Because of this, the generator actually produces less current at high speeds than at the normal operating speed of the automobile. This is an advantage since the car usually is driven for longer periods at its normal speed than at high speeds.

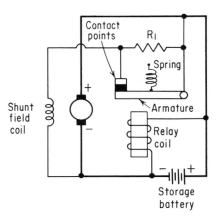

Fig. 13-32. Voltage regulator relay.

There are a number of external devices for regulating and controlling the output of the automobile generator. One such is the **voltage regulator relay,** whose circuit is illustrated in Figure 13-32. The relay (which will be discussed further later in the book) consists of an electromagnet which, when sufficient current flows through its coil, attracts a soft-iron hinged armature. As this armature moves, it brings together or separates (depending upon the type of relay) a set of contact points, thus closing or opening some circuit of which the contact points are a part. The coil of the electromagnet consists of many turns of fine wire, thus producing a high resistance, and is connected across the line leading from the generator to the storage battery.

Normally, not enough current flows through the relay coil to attract the armature. A spring pulls this armature up and, as a result, a contact point on the armature makes contact with another fixed contact point, shorting out resistor R, which is in series with the shunt field coil. Thus, the full line voltage is applied to this field coil.

Should the voltage rise, however, the armature is attracted and contact between the points is broken. This removes the short across resistor R and its resistance appears in series with the field coil. The magnetic field of the generator is reduced and the voltage falls back to its normal value. At this point, the armature is released and the spring pulls it up, restoring contact between the two points. The resistor is shorted out again.

Another device is the **current limiter relay,** illustrated in Figure 13-33. This resembles the voltage regulator relay except that it is wound with fewer turns of heavy wire (hence low resistance) and is placed in series with the generator and the battery. This time, if current increases from normal, the armature is attracted, the contact points are separated, and resistor R_1 is placed in series with the shunt field coil. The output current then drops to normal, at which time resistor R_1 is shorted out again.

While the automobile is in operation, the generator rotates and current is flowing to the storage battery. When the car stops the generator stops as well. However, since it still is connected to the battery, battery current now flows to the generator. This will discharge the battery and

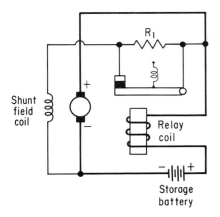

Fig. 13-33. Current limiter relay.

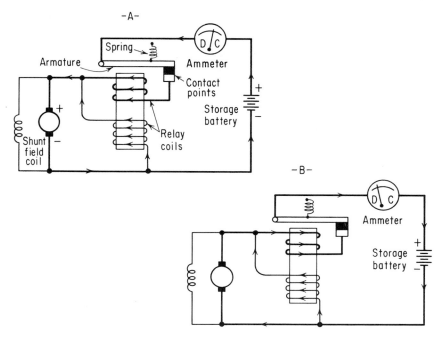

Fig. 13-34. Reverse current, or cutout, relay.

    A. Current flow while generator is rotating and storage battery is being charged.

    B. Current flow the instant generator stops and storage battery starts to discharge.

may ruin the generator. What we need is some device that disconnects the generator when it stops rotating.

This is the function of the **reverse current,** or **cutout, relay** shown in Figure 13-34. This relay has two coils. One is a heavy-wire winding of few turns in series with the generator, the closed contact points, the relay armature, the ammeter, and the storage battery. The other is a fine-wire winding of many turns in parallel with this circuit.

In Figure 13-34A the circuit is shown while the generator is rotating. When the current flows as indicated by the arrows, the magnetic pulls of both windings aid each other and the armature is attracted, closing the contact points. The ammeter shows that the battery is being charged.

Figure 13-34B shows the current flow the instant the vehicle stops. The generator no longer supplies any current. The battery now starts to discharge (as shown by the ammeter) and the current flows in the opposite direction — that is, in the opposite direction in all portions of the circuit except in the fine-wire winding of the relay. The magnetic fields of both windings now oppose each other. The result is an overall weakening of the pull on the armature. The spring now is able to overcome this pull and the contacts are separated, opening the circuit from the battery to the generator.

The cutout relay may be used with other than automobile generators. When two generators are connected with their outputs in parallel, all is well as long as their voltages remain the same. Should the voltage of one drop, current from the other would tend to flow into the first. Thus it becomes necessary to cut that one out of the line until its voltage is restored. To accomplish this a reverse current relay may be placed in series with each generator.

QUESTIONS

1. What are the four essential parts of an a-c generator?

2. What determines (*a*) the voltage of a generator; (*b*) the maximum safe current that can be drawn from a generator; (*c*) the frequency of the output of an a-c generator?

3. What must be the speed (in revolutions per minute) of a 12-pole a-c generator to produce a 400-Hz current?   [*Ans.* 4000 rpm.]

4. What is meant by a **three-phase** a-c generator? What is the phase relationship between currents produced by such a generator?

5. Draw the circuit of the armature of a three-phase generator wound in (*a*) a Y connection; (*b*) a delta connection.

6. What are the four essential parts of a d-c generator?

7. Explain the function and operation of the commutator.

8. What is meant by **armature reaction?**

9. Explain the difference between the **stationary neutral plane** and the **running neutral plane** of the d-c generator.

10. Explain the action of the **commutating poles** of a d-c generator.

11. Explain how the ripple in the output of a d-c generator may be reduced.

12. What are two advantages of the **drum-type** armature over the **ring-wound** type?

13. Draw the schematic circuit of a series field d-c generator. What is the effect of an increase in load upon its voltage?

14. Draw the schematic circuit of a shunt field d-c generator. What is the effect of an increase in load upon its voltage?

15. Draw the schematic circuit of a compound field d-c generator. What is the chief advantage of this type generator?

16. A self-excited d-c generator whose power output is rated at 500 kilowatts is turned by a 750-horsepower diesel engine. What is its efficiency?

[*Ans.* 89.4%.]

17. What are the **mechanical** losses in a generator? What steps are taken to reduce them?

18. What are the **copper** losses in a generator? What steps are taken to reduce them?

19. What are the **iron** losses in a generator? What steps are taken to reduce them?

20. What is meant by the **voltage regulation** of a generator?

21. What is the regulation of a generator whose voltage is 220 volts at no-load and drops to 205 volts at full-load? [*Ans.* 7.3%.]

22. Explain three methods by which the voltage of a generator may be varied. Which method is used most commonly?

23. Explain the use of the **third brush** in controlling the output of an automobile generator.

24. Explain the action of the **voltage regulator relay** of the automobile.

25. Explain the action of the **current limiter relay** of the automobile.

26. Explain the action of the **reverse current relay** of the automobile.

OTHER SOURCES
OF
ELECTRICAL ENERGY

14

In Chapter 13 we learned how the generator converts mechanical energy into electrical energy. You must not get the impression that the generator creates the electrons that constitute the electric current. Actually, the generator creates the electromotive force that sets flowing the billions and trillions of electrons in the circuit itself, just as a water pump creates the pressure that sets flowing the water in the pipes. So we may consider the electrical generator as a sort of electromagnetic "pump."

A. CHEMICAL CELLS

1. The Voltaic Cell

There are other types of "pumps" that can create an electromotive force. In 1798 Alessandro Volta, an Italian physicist, invented a "chemical pump." He noticed that if two dissimilar metal strips are placed in an acid solution, an electromotive force appears between the two metals. If a conductor connects the two metal strips, electrons will flow through it. Such a "chemical pump" is called a **voltaic cell** in honor of its inventor.

Let us perform an experiment that will duplicate Volta's findings. Pour some hydrochloric acid (a compound of hydrogen and chlorine) into a jar of water. We believe that when hydrochloric acid is placed in water the compound breaks up. The chlorine atom seizes an electron from the hydrogen atom and thus becomes a negative chlorine ion (which appears in Figure 14-1 as Cl^-). The hydrogen atom, having lost an electron, becomes a positive ion (H^+). If, when a substance goes into solution ions are formed, we say that the substance has **ionized** and call the solution an **electrolyte.** Thus the water solution of hydrochloric acid is an electrolyte.

Into this solution insert a strip of copper and one of zinc. Connect a piece of copper wire between the two strips and place an ammeter in the circuit. The meter will show that a current is flowing through the wire from the zinc to the copper strip.

In simplified form, this is what happened. When the zinc strip was placed in the acid solution, the zinc started to dissolve — that is, zinc atoms started to leave the strip and enter the solution. As each zinc atom left the strip, however, it left behind two electrons. Thus the zinc atom became a positive zinc ion (Zn^{++}). And the zinc strip, because of the electrons left behind, became negatively charged.

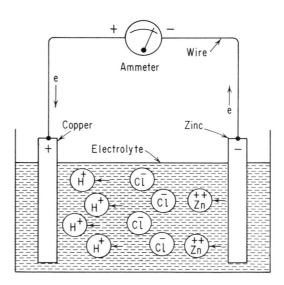

Fig. 14-1. Voltaic cell.

The positive zinc ions repelled the positive hydrogen ions in the solution toward the copper strip. As each positive hydrogen ion reached the copper, it seized an electron from the strip and, becoming in this way a neutral hydrogen atom, bubbled off into the air. The copper strip, having lost electrons, became positively charged.

Thus a difference of potential (electromotive force) was created between the zinc and copper strips. When connected by a conductor, electrons flowed from the zinc to the copper strip, as indicated by the meter. This process will continue until the entire zinc strip is dissolved.

Almost any two dissimilar metals can be used. A carbon rod may be substituted for the copper strip, since the function of the positive strip is merely to supply electrons within the cell. Almost any acid may be used for the electrolyte, and there are a number of other substances, such as sal ammoniac or lye, that may be used as well. An interesting thing about the voltaic cell is the fact that its electromotive force does not depend upon its size. Making the cell larger increases the amount of current that we may draw from it, but it will not increase the electromotive force or voltage. Its voltage depends, mainly, upon the chemical action and this, in turn, depends upon the materials of the strips (**electrodes**) and upon the substance used for the electrolyte. In the cell we have described, the electromotive force is about one volt.

There are a number of disadvantages to the voltaic cell we have described. For example, when the hydrogen ions reach the copper electrode and take away electrons from it, these ions become neutral hydrogen atoms. You can see them as bubbles around the positive electrode when the external circuit is completed. Some of these hydrogen bubbles tend to cling to the positive electrode, forming a sheath completely surrounding it. After a short time, the action of the cell ceases, owing to the insulating action of the hydrogen bubbles which prevent any new hydrogen ions from reaching the positive electrode. We call this effect **polarization.**

In 1866 Georges Leclanché, a French scientist, using a carbon rod as a positive electrode, overcame the effect of polarization by placing this rod in a porous cup containing manganese dioxide. The porous cup and its material did not prevent the positive hydrogen ions from reaching the carbon rod. But after these hydrogen ions became hydrogen atoms, they combined chemically with the manganese dioxide and thus could not form the insulating sheath around the rod. The manganese dioxide is called, appropriately enough, a **depolarizer.**

Two or more cells may be connected together to form a **battery.** (The term "battery" is used rather loosely to indicate a single cell or a

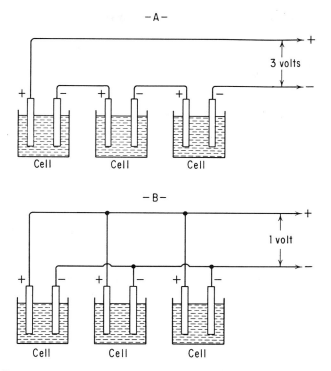

Fig. 14-2. Three cells connected to form a battery.
  A. Series connection.
  B. Parallel connection.

group of cells.) There are several ways in which these cells may be connected. We may connect them in **series** — that is, with the positive electrode of one to the negative electrode of the other. In Figure 14-2A a battery of three similar cells connected in series is illustrated. The electromotive force of each cell is added to that of the others and, since the electromotive force of each is one volt, a total of three volts is obtained.

But note that when the cells are connected in series, so are their internal resistances. Since both the electromotive force and internal resistance are tripled, the current that can be drawn from this battery is the same as that which a single cell is capable of producing.

Also, the three cells may be joined to form a battery by connecting them in **parallel** (see Figure 14-2B). In this method of connection all the positive electrodes are joined together, and in like manner all the negative electrodes. Since all the electromotive forces of the cells are in parallel, the electromotive force of the battery is equal to that of a single cell — one volt.

But note that now the internal resistances of all three cells are connected in parallel. Consequently, the total internal resistance of the battery is equal to one-third of the internal resistance of a single cell. Since the electromotive force of the battery is equal to that of a single cell and its internal resistance is only one-third, the current that may be drawn from the battery is three times that which a single cell may produce.

Where a large voltage is desired, the cells are connected in series. Where a large current is desired, they are connected in parallel. We may connect a number of cells in series to produce a larger voltage, and a number of sets of such series-connected cells in parallel to produce a larger current. Such a connection is called a **series-parallel battery.**

In electrical diagrams the symbol for a cell is ⊣⊢. The small vertical line indicates the negative electrode and the long vertical line the positive electrode. The symbol for a battery of cells connected in series is ⊣∣∣⊢.

2. Primary Cells

By far the greatest amount of electrical power produced and consumed is in the form of alternating current furnished by mechanical generators. Not only are these generators capable of producing tremendous amounts of electrical power, but they do so more cheaply than can be done by any other means. Alternating current is employed because the currents and voltages may be stepped up or down by means of transformers, thus making for greater flexibility to meet various types of requirements. However, these generators are very large and, obviously, must remain fixed in place. Power is supplied to the home and factory by means of feeder lines, or **mains**, from the central generating plant.

Where portable power supplies are required, the need is best met by smaller and lighter chemical cells and batteries. (Sometimes small mechanical generators are also used for portable work. But such generators may be employed only where a source of mechanical energy, such as produced by gasoline or diesel engines, is available to rotate the generator.)

In the voltaic cell we have described, the chemical action is not reversible — that is, the ingredients cannot be restored to their original forms once they have entered into the chemical reaction. Hence, when these ingredients have been used up, the cell ceases to function and, generally, is discarded. Such a type of cell is called a **primary cell.** Note, too, that its output is a direct current.

A serious drawback of the voltaic cell is that its electrolyte is a liquid and, thus, spillable. Hence this cell does not lend itself readily to portable use. However, certain types of primary cells have been devised where the electrolyte is a paste or gel and the entire cell is sealed in a leak-proof container. There are many varieties of such cells. Here we shall consider several types that are commonly employed.

a. The carbon-zinc cell

Because they are cheapest and most easily available, most of the primary cells in use today are of the **carbon-zinc** type illustrated in Figure 14-3. A zinc can, protected by a metal shell with which it makes electrical contact, is the negative element (called the **anode**) and acts as the container for the cell. This can is partially filled with a pastelike mixture

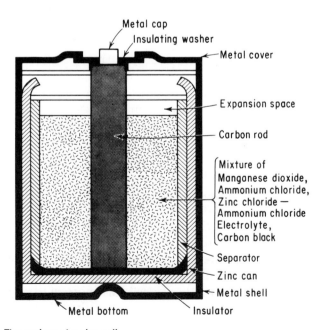

Metal cap
Insulating washer
Metal cover
Expansion space
Carbon rod
Mixture of Manganese dioxide, Ammonium chloride, Zinc chloride — Ammonium chloride Electrolyte, Carbon black
Separator
Zinc can
Metal shell
Metal bottom
Insulator

Fig. 14-3. The carbon-zinc dry cell.

containing manganese dioxide, ammonium chloride, a zinc chloride–ammonium chloride mixture, carbon black, and certain other substances which act as a filler. This mixture is separated from the zinc can by means of a thin plastic film. This separator, while it prevents the mixture from touching the can, nevertheless offers free passage to ions from one to the other.

A carbon rod, which is a conductor of electricity, runs through the center of the cell and makes electrical contact with the mixture. Top and bottom metal covers seal the entire cell and so prevent any of the paste from leaking out. Air space is left near the top of the cell to accommodate the expansion of the ingredients.

The manganese dioxide is the positive element (called the **cathode**) of the cell. It also acts as a depolarizer. The electrolyte is the zinc chloride–ammonium chloride paste. The chemical action of this cell is somewhat more complicated than that of the voltaic cell. But as the cell is connected to an external circuit, the manganese dioxide loses some of its oxygen, thus obtaining a positive charge, and the zinc is oxidized, thus obtaining a negative charge. These charges are transferred to the carbon rod and metal shell, respectively. As the available oxygen from the manganese dioxide lessens, so does the activity of the cell.

The carbon black is used to reduce the internal resistance of the cell. The cell's terminals are its bottom cover ($-$) and the metal cap on the carbon rod ($+$).

The electromotive force that this type of dry cell can produce is approximately 1.5 volts. This voltage remains fairly constant as long as the cell is in good operating condition, regardless of the cell's size. How much current can be drawn from such a cell? The amount of current will depend upon two factors — the resistance of the external circuit and the internal resistance of the cell itself. The greater the external resistance, the less the current drawn from the cell.

If the resistance of the external circuit is reduced to practically zero (we call this a **short circuit**), the amount of current drawn then would be limited by the internal resistance of the cell. (You realize, of course, that the movement of ions within the cell encounters a certain amount of resistance.) This internal resistance depends upon the material and structure of the cell. In general, the larger the size of the cell, the lower the internal resistance, and hence the greater the amount of current that can be drawn. Also, the larger cell has more active material and thus is able to function for a longer period of time than a smaller cell.

Dry cells of this type generally are used for fairly light, intermittent work such as operating door bells or flashlights. If a heavy drain is placed on the cell for an appreciable length of time, its life may be shortened or it may even be ruined. Hydrogen may be formed faster than the depolarizer can consume it and the cell may cease functioning due to polarization. Or else the air space may be too small to contain the excessive amount of gas. As a result, the sides of the cell may split under the pressure, permitting the electrolyte to leak out.

Even when not in use the dry cell has a limited shelf life. Commercial zinc used in these cells contains a small percentage of impurities. When placed in contact with the electrolyte, these impurities form minute cells with the zinc. This is called **local action** and it results in the gradual eating away of the zinc. In time, holes, through which the electrolyte may leak, are formed in the zinc shell. Although local action is greatly inhibited in modern cells by use of a mercury coating on the inside of the zinc shell, enough exists to limit the shelf life of the cell to several years. Also, some of the water of the electrolyte will evaporate in time, thus inhibiting the chemical action of the cell.

Like all other types, these cells come in a great variety of shapes and sizes, depending upon their application. Generally, they are constructed cylindrical or flat. Also, they may be connected to form series, parallel, or series-parallel batteries.

b. The alkaline-manganese cell

Another type of primary cell is the **alkaline-manganese cell.** Here the cathode (positive) element is manganese dioxide, which also acts as a depolarizer. The anode (negative) element is zinc. The electrolyte is a paste of potassium hydroxide. The whole is sealed in a steel can for protection.

Its action is somewhat similar to that of the carbon-zinc type, as is its nominal voltage — approximately 1.5 volts. However, the alkaline-manganese cell is better suited for heavy current drain and for long, continuous action. It has a lower internal resistance and can remain idle for longer periods of time without deterioration. Further, it can operate efficiently at relatively low temperatures better than can the carbon-zinc type.

The alkaline-manganese cell is used to supply electrical power to a great many devices such as toys, portable phonographs, radio and television receivers, and photographic equipment such as the electronic flash. It also is used to supply electrical power to a variety of portable power tools. However, its cost is somewhat higher than for an equivalent carbon-zinc cell.

c. The mercury cell

A third type of primary cell is the **mercury cell.** Here the cathode (positive) element and depolarizer is mercuric oxide mixed with a small amount of graphite. The anode (negative) element is an amalgam of zinc and mercury. The electrolyte is a paste of potassium hydroxide mixed with zinc oxide. The cathode and anode elements are separated by a porous membrane, which permits the ions to migrate through freely. The whole is sealed in a protective metal case.

Its action, too, is somewhat similar to that of the carbon-zinc cell. It has the longest life of the three cells discussed and its chief advantage is that its voltage remains fairly constant for the duration of its useful life. Its nominal voltage, approximately 1.3 volts, is somewhat lower than those of the other types. Also, it is the most expensive of the three.

Its constant voltage makes the mercury cell suitable for a large variety of electronic instruments and other devices, such as heart pacemakers, hearing aids, and electric wrist watches. It also is used in portable radio and television receivers, and communication equipment.

3. Storage Cells

The greatest drawback of the primary cell lies in the fact that its chemical action is not reversible. Once its ingredients are used up in the chemical action that takes place within it, the cell ceases to function and must be discarded. Since its total power output generally is rather limited, this fact makes the cell a rather expensive source of electrical energy.

In 1859 Gaston Planté, a French scientist, discovered he could convert electrical energy into chemical energy which could be stored in a cell. Then, by connecting this cell to an external circuit, he was able to reconvert the chemical energy to electrical energy as he needed it. Here, then, is our reversible chemical action. As electrical energy is fed to the cell, a certain chemical action takes place in one direction. When an external circuit places an electrical load upon the cell, the chemical action reverses itself, and the ingredients are restored to their original conditions, releasing electrical energy to the external circuit in the process. In a sense, electrical energy is stored, to be delivered upon demand. Naturally enough, such a cell is called a **storage cell.**

a. The lead-acid storage cell

In the Planté storage cell two lead plates are placed in an electrolyte of sulfuric acid and water. When electrical energy is "pumped" into this cell by connecting the plates to a source of direct current (we call this process **charging)**, chemical action converts the surface of one of these lead plates to a negative electrode of spongy lead. The surface of the other is converted to a positive electrode of lead peroxide. The cell then is **charged** and ready to deliver electric current on demand.

When we draw current from the cell (this process is called **discharging),** a chemical action takes place within it which ends with both electrodes becoming converted to lead sulfate. The cell then is **discharged** and no more current may be drawn from it. The cell may be charged again, and the electrodes will be transformed to spongy lead and lead peroxide once more. The charge-discharge cycle can be repeated many times.

In modern storage cells the plates are constructed in the form of open metallic grids instead of solid lead sheets. The openings in the grids are filled with the active materials. Thus the spaces of the positive grid are packed with a paste of dark-brown lead peroxide. The spaces of the negative grid are packed with gray spongy lead. (See Figure 14-4.) The advantage of grid-type electrodes is that they are easier to manufacture than those started from solid lead plates.

This kind of storage cell is known as the **lead-acid** type. Its elec-

–A–   –B–

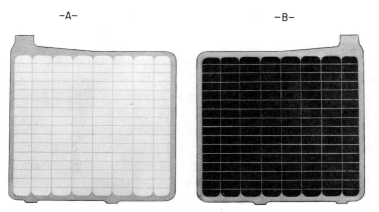

Willard Storage Battery Co.

Fig. 14-4. A. Negative plate of lead-acid storage cell.
B. Positive plate.

General Motors Corp.

Fig. 14-5. How plates and separators are meshed.

tromotive force when fully charged generally is considered to be two volts, though it is slightly higher when measured at **open circuit** — that is, when it is not connected to an external circuit. When a cell is connected into a circuit and current flows through it, its voltage becomes somewhat lower than its open-circuit voltage because of the IR drop due to the cell's internal resistance.

As is true of primary cells, the voltage of the storage cell does not depend upon the size of its plates. However, the larger the plates, the more chemical energy that can be stored in the cell and, hence, the greater the current that can be drawn from it. To increase the effective size of the plates and yet not make the cell too bulky, alternate positive and negative plates are sandwiched together with insulators (called **separators**) of wood, rubber, or other materials between plates (Figure 14-5). The negative plates are connected together, as are the positive plates, producing the effect of a single cell with very large plates. Commercial cells may contain 13, 15, 17, or more, plates. Generally, there is one more negative plate than there are positives. Hence the two outside plates of the stack are negatives.

Storage cells generally are connected in batteries, the chief use for which is in the automobile. Such batteries usually are of the 6-volt and 12-volt types. Thus, three similar cells are connected in series to form the 6-volt type and six cells are series-connected for the 12-volt type.

The battery container usually is made of hard rubber or some other material that is able to withstand mechanical shock, extremes of heat and cold, and is resistant to the action of the acid electrolyte. A separate compartment is provided for each cell and each compartment has space at its bottom for any sediment that may drop from the plates. Each set of plates of the cells has a heavy lead terminal post and these posts are connected in series (that is, the positive terminal of one cell to the negative terminal of its adjacent cell) by means of heavy lead connectors. (See Figure 14-6.)

Each cell compartment has a cover, usually of molded hard rubber. Openings are provided in these covers for the two terminal posts and for a vent. Each vent has a plug so constructed that the electrolyte cannot splash out, although gases may escape from the cell. The joints between covers and containers are sealed with an acid-resistant compound.

The capacity of a storage battery is rated in **ampere-hours.** This means that a 120 ampere-hour battery, for example, theoretically can deliver one ampere of current for 120 hours, 120 amperes for one hour, or any other combination of amperes and hours that, when multiplied together, gives 120. However, if the battery is discharged slowly, it may show a capacity greater than the rated 120 ampere-hours. On the other hand, if the battery is discharged rapidly, its capacity is reduced.

DISCHARGING THE BATTERY. Let us see what happens as we draw current from a fully charged battery. The positive plates consist of grids containing lead peroxide. The negative plates contain spongy lead. The electrolyte is a solution of sulfuric acid and water.

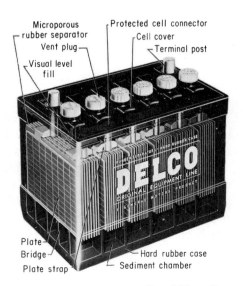

Microporous
rubber separator
Vent plug
Visual level
fill
Protected cell connector
Cell cover
Terminal post
Plate
Bridge
Plate strap
Hard rubber case
Sediment chamber

Fig. 14-6. "Phantom" view of the structure of lead-acid storage battery with six cells.

General Motors Corp.

If we compare concentrated sulfuric acid with an equal volume of water, we find the acid is 1.834 times as heavy as the water. We say that the **specific gravity** of the concentrated sulfuric acid is 1.834. As we add water to the sulfuric acid, the specific gravity of the solution goes down. The specific gravity of the electrolyte of a fully charged storage battery should be approximately 1.280.

We may measure the specific gravity of a liquid quickly by means of a **hydrometer.** This consists of a glass barrel and bulb syringe for sucking up a sample of the liquid. Inside the barrel is a sealed glass float weighted at one end to make it float upright. The depth to which the float sinks in the liquid indicates the relative weight of the liquid compared to water and thus gives us a measure of the specific gravity of the liquid. If the float sinks low in the liquid, the specific gravity is low. If it floats high, the specific gravity is high. A paper scale inside the float indicates the specific gravity of the solution if we note the mark that is level with the surface of the liquid.

So in the fully charged storage battery we have positive plates of lead peroxide and negative plates of spongy lead immersed in an electrolyte of sulfuric acid and water. In this solution some of the sulfuric acid ionizes into positive hydrogen ions and negative sulfate ions. As the battery is discharged by connecting it to an external circuit, the following chemical reaction takes place:

Positive plate:
Lead peroxide + sulfuric acid + hydrogen ions →
lead sulfate + water.

Negative plate:
Lead + sulfate ions → lead sulfate.

You see that on discharge both plates tend to turn to lead sulfate. The more current we draw from the battery, the more lead sulfate is formed. When enough lead sulfate forms on the plates, the chemical action ceases and the battery goes dead. It is completely discharged.

Note, too, that water is produced on discharge. The more we discharge the battery, the more water is formed and the lower the specific gravity of the electrolyte becomes. By means of a hydrometer we can measure the specific gravity of the electrolyte and thus determine how much charge is left in the battery. The following table may be used:

1.280 specific gravity............... 100% charged
1.250 specific gravity............... 75% charged
1.220 specific gravity............... 50% charged

1.190 specific gravity................ 25% charged
1.160 specific gravity................ Very little useful capacity left
1.130 specific gravity................ Discharged

Now, how much current can we draw from a battery? As much as the battery can furnish without dangerous overheating. For example, the starting motor of an automobile may require more than 300 amperes! Because of its low internal resistance, the lead-acid storage battery can furnish such currents — but only for a few seconds at a time; then it must be recharged.

If the battery is discharged excessively, too much lead sulfate is formed, clogging the pores of the plates, possibly damaging them, and making recharging increasingly difficult. Also, the excess lead sulfate in the plates raises the internal resistance of the cell, thus lowering its voltage. The specific gravity of the electrolyte, too, is reduced by the water that is formed, and this lowers the cell's voltage even more. A lead-acid storage cell must not be discharged beyond the point where its voltage falls to 1.75 volts.

The lead-acid storage battery should never be permitted to remain in a low-charge state for any length of time. The specific gravity of the electrolyte should not be permitted to go below about 1.225 (for standard electrolytes). If it remains in this low-charge state, crystalline lead sulfate will form. Since this lead sulfate occupies a larger volume than the material from which it is formed, the plates may buckle, the separators may be damaged, and active material may be dislodged from the grids. Besides, if enough of this lead sulfate forms on the plates, the battery cannot be recharged. If the battery is to remain idle for an extended period, it will slowly discharge itself and, therefore, it must be recharged from time to time.

CHARGING THE BATTERY. The storage battery is a direct-current device. When it runs down and must be recharged, the recharging current must come from a direct-current source, such as a d-c generator. If only alternating current is available, it must be rectified to a direct current before being applied to the battery. During the charging period the positive terminal of the source must be connected to the positive post of the battery and the negative terminal of the source to the negative post.

You will recall that as the battery discharges, both the positive and negative electrodes tend to turn to lead sulfate and the specific gravity of the electrolyte is lowered. As the battery is recharged, the following chemical reaction takes place:

Positive plate:
Lead sulfate + water + sulfate ions →
lead peroxide + sulfuric acid.

Negative plate:
Lead sulfate + hydrogen ions → lead + sulfuric acid.

Note that the positive electrode is reconverted to lead peroxide and the negative electrode to lead. Because sulfuric acid is produced on charging, the specific gravity of the electrolyte rises. When this specific gravity reaches its maximum and will go no higher, the battery is fully charged.

At what rate can the storage battery be recharged? At as high a rate as the battery can take without excessive "gassing" or heat. When the battery is being recharged, only a portion of the current goes to re-forming the electrodes. The rest acts to break up the water of the electrolyte into hydrogen and oxygen gases. This is called **electrolysis.** During recharge a certain amount of "gassing" is normal. But when the charging rate becomes too high, the "gassing" becomes excessive. The vigorous bubbling action loosens the active material of the grids (especially at the positive plates). The loose particles fall to the space at the bottom of the compartments. If enough sediment collects there, it may short-circuit the plates of the cells, making the battery inoperative.

Too great a charging rate may produce excessive heat. This will cause plates to buckle and a short-circuit may occur. Also, the water of the electrolyte will be dissipated by electrolysis and evaporation, raising the concentration of the sulfuric acid left behind. The strong acid then may char the separators, especially if they are made of wood. The temperature of the electrolyte should not be permitted to rise above 110°F.

There are a number of methods for charging a storage battery. (See Figure 14-7.) A d-c generator is connected in series with a rheostat, ammeter, and the battery to be charged. Note that the positive brush of the generator goes to the positive post of the battery and the negative brush is connected to the negative post. The generator and battery are in opposition. Thus, if current is to be fed to the battery, the voltage of the generator must be high enough to overcome the opposing electromotive force of the battery, its internal resistance, and the resistance of the rheostat and ammeter.

A rule-of-thumb method for determining the rate of charge is to

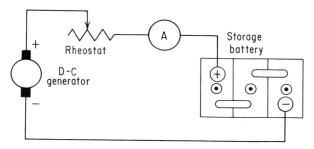

Fig. 14-7. Charging circuit.

allow one ampere for every positive plate of a single cell. Thus, if the cell contains 13 plates, six of them will be positives. The rheostat is adjusted until the ammeter shows six amperes flowing into the battery.

When the battery is in its most discharged state, most of the charging current will go to re-form the plates and very little "gassing" will occur. But as the battery acquires a charge, the "gassing" increases. The rate of charge then must be reduced by introducing more resistance into the circuit by means of the rheostat. When the "gassing" becomes quite heavy and cannot be reduced, and the specific gravity of the electrolyte shows no further increase, the battery is fully charged.

When the battery is being charged, the vents must be open so that the gases may be able to escape. Otherwise, the case may be cracked or the battery otherwise damaged. The mixture of hydrogen and oxygen gases is explosive and care must be taken not to ignite it.

After charging is completed, water should be added to compensate for that lost by "gassing" and evaporation. The level of the electrolyte should be about one-half inch above the tops of the plates. If water is not added, the excessive concentration of sulfuric acid may char the separators, or the battery may be otherwise damaged. Only distilled water or pure water that is free of minerals or sediment should be used. Otherwise, a coating may be deposited on the plates, ruining the battery.

Storage batteries are employed wherever a low-voltage, high-current, direct-current source is required. As stated, the lead-acid storage battery is used extensively in automobiles since its low internal resistance permits the very large output current required for starting. In addition, the battery is called upon to furnish the power for the lights, radio, heater, and so forth, when the engine is stopped. When the engine is running it operates a direct-current generator that takes over the duties of the battery. This generator also charges up the battery so that it will be ready when needed. (See Chapter 13, subdivision D.) However, the plates of the automobile storage battery wear out rather rapidly. The normal life of such a battery is about two years.

Storage batteries are used for lighting purposes in remote rural areas where there are no power lines. The batteries are kept in charge by means of a d-c generator usually driven by a gasoline engine. In d-c areas, storage batteries are used to "iron out" fluctuations in the power lines. Batteries, whose voltages are equal to that of the line, are connected with their positive terminals to the positive side of the line and their negative terminals to the negative side. This is called "floating." So long as the voltage of the line is maintained, nothing happens. But if the line voltage drops, the batteries discharge into the line, bringing its voltage back to normal. When, as a result of discharge, the battery voltage falls, the line current recharges it. Storage batteries are also used for standby service if power lines fail.

b. The Edison storage cell

In 1908 the American inventor, Thomas A. Edison, invented another type of storage cell. In this cell the positive plate consists of a number of perforated steel tubes welded together. Each tube is filled with nickel peroxide. The negative plate is constructed of many small, perforated steel pockets, each filled with pure powdered iron. (See Figure 14-8.) As in the lead-acid cell, alternate positive and negative plates, separated by rubber insulators, are stacked together. All the positive plates are connected together, as are all the negative plates. The electrolyte is a solution of potassium hydroxide in water. The perforations in the tubes and pockets of the plates are to permit the electrolyte to get at the active ingredients.

The assembled plates are housed in a nickel-plated steel container. A steel container is used because there is no acid to attack it. Indeed, the potassium hydroxide of the electrolyte helps protect the container from corrosion. The positive and negative posts extend from the top of the container through insulated bushings. A vent is provided through which gases may escape and through which distilled water may be added as needed. No sediment space is required at the bottom of the container because none of the active material can flake off the plates.

The chemical action that takes place in the cell is somewhat complicated. In brief, as the cell is discharged, the nickel peroxide of the positive plate is changed to nickel oxide and the iron of the negative plate is converted to iron oxide. When the cell is charged, the nickel

-A-   -B-

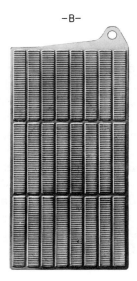

Thomas A. Edison, Inc.

Fig. 14-8. A. Negative plate of Edison storage cell.
B. Positive plate.

oxide is reconverted to nickel peroxide, and the iron oxide to iron. The electrolyte undergoes no change. We may illustrate the chemical process as follows:

CELL CHARGED

Nickel peroxide + iron + potassium hydroxide + water $\rightleftharpoons$

nickel oxide + iron oxide + potassium hydroxide + water.

CELL DISCHARGED

(The arrows $\rightleftharpoons$ indicate that the chemical action is reversible.)

The Edison cell produces an electromotive force of approximately 1.2 volts when fully charged and falls to 0.9 volt when discharged. It has a higher internal resistance than the lead-acid cell and, hence, cannot furnish as high a current. For this reason the Edison battery is not used to operate the starting motor of the automobile. Because of its lower voltage, five cells are connected in series to form a 6-volt battery.

Charging an Edison battery is like charging a lead-acid type, except that an Edison battery can stand an overcharge without damage, provided that there is no frothing and that the temperature of the electrolyte does not rise above 115°F. For ordinary purposes it is best to charge the Edision battery at its **normal** rate, which is about one-fifth of its rated capacity. Thus, if the battery is rated at 150 ampere-hours, then its normal charging rate will be about 30 amperes.

Since the electrolyte undergoes very little change in its specific gravity, a hydrometer is not suited for checking the battery for charge. Instead, a voltmeter may be employed to determine whether the cells are up to their rated voltages.

The Edison storage battery has a number of advantages over the lead-acid type. It can be discharged at a high rate for long periods without danger to the battery and can be recharged safely at a rapid rate. It can stand idle for long periods of time, regardless of its charge, and its life is much longer than that of the lead-acid type.

The Edison battery is much more rugged than the lead-acid type and is lighter in weight than an equivalent lead-acid battery. It cannot produce acid fumes that corrode nearby objects. It can withstand high temperatures better than the lead-acid type and cannot freeze in cold weather. The lead-acid battery will not freeze only if it is at least three-quarters charged.

On the other hand, the lead-acid storage battery has a number of advantages over the Edison type. Chief of these is its lower internal resistance, enabling it to furnish a higher current on demand. Also, the lead-acid battery is cheaper.

Edison storage batteries are used to furnish power to electric motors that operate electric trucks and other types of vehicles, such as mine locomotives and submarines. When the submarine is submerged, it cannot use engines that consume air. Hence it runs on electric motors operated by storage batteries. The lead-acid battery is not suitable because it produces acid-laden fumes which may contaminate the air and corrode the equipment.

On the surface, the submarine is run by diesel engines. These engines also operate a d-c generator that recharges the batteries. (Atomic-powered submarines do not use batteries for motive power since the reactor that furnishes the heat for the turbines does not exhaust the air.)

c. The nickel-cadmium storage cell

In 1899 Waldemar Jungner, a Swede, developed another type of storage cell — the **nickel-cadmium storage cell.** In appearance this cell is very similar to the Edison cell. The plates, of nickel-plated steel, are flat and contain parallel rows of small pockets. Within these pockets are the active ingredients, which consist of nickel hydroxide for the positive element and cadmium hydroxide for the negative element. The pockets contain many fine perforations, which permit the electrolyte to enter but keep the ingredients from escaping. The electrolyte is a solution of potassium hydroxide.

As in the other types of cells, the plates may be stacked for greater capacity. Polystyrene or glass rods are used as separators between plates. And, as with the other types of cells, a number of cells may be arranged to form a battery, cased in steel or plastic.

When the cell is discharged, the positive element becomes nickelous hydroxide and the negative element is cadmium hydroxide. When the cell is charged, the positive element changes to nickelic hydroxide and the negative element to metallic cadmium. After the cell is fully charged and the charging process continues, oxygen gas is generated at the positive element and hydrogen gas is generated at the negative element. This requires that the cell be vented to permit the escape of gas.

The average operating voltage for a fully charged nickel-cadmium cell is approximately 1.2 volts. Its internal resistance is very low, which means that it can deliver very large currents with little loss of voltage.

The nickel-cadmium storage battery combines the best features of both the Edison and the lead-acid types. It is rugged and has a very long active life. It can be stored indefinitely without suffering any ill effects. Any rechargeable storage battery loses some of its charge when it stands idle, but the nickel-cadmium battery has the lowest self-discharge rate of any battery. When it has to be recharged, it is safe to charge it rapidly. And because of its low internal resistance, it can furnish the large current required for starting the automobile and similar vehicles. Its chief disadvantage compared to the lead-acid battery is its greater cost.

The obvious advantages of a rechargeable cell have enabled the nickel-cadmium storage cell to invade the field formerly dominated by primary cells. This is the field of small cells and batteries for portable use where cell and battery maintenance is at a minimum.

The ingredients of the small nickel-cadmium cell are the same as for the larger type. The plates are formed of a woven nickel screen, and a paste consisting of the active ingredients is pressed into the spaces within the screen. Another method is to make the plates in **sintered** form. Here the screen is covered with powdered nickel and heated until the powder forms into a solid mass. Then the plate is impregnated with the nickel salt, to form the positive plate, and with the cadmium salt, to form the negative plate. The electrolyte is a potassium hydroxide solution.

A separator, consisting of a porous insulating membrane, is placed between a positive and negative plate and the whole rolled to form a cylinder which is sealed into a small steel can for protection. The metal can forms the negative terminal of the cell and an insulated metal button at the top is the positive terminal. In some cells the connections are reversed so that the can is the positive terminal and the button at the top is the negative terminal.

In order that the liberated gases may escape, the cell is provided with a spring-loaded, one-way valve. When the pressure within the cell becomes too great, the valve is forced open and the gases are permitted to escape. As the pressure drops, the valve closes, resealing the cell.

The **sealed** nickel-cadmium storage cell is an improvement over the vented type. Here no valve is required. Instead, the gases formed on charging are not permitted to create pressures great enough to damage the cell.

This is accomplished by constructing the cadmium element with a larger capacity than the nickel element. Starting with both elements fully discharged, charging the cell causes the nickel element (which has the lower capacity) to reach its full-charge state first. In this state, it starts giving off oxygen. Since the cadmium element has not yet reached its full-charge state, it produces no hydrogen.

The oxygen is used up as it oxidizes the cadmium element, forming cadmium oxide. As the charging continues, the cadmium is oxidized at a rate just sufficient to offset the input energy, thus keeping the cell in equilibrium at full charge. In this way, no dangerous gas pressures can be built up within the cell.

The sealed nickel-cadmium cell has made possible a host of "cordless" electrical devices, such as electric shavers, toothbrushes, and power tools, which formerly had to be connected to the power mains. These cells are used in many electronic devices such as hearing aids, portable radio and television receivers, phonographs, tape recorders, communication equipment, and photographic equipment. Such cells are used extensively in space exploration where the energy of the sun is transformed into electrical energy by means of **solar cells** (which will be described later). The electrical energy so generated is used to charge the nickel-cadmium cells, which supply the current to the various devices as needed.

The chief disadvantage of the nickel-cadmium cell, compared to the carbon-zinc primary cell, is its higher initial cost. But, because it is rechargeable, the nickel-cadmium cell offers a lower total cost, since it does not have to be discarded after it has been exhausted.

4. Fuel Cells

A major portion of the energy we consume for useful work comes from the breaking of the chemical bonds that bind the atoms of fuels, such as coal, oil, and natural gas, to form molecules. As we burn these fuels in the oxygen of air, the bonds are broken and their energies are converted to heat energy. In the power plant, this heat is used to change water to steam, which causes some device, such as a steam turbine, to rotate. The mechanical energy of the rotating turbine then may turn an electrical generator, producing electrical energy.

In such an indirect transformation from chemical to electrical energy the major portion of the energy is lost — mainly in the form of wasted heat energy. Some additional energy is lost overcoming the friction of the various rotating devices. It has been estimated that less than half of the original energy inherent in the fuel is made available as useful electrical energy.

Obviously, a more efficient transformation would take place if we could go directly from chemical energy to electrical energy, thereby eliminating the losses in the heat cycle and the friction in the turbine and generator. Such a transformation takes place in the voltaic cell and the other cells we have discussed. The difficulty with these cells, however, is that their ingredients (lead, zinc, mercury, etc.) are expensive, requiring about as much electrical energy to produce them as the energy they can furnish.

In 1839 an English scientist, William Grove, constructed a chemical cell using less expensive ingredients. Hydrogen gas (the **fuel**) and oxygen gas (the **oxidant**) were combined to form water, generating an electric current in the process. Unfortunately, the mechanical electric generator was being developed at about this time and very little was ·done to investigate Grove's idea further. In recent days, however, the need for an efficient, portable power supply has revived interest in such a project.

A chemical cell using a fuel such as hydrogen to produce electrical energy is called, appropriately enough, a **fuel cell.** We may understand

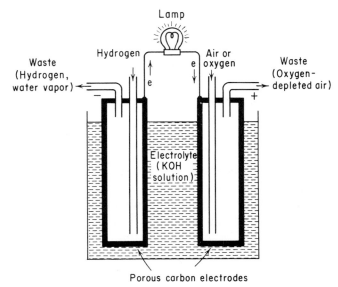

Fig. 14-9. Simple oxygen-hydrogen fuel cell.

its operation from an examination of Figure 14-9. Two hollow, porous carbon electrodes are suspended in a solution of potassium hydroxide (KOH) in water. Hydrogen gas is pumped through a tube into one of these electrodes (the negative electrode) and the waste (consisting of excess hydrogen gas and water vapor) exits through another tube. Oxygen gas (or air, which contains oxygen) is pumped through a tube into the other electrode (the positive electrode). The waste, consisting of the oxygen-depleted air, exits through another tube.

The potassium hydroxide solution forms the electrolyte of the cell. As the potassium hydroxide dissolves in water, it ionizes, forming positive potassium (K^+) ions and negative hydroxyl (OH^-) ions. (A hydroxyl ion is a negative ion composed of one hydrogen atom and one oxygen atom.)

At the negative side of the cell the molecules of hydrogen gas diffuse through the pores of the carbon electrode. Each molecule consists of two atoms of hydrogen held together by means of a chemical bond. As the molecules diffuse through the pores of the electrode, a **catalyst,** embedded in the electrode surface, causes them to break up into hydrogen atoms. (A catalyst is a substance, such as powdered platinum, that assists a chemical reaction without entering the reaction directly.)

The hydrogen atoms react with the hydroxyl ions in the electrolyte to form water. In the process, they give up electrons to the electrode. The water thus formed goes into the electrolyte. These electrons flow around the external circuit to the positive electrode, doing useful work on the way. This electron flow is the output of the cell.

At the positive side of the cell the oxygen molecules diffuse through the pores of their electrode. In a rather complicated process, the oxygen, assisted by the electrons from the negative electrode, reacts with the water of the electrolyte to form hydroxyl ions. Here, too, a catalyst is used to assist the reaction. The hydroxyl ions migrate through the electrolyte to the negative electrode, thus completing the cycle.

If the external circuit is open, the electrons released by the hydrogen atoms accumulate at the surface of their electrode. This negative charge attracts the positive potassium ions of the electrolyte, thus forming a sort of layer which balances the negative charge at the surface of the electrode. In an equivalent manner the oxygen electrode acquires a layer of negative hydroxyl ions which balances its positive charge. These layers prevent further reaction between the gases and the electrolyte.

When the circuit is closed and the electrons at the hydrogen electrode can flow through the external circuit to the oxygen electrode, the process is resumed. Thus the cell remains idle as long as no drain is put on it. (This is an advantage over the other chemical cells we have discussed, in which a certain small activity goes on, even with the external circuit open.)

The fuel cell is still in its developmental stage. Experimentation is continuing with other types of fuels — gases, such as methane, propane, vaporized gasoline and kerosene, and gases formed by the gasification of coal, and liquids, such as alcohol and formaldehyde. The simple hydrogen-oxygen cell we have described will operate at ordinary room temperatures and pressures. Higher output may be obtained by using elevated temperatures and pressures. At temperatures high enough to evaporate the water of the electrolyte, molten electrolytes, such as molten potassium carbonate, may be employed. Cells have been constructed that use a solid electrolyte — an ion-exchange membrane through which the ions may migrate from one electrode to the other. Other types of electrodes, too, have been employed, such as electrodes made of porous nickel.

The future for the fuel cell is bright, indeed. We have the possibility of a small fuel cell, using cheap fuels and requiring practically no maintenance, operating quietly in a closet in the home and supplying the electrical power to heat and light the house and to operate a host of home appliances and devices. In the field of transportation, electric motors will replace gasoline and diesel engines, the power being furnished by fuel cells. Such motors will make the vehicles quieter and more efficient and will produce no noxious fumes to pollute the air. In space travel the fuel cell will provide the electrical power for the various devices of the craft, and its by-product, water, could be used by the passengers for drinking and other purposes.

B. THE SOLAR CELL

Our sun constantly radiates enormous quantities of light and heat energy to the earth. Large quantities of this energy have been stored as coal, oil, and natural gas, which are used as fuels in our thermal power plants. Here the fuels are burned to produce heat energy. The heat energy is converted to mechanical energy which, in turn, is converted to electrical energy.

For many years, scientists have been seeking methods by which the heat and light energy of the sun may be converted directly to electrical energy. To date, they have not been too successful; their best devices have been able to convert only a very small percentage of the available energy. But a start has been made.

The **solar cell** is a device that converts light energy directly to electrical energy. (See Figure 14-10.) A wafer consisting of a crystal of pure silicon, a semiconductor, is "doped" with a slight amount of arsenic. As a result of this doping, the wafer contains an excess of free electrons which, you will recall, are negative current carriers. (See Chapter 4, subdivision A.) This wafer is coated at its top surface with a very thin layer of silicon doped with a slight amount of boron. This

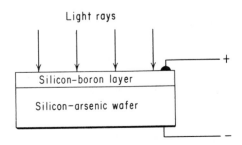

Fig. 14-10. The solar cell.

doping produces a layer containing an excess of holes, which are positive carriers.

Some of the free electrons of the wafer drift toward the layer and some of the holes of the layer drift toward the wafer. At the junction between the two, an electrical barrier is produced which prevents the further migration of the electrons and holes from one to the other.

If the cell now is exposed to light, the silicon absorbs some of the light energy. This energy breaks some of the bonds between the atoms, freeing a quantity of electrical charge carriers (electrons and holes). Some of these are collected and separated by the junction, giving the wafer a negative charge and the layer a positive charge. Thus an electrical potential difference (electromotive force) appears between the wafer and the layer.

The bottom of the wafer and a spot on the layer are tinned for the connection of leads. If an external circuit is connected to these leads, the electromotive force will cause a current to flow through this circuit.

The output of the cell depends, among other factors, upon its exposed area and the intensity of the light falling upon it. The area of the cell cannot be very large, since it is extremely difficult to produce large silicon crystals. A typical cell, such as used in the Telstar communication satellite, has an effective area of about two square centimeters, or about one third of a square inch. Such a cell can produce a maximum output of about 0.45 volt at about 50 milliamperes (0.05 ampere). This indicates a maximum power output of about 0.0225 watt. Of course, such cells may be connected in series (for higher voltages) or in parallel (for higher currents). In the original Telstar satellite 3600 such cells were connected to form a series-parallel battery, which was capable of generating about 15 watts at about 20 to 29 volts (depending upon the intensity of the sunlight to which it was exposed at a particular time). To date, solar batteries designed to produce peak-power outputs to about 250 watts have been manufactured.

The solar cell is simple, light, and reliable. It is rugged and can operate in an environment where temperature and humidity vary over a wide range. Its operating life is practically unlimited. On the other hand, it is quite expensive.

As indicated, it may be employed in space vehicles to furnish the electrical power required by the various instruments. It generally is used to charge up storage cells, such as the nickel-cadmium type, so that power may always be available, even when the cells are not exposed to the sun.

Hoffman Electronics Corp.

Fig. 14-11. Explorer VII weather satellite, showing solar cells on surface.

The solar battery can also be used in sunny areas for remote and unattended radio stations, lighthouses, beacons, and telephone repeaters. For example, in a recent test, the solar battery was used to supply electricity to a telephone system. A battery consisting of 432 individual cells was mounted on a pole and exposed to sunlight. On a bright day it generated 10 watts of electrical power. Part of the electricity went to the telephone system and the rest was used to charge up a storage battery. This storage battery was a reserve to be used at night or on cloudy days when the sun was obscured.

Of course, the solar cell may be used with light sources other than the sun. For example, light from a lamp, shining through holes punched in cards or tape and falling on such a cell, generates electrical pulses which operate devices such as data-processing machines. In addition, many other industrial devices and systems utilize the electrical output of the cell for light-detection and light-response purposes.

C. THE ATOMIC CELL

Atomic energy can be used to generate electricity. In the atomic power plant the atomic reactor produces a tremendous amount of heat resulting from the fissioning atoms of the nuclear fuel. This heat, as in the conventional thermal power plant, is used to change water to steam to rotate steam turbines which, in turn, rotate the electric generators. Here, too, scientists are seeking ways to bypass the wasteful heat cycle by transforming atomic energy directly to electrical energy.

Steps in this direction are being taken by utilizing the radioactive isotopes that are a by-product of the fissioning atoms of the nuclear fuel. For example, in the cell illustrated in Figure 14-12 a small amount of radioactive material is enclosed in a capsule of polystyrene, an insulator. As electrons emitted by the radioactive source penetrate the walls of the capsule, they are absorbed by the aluminum collector surrounding it and flow through the lead shield around the aluminum collector to the negative terminal.

At the same time, the radioactive source, having lost electrons, acquires a positive charge. This charge is carried to the positive terminal by means of the monel wire. Hence the negative terminal obtains a negative charge and the positive terminal a positive charge. If an external circuit is connected between these terminals, current will flow.

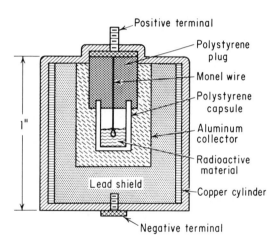

Fig. 14-12. Atomic cell.

The power output from such a cell is quite small. However, since the number of atomic reactors is increasing steadily, very large quantities of radioactive by-products will become available. And since some of these are capable of emitting electrons for thousands, and even millions, of years, the future of the atomic cell is bright indeed.

D. MISCELLANEOUS SOURCES OF ELECTRIC ENERGY

Besides those we have already discussed, there are several other sources of electric energy based upon various types of physical effects. For the most part the amount of electrical power generated by these sources is very small, hence generally they are not employed as power supplies. Instead, their outputs are used in certain devices having special applications. It may well be, however, that these sources will someday be developed to the point where they are able to generate appreciable electric power.

1. Thermoelectric Effect

In the eighteenth century Alessandro Volta, the inventor of the voltaic cell, discovered a curious phenomenon. He found that if two dissimilar metals are placed in contact with each other, one of the metals will become slightly negative and the other slightly positive. In other words, a potential difference appears between the two. We call this **contact potential.** He further found that the contact potential is affected by the metals used and the temperature of the junction between them.

Today we believe that the contact potential is produced because of the free electrons present in the metals. These free electrons move from one metal to the other but, depending upon the metals used, can cross more readily in one direction than in the other. The metal receiving the most free electrons will then become negative. The other, because of its deficiency of electrons, becomes positive.

In 1822 Thomas J. Seebeck, a German physicist, while investigating the phenomenon discovered by Volta, found that if two dissimilar metal strips are joined at one end and this joint is heated, a small direct voltage, of the order of millivolts, appears between the cool unjoined ends of the strips. And the magnitude of this voltage depends directly upon the difference in temperature between the heated and cool ends of the strips.

Note that here is a method for converting heat energy directly to electrical energy. The effect discovered by Seebeck is known as the **Seebeck,** or **thermoelectric, effect.** We believe this effect is produced by the heat, which causes the electrons of the strips to move away from the hot end of each strip and concentrate at the cool end. But the degree of such concentration is different for different metals. Thus, if we use two different metal strips, there is a different concentration of electrons at the cool end of each. This difference of concentration accounts for the difference of potential (electromotive force) between the two cool ends.

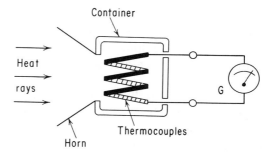

Fig. 14-13. Thermopile.

The combination of the two dissimilar metal strips is called a **thermocouple.** You will recall we discussed the use of the thermocouple for electrical and temperature measurement in Chapter 12, subdivision B, 2.

A number of thermocouples may be joined to produce a **thermopile,** which is an extremely sensitive detector of heat rays. (See Figure 14-13.) As many as several hundred thermocouples may be joined in series (only three sets are shown here) and encased in a container that is open at one side. One set of junctions is kept cool by being placed at the back of the container. The other set of junctions is exposed to heat rays entering through the opening. A horn, mounted at the open side, acts as a sort of funnel to catch more of the heat rays and thus increase the voltage generated across the open ends of the thermopile. A sensitive galvanometer is used to measure this voltage.

Similarly, a number of thermocouples may be joined in series to produce a voltage sufficient to operate a radio receiver or some similar device. If one set of junctions is heated and the other set kept cool, the required voltage may be obtained from across the two free ends.

It is interesting to note that the Seebeck effect is reversible, as was discovered by Jean Peltier, a French physicist, in 1834. If two dissimilar metal strips are joined together to form a complete loop and electric current is set flowing through this loop, one junction will become heated and the other cooled. (The heating effect is distinct from the heating of both strips because of their resistances.)

The **Peltier effect** makes possible the construction of devices that may be used for both heating and cooling. Small refrigerators have been made that employ no moving parts or compressors. A portable cart has been manufactured with both refrigeration and oven compartments. These are but a few of the various devices employing the Peltier effect that have been constructed or are being contemplated. It may well be that the homes of the future will be electrically heated in the winter and cooled in the summer by small devices that operate quietly and require very little maintenance.

There is a second method for converting heat energy directly to electrical energy. In 1883 Thomas A. Edison, while experimenting with his incandescent lamp, noticed a peculiar effect. His lamp consisted of a filament within a glass bulb from which all the air had been removed. When an electric current was passed through this filament, the resistance it encountered caused the filament to glow, producing light.

In this particular lamp, Edison had sealed in a metal plate. (See Figure 14-14.) When he connected a sensitive galvanometer between the plate and one side of the filament, the instrument showed that a small current was flowing through it from the plate to the filament. Edison was unable to explain the presence of this current.

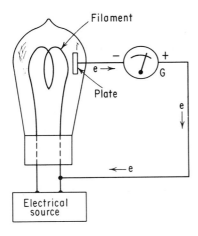

Fig. 14-14. Circuit used by Edison.

Today we believe that, as the filament was heated, the normal activity of the free electrons was speeded up until some of them were shot out into space. A number of these free electrons landed upon the metal plate, making it negative. The filament, having lost electrons, became positive. When the external circuit was completed by connecting the galvanometer between the plate and the filament, current flowed through the circuit as indicated.

Note that the electrical source in Edison's experiment was used only to heat the filament. It makes no difference how the filament is heated. The result is the same — electrons are emitted. The giving off of electrons by a heated body is called **thermionic emission.** Most of the electron tubes used in radio, television, and other electronic devices operate on this principle.

2. Photoelectric Effect

Electrical energy is transformed into light energy in the electric lamp. Can light energy be changed back to electrical energy? The first clue came in 1887 when Heinrich Hertz, the German scientist who is known as the "father of radio," discovered that, for a given electromotive force, an electric spark will jump across a larger gap if this gap is illuminated by ultraviolet light than if the gap is left in the dark.

The second clue came about a year later when Wilhelm Hallwachs, another German scientist, found that ultraviolet light falling upon a negatively charged metal plate caused it to lose its charge. If the plate was charged positively, there was no apparent loss of charge. The final clue came about ten years later when Joseph J. Thomson, the famous English scientist, discovered that ultraviolet light falling upon a metallic surface caused it to emit electrons.

Here was the reason for the behavior of Hertz's spark gap. The presence of the emitted electrons (from the metal balls between which the spark jumped) reduced the effective resistance of the gap and thus the spark was able to jump across it more easily. Also, if a negatively charged plate emitted electrons, it lost its negative charge. If, on the other hand, the plate was charged positively, the loss of electrons would merely increase the charge.

The emission of electrons under the impact of light energy is called **photoelectric emission.** The more intense the light, the more electrons are emitted by the exposed metal. Although most metals will emit electrons when their surfaces are exposed to ultraviolet light, two other German scientists, Julius Elster and Hans Friedrich Geitel, discovered that some metals, such as sodium, potassium, and certain others, will emit electrons when exposed to ordinary visible light rays and infrared rays as well.

This is the principle of the **phototube** illustrated in Figure 14-15. A half-cylinder of metal (called the **cathode**) is coated on its inner surface with some emissive substance such as potassium or cesium. A thin metal rod (called the **anode**) is placed along the central axis of the cathode. Both the cathode and anode are enclosed in a glass envelope from which the

−A−

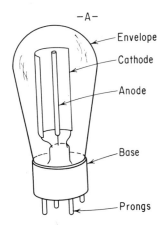

Envelope

Cathode

Anode

Base

Prongs

−B−

Radio Corporation of America

Fig. 14-15. A. A phototube, showing its parts.
B. Commercial phototube.

air has been evacuated and which is set in a bakelite base. Connections are made to the anode and cathode by means of prongs mounted in the base.

There are two reasons for evacuating the air from the envelope. First, we wish the emitted electrons to have an unimpeded path from the cathode to anode without colliding with air molecules. Second, the emitting surface must be absolutely clean. Air would soon corrode the surface and impair the action of the tube.

To keep out unwanted light, the inner surface of the envelope may be masked, except for a small, circular window facing the inner surface of the cathode through which the light may enter. Or else a shield with a similar window may be placed over the entire tube.

As light strikes the inner surface of the cathode, electrons are emitted; and the greater the intensity of the light, the stronger is the emission. Some of these electrons strike the anode, making it negative. The cathode, having lost electrons, becomes positive. If a microammeter is connected between the two, current will flow through the meter from anode to cathode.

If a battery is placed in the external circuit as illustrated in Figure 14-16, a positive charge will be placed on the anode. This will attract more of the emitted electrons and a larger current will flow through the external circuit. (The symbol for the phototube is ⊖ . The half-circle stands for the cathode; the bar is the anode.)

Here, then, is a device that converts light energy directly into electrical energy. The numerous applications of the phototube are as simple as they are ingenious. For example, a beam of light at one side of a room may be focused to strike the cathode of a phototube at the opposite side. Electrons are emitted and the flow of current is amplified until it is large enough to close an electromagnetic relay. If a fire breaks out and smoke fills the room, the light beam is obscured and the current in the phototube circuit drops. This causes the relay to open, and an alarm is sounded.

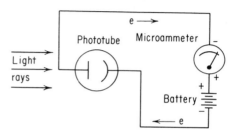

Fig. 14-16. Phototube circuit.

A similar setup may be used as a burglar alarm, except that an invisible ultraviolet ray will be used instead of a visible light ray. When the beam is interrupted by the body of the burglar passing through it, an alarm will be given. Similar circuits may be employed for counting and sorting purposes where the interruption of the light beam operates a counter or some other device. Since the current flowing in the external circuit of the phototube will vary directly as the intensity of the light striking the tube, a meter in that circuit will indicate the intensity of the light.

Perhaps the most ingenious application of the phototube is to reproduce the sound from sound-on-film motion pictures. The original sound enters a microphone, producing an electric current that varies directly with the frequency and loudness of the sound. This current is sent to a recorder where the illumination produced by an electric lamp is made to vary in step with the variations of the current.

The motion picture film is divided into two unequal parts. One part, which occupies most of the width of the film, contains the picture. The rest of the width of the film is reserved for the recording of the sound that corresponds to the picture. This portion is called the **sound track.**

As the varying light from the lamp strikes the unexposed sound track,

Weston Instruments Div., Daystrom, Inc.

Fig. 14-17. Photronic cell.

it produces, upon development, a series of dark bands whose number corresponds to the frequency of the original sound. The density of these bands depends upon the loudness of the sound.

In the projector, a light shines through the pictures on the film, projecting them on the screen. At the same time, another light shines through the sound track of dark bands and falls upon a phototube. Since the film is moving, the light entering the tube will be interrupted at a frequency corresponding to the number of bands on the film, which, as you know, corresponds to the frequency of the original sound. The intensity of the light reaching the tube will vary with the density of the bands, which corresponds to the loudness of the original sound. Consequently, a varying beam of light, whose variations correspond to the frequency and to the loudness of the original sound, enters the phototube. As a result, a varying current will flow in the tube. The current is amplified and passed into a loudspeaker, which reproduces the original sound that entered the microphone.

Another device for converting light to electric energy is the **photronic cell** illustrated in Figure 14-17. A thin layer of selenium is deposited upon a metal base plate. A special metallic grid is placed over the selenium layer. Between the selenium and grid a sort of **barrier layer** is formed. This layer permits electrons to pass from the selenium to the grid, but not in the opposite direction.

A metal ring, which acts as the negative electrode of the cell, is placed over the grid. The metal base plate in contact with the selenium is the positive electrode. The whole is enclosed in a case of glass and plastic.

Weston Instruments Div., Daystrom, Inc.

Fig. 14-18. Front and rear views of exposure meter.

When light, passing through the glass of the case and the openings of the grid, strikes the selenium, some of its electrons will absorb sufficient energy to break away from their parent atoms. These high-energy electrons will be able to surmount the barrier layer and, reaching the metallic grid, will give it a negative charge. The selenium layer, having lost electrons, acquires a positive charge.

If an external load is connected between the metal ring (which is in contact with the grid) and the base plate (which is in contact with the selenium), the difference in potential between the two will cause current to flow through the load. This load may be a sensitive meter and, since the amount of current depends upon the intensity of the light shining upon the cell, this light intensity can be measured. Such a cell and meter constitute the **exposure meter** used in photography to measure the intensity of illumination to determine the proper exposure for the film. (See Figure 14-18.)

3. Piezoelectric Effect

We have seen how the generator converts mechanical energy into electrical energy. But this is an indirect action. It requires magnetic energy as an intermediate step between the mechanical energy applied to the generator and the electric current flowing from it.

Webster Electric Co.

Fig. 14-19. Phonograph pickup using a piezoelectric crystal.

We have, however, a method for converting mechanical energy directly into electrical energy. Certain crystals, such as Rochelle salts and quartz, have the property of generating an electromotive force when they are compressed. The voltage generated depends upon the degree of compression. This is known as the **piezoelectric effect.**

For example, the Rochelle salt crystal is often employed in **phonograph pickups** to convert the variations in the grooves of a phonograph record into a varying electrical voltage. (See Figure 14-19.) The phonograph needle is held firmly against the crystal. As it passes through the grooves of the record, the needle is vibrated from side to side by the variations in the grooves. These vibrations are transmitted to the crystal as variations in pressure. As a result, a varying voltage is generated by the crystal, which, when it is amplified and fed into a loudspeaker, produces sound.

Another use of the piezoelectric crystal is in the **pressure sensor,** where a mechanical force or pressure is to be measured. Here the force is applied to the face of the crystal. Since the resulting voltage is proportional to the force producing it, the amount of this force can be determined by measuring the voltage.

The interesting thing about the piezoelectric effect is that it is reversible. A mechanical strain applied to opposite faces of the crystal will generate a voltage — that is, it will set up an electrostatic field between the two faces. Conversely, if a voltage is applied to electrodes on two parallel faces of the crystal, a mechanical strain occurs in the crystal.

This property of a quartz crystal may be employed to generate the high-frequency alternating currents that are used in radio transmitters. Flat metal electrodes are placed against two opposite parallel faces of the crystal. A voltage placed across these electrodes produces a mechanical strain in the crystal. This strain, in turn, produces an electrostatic field which, in turn, again produces a strain. This process goes on.

You may understand this better, perhaps, if you review the action in a parallel-resonant circuit (see Chapter 11, subdivision C, 2). There an electrostatic field alternately changes to a magnetic field and back again. In the quartz crystal the electrostatic field alternately changes to a mechanical strain and back again.

At the natural frequency of the mechanical vibrations of the crystal, we may make the two actions mutually self-sustaining by bringing in sufficient electrical energy to replenish the energy that is lost as heat during each cycle. The effect of the crystal, then, is to produce an oscillating voltage whose frequency is determined by the natural frequency of the crystal. This frequency, in turn, is determined by the mechanical structure of the crystal. Quartz crystals can be cut whose natural frequency may be thousands, and even millions, of cycles per second. When used in this way, the crystal, with its associated circuit, is known as a **crystal oscillator.**

4. The Magnetohydrodynamic Generator

In the mechanical generator discussed in the previous chapter, a conductor (consisting of a coil or coils of copper wire) moving through a magnetic field generates an electromotive force. It makes no difference whether the conductor consists of a copper wire or a conductive gas; the result would be the same. The flow of the gas through the magnetic field would induce an electromotive force which would be at right angles to both the flow of the gas and the magnetic field.

This is the principle upon which the **magnetohydrodynamic** (MHD) **generator** operates. The amount of electric power generated depends (among other factors) upon the strength of the magnetic field, the velocity of the gas through it, and the resistivity of the gas. The stronger the magnetic field, the greater is the amount of power generated. Similarly, the greater the velocity of the gas, the greater will be the power. On the other hand, the larger the resistivity of the gas (or, what is the same thing, the lower its conductivity), the smaller will be the power.

The magnetohydrodynamic generator is still in its developmental stage, and many serious problems must be solved before it becomes a practical supplier of appreciable quantities of electric power. The

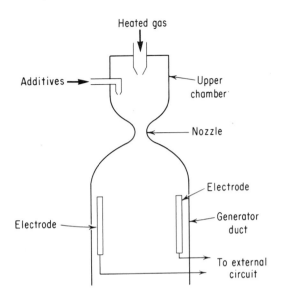

Heated gas

Additives →

Upper chamber

Nozzle

Electrode

Electrode

Generator duct

To external circuit

Fig. 14-20. Model of a magnetohydrodynamic generator.

diagram of one experimental model is illustrated in Figure 14-20. Hot helium gas (at a temperature of 2500°C or more) is introduced under pressure into the upper chamber. This gas may be heated by conventional means or by the heat of an atomic reactor. At this heat the gas is partially ionized — that is, it is a conductor of electricity. To further increase its conductivity (or, what is the same thing, to reduce its resistivity), additives, consisting of ions of sodium, potassium, or cesium, are introduced into the same chamber.

The gas then flows through a nozzle into the generator duct, where it expands and its velocity is increased to supersonic speeds. At this high speed the gas passes through a strong magnetic field (not shown in the diagram but assumed to be at right angles to the plane of the page). As it does so, an electromotive force is generated between the two electrodes, which are connected to the external electrical circuit. The gas then is recovered, reheated, and passed through the generator again.

There are a number of variants of this system. For example, instead of hot helium gas, a fuel and oxidant may be burned in the upper chamber, which thus becomes a combustion chamber. The hot gases produced by the combustion then pass through the nozzle into the generator duct.

The electromotive force generated by the above systems is a direct voltage. If an alternating voltage is desired, the hot gas may be passed through the magnetic field in a series of pulses, and the output would then be a series of electrical pulses. By transformer action these pulses may be converted to alternating voltage.

To date, relatively small amounts of electric power have been so generated. However, the magnetohydrodynamic generator gives promise of becoming a major source of electric power in the near future.

QUESTIONS

1. Explain the action of the voltaic cell.

2. Describe and explain the action of the carbon-zinc dry cell.

3. Describe the structure of the alkaline-manganese dry cell.

4. Describe the structure of the mercury dry cell.

5. Explain the difference between primary and storage cells.

6. Describe and explain the action of the lead-acid storage cell.

7. What happens in the lead-acid storage cell (*a*) as it is charged; (*b*) as it is discharged?

8. Describe and explain the action of the Edison storage cell.

9. Describe and explain what happens in the nickel-cadmium storage cell as it is charged and discharged.

10. Explain how the nickel-cadmium storage cell may be constructed as a sealed type.

11. Explain the principle of operation of the fuel cell.

12. Explain the principle of operation of the solar cell.

13. Explain the principle of operation of the atomic cell.

14. Explain what is meant by the **thermoelectric effect; thermionic emission.**

15. Explain what is meant by the **photoelectric effect.**

16. Explain what is meant by the **piezoelectric effect.**

17. Explain the principle of operation of the **magnetohydrodynamic generator.**

TRANSMISSION AND CONTROL OF ELECTRIC POWER

V

TRANSMISSION
OF
ELECTRIC POWER

15

In the United States electric power generally is generated and distributed in the form of 60-Hz alternating current. Where, as in certain industrial applications, direct current is required, direct-current generators may be employed, although usually the alternating current from the power mains is rectified to produce the direct current.

Single-phase alternating current generally will suffice for lamps, heating devices, and appliances employing low-power motors. Hence most households are supplied with this type of current. For industrial plants, however, where large quantities of electric power are consumed, it is customary to supply three-phase alternating current to the power mains.

A. SINGLE-PHASE ALTERNATING CURRENT

Most household appliances and devices are constructed to operate on 120-volt, 60-Hz, single-phase alternating current. (Although we usually refer to the household mains as 120-volt lines, the electricity may be supplied at 110, 115, or 117 volts, depending upon the locality.)

311

Simplest, perhaps, is the arrangement illustrated in Figure 15-1. Here a single-phase, 120-volt, 60-Hz alternator supplies the power for the load. (The symbol ϕ, the Greek letter **phi**, is frequently used to indicate "phase." Thus, single phase may appear as 1ϕ.) As illustrated, the load is shown to be resistive in nature; hence the voltage and current of the alternator would be in phase. Should the load be reactive, the voltage and current would be out of phase. (See Chapter 10.)

(The arrows in Figure 15-1 are merely a convention to show the flow of current from the alternator to the load and back to the alternator. Actually, the current reverses the direction of its flow with every alternation.)

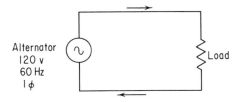

Alternator
120 v
60 Hz
1ϕ

Load

Fig. 15-1. Simple single-phase circuit.

Electrical appliances are connected in parallel across the line. Thus the voltage across each appliance remains the same — 120 volts. The current flowing through the line will depend, of course, upon the sum of the currents drawn by each appliance. If this current exceeds the rating of the fuse, the latter "blows," opening the circuit and cutting off all the appliances on that particular line.

You can see, therefore, that it is not wise to connect all the appliances to a single line. What we need are two or more lines. Of course, we may bring these extra lines into the house, using two wires for each line. Thomas Edison invented a **three-wire system** whereby two lines can be brought in, using only three wires instead of four.

Look at Figure 15-2. Here two identical alternators (each 120 v, 60 Hz, 1ϕ) are connected in series. Line #1 is connected to the free end of alternator #1 and line #2 is connected to the free end of alternator #2. The **neutral** line is connected to the junction between the two alternators. Current flows (using the previously mentioned conventional indication of current flow) from each alternator to its respective load and back to the alternator through the neutral line.

The voltage between the neutral line and either line #1 or line #2 is 120 volts. Thus, with only three wires we have two 120-volt lines. Further, should a 240-volt source be required, it may be obtained by connecting the load between lines #1 and #2.

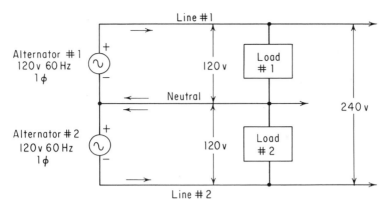

Fig. 15-2. Edison three-wire system.

Should load #1 be equal to load #2, the system would be **balanced** and no current would flow through the neutral line. (The effect then would be as if the neutral line were removed.) Unfortunately, such a balance is rarely achieved in practice; hence some current usually flows through the neutral line.

Since the neutral line of this system is at a potential which is halfway between +120 volts and −120 volts, it must always be at zero potential. The ground, too, is at zero potential. Hence, the neutral line may be connected to ground and no current will flow from that line to the ground connection. (Grounding the neutral line provides a safety factor, as we shall see later.)

Note, however, that the current (whether at 120 volts or 240 volts) is still of the single-phase type.

The same effect can be obtained by means of a center-tapped transformer, as illustrated in Figure 15-3. The primary of the transformer is

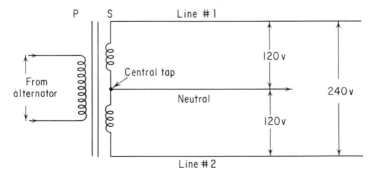

Fig. 15-3. Edison three-wire system using a center-tapped transformer.

313

connected to a suitable source and the center-tapped secondary supplies the two outer lines (known as the "hot," or "ungrounded," lines) and the neutral (or grounded) line.

Figure 15-4 shows a typical household installation. The connection from the neutral line to ground generally is made just before or just after the conductors enter the side of the house. After the three conductors enter the house, a fuse (whose symbol is ⌇) is placed in series with each ungrounded conductor. A **two-pole switch,** to be used to break the circuit, if desired, is placed in series with the ungrounded conductors. (This switch is called the **main switch,** and the fuses that precede it are known as the **main fuses.**) Then all three conductors are attached to the **watthour meter,** which indicates the electrical power consumed in the house.

As the three conductors emerge from the watthour meter, one 120-volt circuit is taken off between one of the ungrounded conductors and the neutral conductor. A second 120-volt circuit is taken off between the other ungrounded conductor and the neutral conductor. If a 240-volt

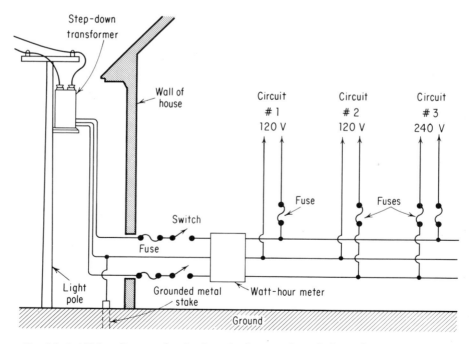

Fig. 15-4. Wiring diagram showing how the lines are brought into a house.

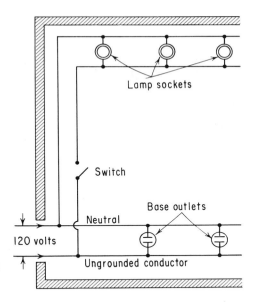

Lamp sockets

Switch

Base outlets

Neutral

120 volts

Ungrounded conductor

Fig. 15-5. Wiring diagram showing how current is brought into a room.

circuit is required — as, for example, if an electric stove is to be operated — a third circuit is taken off between the two ungrounded conductors.

If any connection to the neutral conductor should accidentally be touched to ground **(grounded),** no current will flow because both points are at the same potential. But should one of the ungrounded conductors, or any connection to it, become grounded, there is a 120-volt difference of potential that will cause current to flow. If the resistance of the grounded circuit is very low, sufficient current may flow to heat the conductors and set the house afire. For this reason, fuses must be placed in all the ungrounded conductors.

A typical wiring circuit is shown in Figure 15-5. As the 120-volt line enters the room, it divides into two parallel circuits. One circuit runs along the bottom of the room and connects to a number of base outlets. Note that all these outlets, and hence all appliances connected to them, are in parallel.

The other circuit runs up the wall. About four feet above the floor, the ungrounded conductor is broken and a switch is inserted in series with it. The circuit then is connected with a number of lamp sockets. These lamp sockets, too, are connected in parallel. Closing the switch connects all the lamps; opening it turns them all off. Note, however, that the switch does not affect the base outlets.

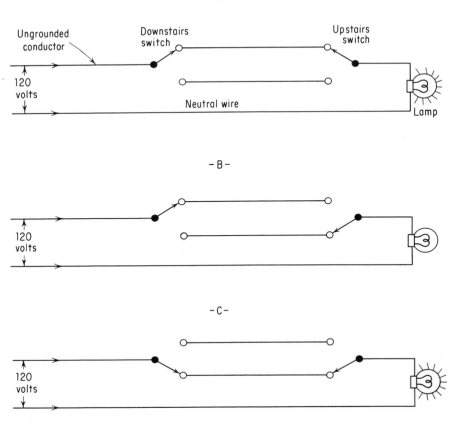

Fig. 15-6. Two-way light circuit.

An interesting lighting circuit is illustrated in Figure 15-6. A flight of stairs may be illuminated by a lamp located at the top of the stairs. Suppose that you wish to be able to turn the lamp on and off at both the top and the bottom of the stairs. The ungrounded conductor is broken at both ends of the stairway. Two **single-pole, double-throw switches** are inserted, one at each break. Note that a third conductor must be connected between the switches. With the switches in the position shown in Figure 15-6A the lamp is lit. Throwing either switch to its opposite position turns the lamp off (Figure 15-6B). When the lamp is off, throwing either switch to its opposite position causes the lamp to light again (Figure 15-6C). In this way, the lamp may be turned on and off from either the top or the bottom of the stairs.

Because of the danger of fire from faulty installation, house wiring is carefully regulated by a code drawn up by a National Board of Fire Underwriters and by local ordinances. The conductors must be heavy enough to carry the current without undue heating, they must be insulated adequately, and they must be installed properly so that there is no danger of insulation failure or of conductors breaking.

The permanent conductors in the home generally are solid, made of copper, and lightly coated with tin to facilitate soldering. They are insulated with a coating of rubber or some plastic material, and the whole wrapped in cotton braid that has been impregnated in some fire-resistant material. Where heavy currents are to be carried, the insulation may be of asbestos.

House conductors are color-coded for easy identification. The neutral conductor always is colored white. The ungrounded conductor may be black or red. Where a three-wire system is used, the neutral conductor is white, one ungrounded conductor is black, and the other is red.

When the wiring is exposed, the conductors may be held in place by means of porcelain knobs or cleats. Many localities require that such exposed conductors be enclosed in thin steel tubing, called **conduit.** Where the wiring must pass through the narrow space inside walls or between a ceiling and the floor above it, flexibility is required. Accordingly, a flexible, steel-armored cable, called **BX,** may be employed. This cable consists of the insulated conductors wrapped in fire-resistant paper and the whole enclosed in metal strip wrapped spiral-fashion. (See Figure 15-7A.) Sometimes a **nonmetallic cable,** consisting of the insulated conductors in braided fabric tubing, is used.

The conductor connecting the various appliances to the electrical outlets generally is flexible and is made of a number of thin copper strands twisted together so that the equivalent of about a #18 wire is formed. It then is covered with a rubber or plastic insulation. Two such

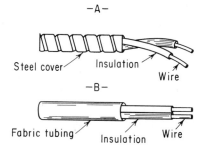

−A−

Steel cover   Insulation   Wire

−B−

Fig. 15-7. A. BX cable.
      B. Nonmetallic sheathed
        cable.

Fabric tubing   Insulation   Wire

insulated conductors are wrapped with cotton or silk to form a single lead, one end of which is attached to the appliance, and the other end terminates in a two-pronged plug that fits into the electrical outlet. Sometimes two rubber-covered conductors are merely attached to each other to form a lead that is known as "zip cord." Where the appliances are heat-producers, such as flatirons or toasters, asbestos insulation is employed.

B. POLYPHASE ALTERNATING CURRENT

Where large quantities of electric power are required, as in most industrial installations, three-phase alternating current (usually at 60 Hz) generally is supplied to the power mains. As previously indicated, such current may be generated by an alternator with three identical sets of armature coils, each set separated by 120 electrical degrees (see Chapter 13, subdivision A, 3). Also, these armature coils may be connected in either a delta or a wye configuration.

1. The Delta Configuration

The delta configuration is shown in Figure 15-8A. Note that the ends of each armature coil are joined to those of its two neighbors. Note, too, that these coils are so joined that all the generated voltages are in the same direction around the triangular configuration. Since these voltages are out of phase with each other, they must be added vectorially. The vector diagram is illustrated in Figure 15-8B. Since the net result of such addition is zero, there will be no current flow around the triangle.

The three-phase lines are taken from each of the junctions between the coils. These lines connect to a three-phase load. (In this illustration the load is shown as consisting of three separate arms connected in a delta configuration.) The voltage produced by an armature coil is known as the **phase voltage.** The voltage produced across the lines as a result of the phase voltage is known as the **line voltage.** In a delta configuration the phase voltage is the same as the line voltage.

The current flowing through an armature coil (when it is connected to a load) is called the **phase current.** The corresponding current flowing through the lines is called the **line current.** It can be proved mathematically that in this configuration the line current is equal to 1.73 times the phase current. At any one instant, current flows from the generator

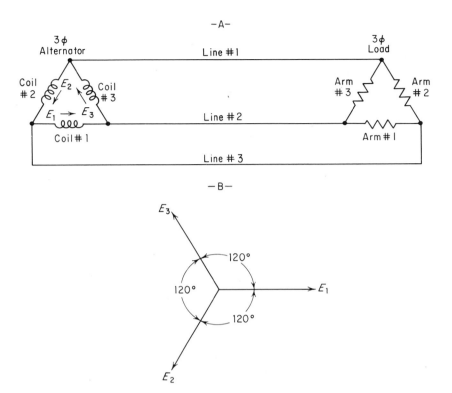

Fig. 15-8. A. Delta configuration.
   B. Vector diagram of the voltages generated in the armature coils.

through one of the lines to the load and returns through the other two lines. Or else current flows from the generator through two of the lines and returns by means of the third.

(In our discussion we have indicated that the load is resistive. It may be resistive, reactive, or a combination of the two. Further, we have assumed that the load on the lines is **balanced** — that is, we have assumed that the load is such that all the line currents are equal and have the same phase angle with respect to the line voltages. This condition is desirable, since it reduces power losses to a minimum, and power companies constantly seek to maintain this balance. In all further discussion we will assume a balanced load.)

If it becomes necessary to step up or step down the voltage of three-phase alternating current, three identical single-phase transformers, one

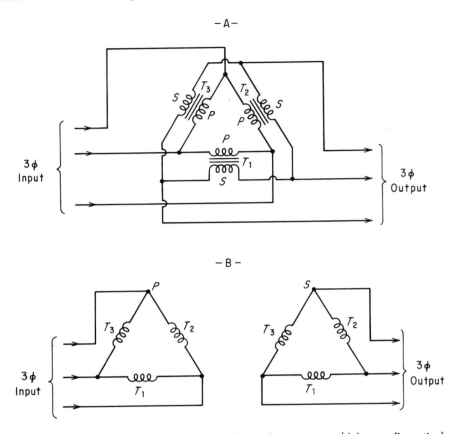

Fig. 15-9. A. Transformer connections for three-phase current (delta configuration). B. Conventional diagram of same.

for each phase, may be employed. (See Figure 15-9A.) Note that all the primary windings are connected in the delta configuration, as are all the secondary windings. (Since the various cross-over wires of the illustration are rather confusing, it is conventional to represent the three-phase transformer as illustrated in Figure 15-9B.) Sometimes, instead of three single-phase transformers, a single **three-phase transformer** may be employed for the same purpose. This transformer has all the windings on a single core.

If a single-phase current is required from a three-phase line, it may be obtained by connecting across any two of the lines. The voltage of this single-phase current may be stepped up or stepped down by means of a single-phase transformer, just as for any other single-phase current.

2. The Wye Configuration

The wye configuration is shown in Figure 15-10. One end of each armature coil is joined to one end of both the others. The free ends of the coils are connected to the three-phase lines. Note that the coils must be connected with the proper polarities, as indicated. As in the case of the delta configuration, the three phase voltages are 120 degrees apart. But the voltage between any two lines is the vector sum of two of the phase voltages. It can be shown mathematically that, in the wye configuration, the line voltage is equal to 1.73 times the phase voltage. The line current, however, is the same as the phase current.

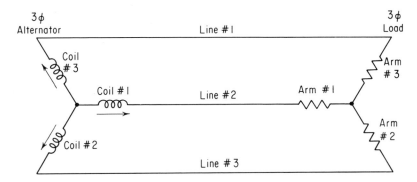

Fig. 15-10. Wye configuration.

As is true for the delta configuration, the voltage of the wye-configuration, three-phase alternating current may be stepped up or stepped down, using suitable transformers. (See Figure 15-11.) Also, single-phase current may be obtained from across any two of the three lines.

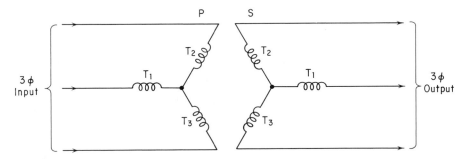

Fig. 15-11. Transformer connections for three-phase current (wye configuration).

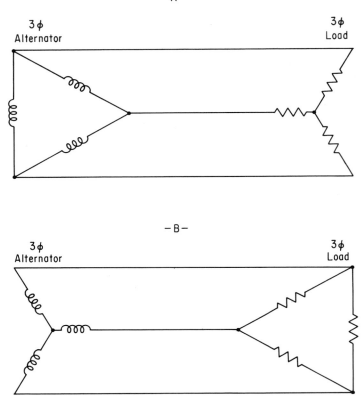

Fig. 15-12. A. How a three-phase current in a delta configuration is applied to a three-phase load in a wye configuration.

B. How a three-phase current in a wye configuration is applied to a three-phase load in a delta configuration.

Not only may a three-phase current in a delta configuration be applied to a three-phase load in a similar configuration, but the delta three-phase current may also be applied to a three-phase load in a wye configuration. (See Figure 15-12A.) Similarly, a three-phase wye current may be applied to a three-phase load in a delta configuration (Figure 15-12B).

Further, a three-phase, delta-configuration current may be changed by transformer action to a three-phase wye-configuration current. (See Figure 15-13A.) Here the primary windings of the transformers are connected in the delta configuration and the secondary windings in the wye

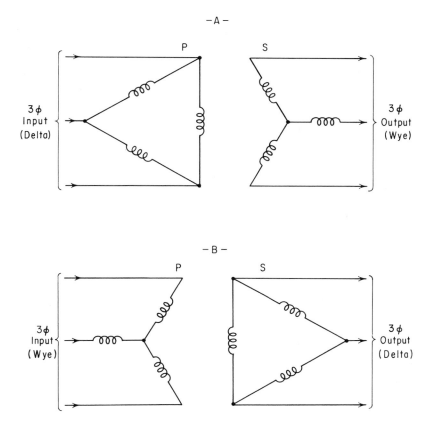

Fig. 15-13. A. Three-phase transformer with a delta input and a wye output.
   B. Three-phase transformer with a wye input and a delta output.

configuration. If the primary windings are in the wye configuration and the secondary windings are in the delta configuration (Figure 15-13B), the three-phase wye current is changed to a three-phase delta current.

A variation of the wye configuration that is widely employed is the **three-phase, four-wire system** illustrated in Figure 15-14. Here a **neutral** line is brought from the junction of the three armature coils to the junction of the arms of the load. This configuration has certain advantages.

In the three-wire system a three-phase current may be obtained by connecting the load to the three wires. Should a single-phase current be desired, the single-phase load is connected to any two of the three wires. In the four-wire system a single-phase current may also be obtained by

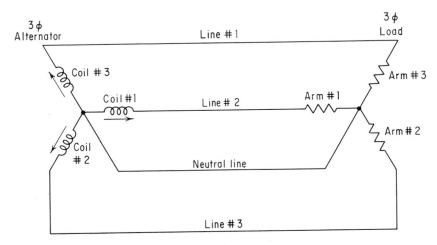

Fig. 15-14. Three-phase, four-wire, wye configuration.

connecting the single-phase load to the neutral and any one of the other wires. (See Figure 15-15.)

In the three-wire system the voltage of the single-phase current obtained by connecting to two of the three wires is the same as that of the three-phase current (usually at 208 v). In the four-wire system if the single-phase current is obtained by connecting to the neutral and one of the other three wires, and if the three-phase voltage is 208 volts, the

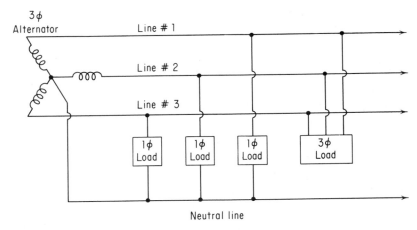

Fig. 15-15. How single-phase and three-phase loads may be connected to lines in a three-phase, four-wire, wye configuration.

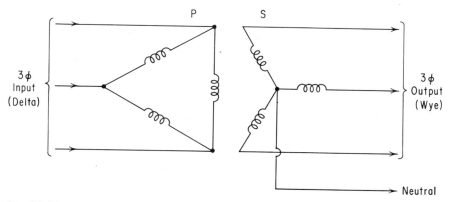

Fig. 15-16. How the four-wire wye configuration is obtained from a delta source.

single-phase voltage will be 120 volts. Since most single-phase devices operate at 120 volts, we now have a method for obtaining this desired voltage without using a transformer.

Because the delta configuration has no common junction, such a configuration usually is not used in the four-wire system. If the source is of the delta configuration, the four-wire wye configuration may be obtained by transformer action. (See Figure 15-16.) The primary windings of the transformers are arranged in the delta configuration to match the source. The secondary windings are arranged in a wye configuration. The neutral wire is connected to the junction of these windings.

3. Advantages of polyphase current

Polyphase current has a number of advantages over single-phase current. For the same line voltage and current a three-wire, three-phase system will deliver 1.73 times as much power as a two-wire, single-phase system. Or, what amounts to the same thing, less copper is needed for the transmission of three-phase power than for an equivalent single-phase power.

Further, as we shall see later in the book, a polyphase current is able to produce a rotating magnetic field. This cannot be done with single-phase current. Such rotating magnetic fields are essential in certain types of control circuits. Also, because of such rotating magnetic fields, polyphase motors are simpler in construction, have higher efficiencies, and operate more smoothly than equivalent single-phase motors, particularly in the larger sizes.

In addition, the ripple component of polyphase current is smaller than that of a single-phase current of the same frequency. As a result, it is easier to filter out the ripple component of polyphase current than that of single-phase current. This is especially useful where the alternating current must be converted to a steady direct current.

C. LONG-DISTANCE TRANSMISSION OF ELECTRIC POWER

In a large country such as the United States it generally is feasible to locate the electric power generating plant at some strategic spot (such as a river dam) and to distribute the generated power over a wide area to the ultimate consumers. This raises serious problems of transmission. Transmission lines, which may extend hundreds of miles, are costly. Also, ohmic and radiation losses cause serious losses of power. These losses may be partially reduced by increasing the size of the wires carrying the current. But this, in turn, increases the cost of the transmission lines (besides introducing other complications such as greater weight, and so on).

Since electric power ($P = I \times E$) is the product of current and voltage, the same amount of power may be transmitted using a high voltage and low current or a high current and low voltage. Since the ohmic power loss is the product of the resistance of the line and the square of the current ($P = I^2 \times R$), it becomes obvious that it is advantageous to transmit the power at the lowest possible current. As a result, power is transmitted at the highest practical voltage.

Transmission at high voltage, too, presents certain problems. For example, the problem of suitable insulation becomes more serious with higher voltages. Practical and economically acceptable compromises must be made. As a result, most long-distance transmission of power in the United States is at about 345,000 volts (although some lines carry considerably higher voltages).

The power is usually generated as three-phase, alternating current at voltages up to about 18,000 volts. Here, too, compromises must be made. The lower the generated voltage, the more it must be stepped up to reach the transmission-line voltage. This means more turns for the step-up transformers which, in turn, means more copper and bulkier equipment. On the other hand, the higher the generated voltage, the more and heavier must be the insulation for the wires of the generator. If the insulation of the rotating wires of the generator becomes too

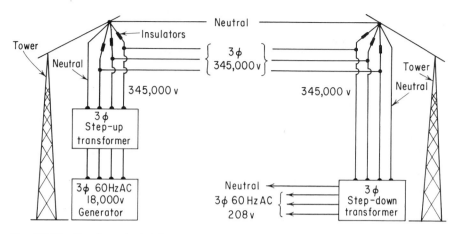

Fig. 15-17. Plan for a high-voltage, alternating-current transmission system.

heavy, there is danger that the centrifugal forces thus generated may become large enough to damage the generator.

The plan for a typical high-voltage transmission system is illustrated in Figure 15-17. At the generating plant 60-Hz alternating current is generated at about 18,000 volts. The armature coils of the generator are usually connected in a three-phase, four-wire, wye configuration. The current is led from the generator to a step-up transformer which raises the voltage to 345,000 volts.

The current may then be transmitted for hundreds of miles, using steel towers suitably spaced to support the wires. The three-phase, four-wire configuration is retained, the neutral line serving to ground the steel towers and thus protect them from lightning. Large insulators are required to insulate the ungrounded lines from the metal towers supporting them. At the receiving end of the line the high-voltage current is stepped down to the 208-volt current that is delivered to the consumer.

The 345,000-volt current generally is not stepped down in a single step. A substation transformer usually steps the high voltage down to some intermediate voltage — say , 13,800 volts. At this lower voltage the current is distributed throughout the locality (by means of either overhead or underground cables) to another transformer located near the ultimate consumer; this latter transformer steps the voltage down to its 208-volt value.

In recent years there has been considerable interest in the practicability of transmitting electric power as high-voltage direct current. Such transmission offers a number of advantages over alternating-current transmission. First of all, it is cheaper because only two wires are required instead of the four wires for three-phase, four-wire, alternating-current transmission. Hence, although it can be shown that polyphase a-c transmission requires wires with a smaller total cross section to supply a given power load at a given voltage, a larger number of costly high-voltage insulators are needed. Also, there are no radiation and certain other a-c losses with direct-current transmission.

On the other hand, direct current has certain disadvantages. For one, alternating-current generators are simpler than their direct-current counterparts. More important, alternating current, which can be stepped up or down by means of ordinary transformers, is therefore much more flexible than direct current.

Accordingly, to overcome these disadvantages, the current is therefore generated as alternating current and stepped up to the high transmission voltage by means of a transformer. Then the high-voltage alternating current is rectified electronically to produce high-voltage direct current, which is transmitted.

At the other end of the transmission line the high-voltage direct current is converted electronically to a high-voltage alternating current. This current then is stepped down by a transformer to the low-voltage alternating current for distribution to the consumer.

Note that such a system requires electronic rectifiers and converters, which are not needed for the alternating-current system. If the savings brought about by smaller transmission costs and power losses are large enough to overcome the costs of the additional equipment required, direct-current transmission may become economically feasible. In the Soviet Union, for example, a direct-current transmission line, extending for several hundred miles and transmitting power at 800,000 volts, is in present-day use.

QUESTIONS

1. Describe the Edison three-wire system for the transmission of current. Why may the neutral line of this system be grounded?

2. Draw and explain a two-way light circuit whereby a lamp may be turned on or off from two different places.

3. Draw the delta configuration for three-phase alternating current. What is the relationship between phase current and line current in such a configuration; the relationship between the phase voltage and the line voltage?

4. Draw the wye configuration for three-phase alternating current. What is the relationship between phase current and line current in such a configuration; the relationship between phase voltage and the line voltage?

5. Draw the four-wire, wye configuration for three-phase alternating current. Describe six methods for obtaining a single-phase current from such a configuration.

6. What are the advantages of three-phase alternating current over single-phase alternating current?

7. Describe the system used in the United States for the long-distance, high-voltage transmission of alternating current. What is the advantage of using high-voltage current?

CONTROL
OF
ELECTRIC POWER

16

The control of power is the process of controlling the transfer of energy from a source to a load. This applies, of course, to all kinds of power. We shall deal here with the control of electric power.

Basically there are three types of control. First of all, there is **off-on** control. This type of control is evidenced by a switch that opens or closes a circuit. A second type is the **stepwise** control evidenced by a tapped transformer. By varying the taps of the windings, we may vary the output voltage of the transformer. The third type is **continuous** control, whereby the power output is varied from minimum to maximum in a continuous, stepless change. A rheostat is an example of a device that can exercise this type of control.

There are many different devices for controlling electric power. We shall discuss a few of the most common.

A. VARIABLE RESISTORS AND TRANSFORMERS

Power, as you know, is the product of voltage and current. So if we wish to control the power supplied to some electrical device, we may vary the voltage, the current, or both. One of the simplest methods for accomplishing this is to insert a variable resistor, called a **rheostat,** in series with the electrical device (the **load**) and the power mains. (See Figure 16-1A.)

Placing the rheostat in series increases the total impedance of the circuit to the power mains. Since the voltage at the mains generally is kept at a constant value, increasing the impedance reduces the amount of current that will flow through the circuit ($I = E/Z$). Hence the power to the load is reduced. By varying the resistance of the rheostat, we may vary the amount of power.

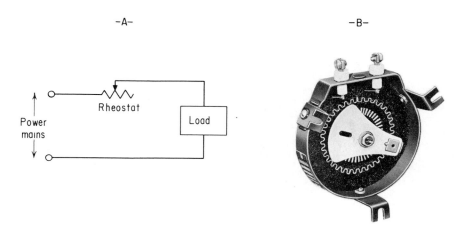

–A–　　　　　　　　　　　　　　　　　　　–B–

Fig. 16-1. A. How a rheostat may be used to control the electric power supplied to a load
　　　　　B. Rheostat.

As an example, assume we wish to be able to regulate the lamps of a theater. A suitable rheostat is inserted in series with the banks of lamps. Then, by varying the resistance of the rheostat, we can vary the current flowing through the lamps, making them glow brighter or dimmer at will.

In the example above we have varied the power by varying the amount of current that may flow through the load. We may also vary the power by varying the voltage applied to the load. This may be done by means of a **variable transformer.** (See Figure 16-2A.)

The primary winding of the transformer is connected to the power mains. The secondary winding, which generally has a few more turns than the primary, is tapped at various points and a switch is used to

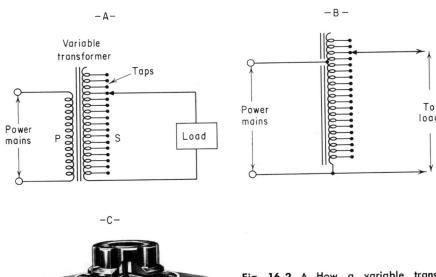

General Radio Company

Fig. 16-2. A. How a variable transformer may be used to
control the electric power
supplied to a load.
   B. Schematic diagram of
the variable autotrans-
former.
   C. Cutaway view of the
variable transformer.

connect to these taps. Thus, by varying the position of this switch, we may vary the number of turns of the secondary winding of the transformer and, hence, the voltage applied to the load.

Usually an autotransformer (Chapter 11, subdivision E, 1) with a single tapped winding is employed. (See Figure 16-2B.) The portion of the winding across the power mains is the primary, and the portion encompassed by the switch is the secondary.

A cutaway view of a commercial variable transformer is shown in Figure 16-2C. The single toroidal winding is wound on an iron core. Instead of a switch, a sliding contact arm carrying a carbon brush is mounted over the transformer, and as the shaft is rotated, the brush slides over the top of the winding. The insulation of the wires over which the brush slides is removed so that an electrical contact can be made. Depending upon the position of the slider, an output-voltage range from zero to a value about 10 or 15 percent higher than the mains voltage may be obtained. The shaft that controls the slider may be rotated by means of a knob or handle.

The variable transformer may be of open construction or cased. Where the power handled is small, ordinary air convection is sufficient to dissipate the heat produced. Where larger power is involved, the transformer may be cooled by a fan or by oil immersion.

Where large amounts of power are to be controlled, both the rheostat and variable transformer offer certain drawbacks. In the case of the rheostat, since it is in series with the load, the same current flows through both. As this current flows through the resistance wires of the rheostat, an appreciable amount of power is lost, being converted into heat ($P = I^2 \times R$). Hence, not only is the rheostat wasteful of power, but it is quite bulky and requires an elaborate cooling system to dissipate this heat.

The variable transformer is more efficient, since the power drawn by the primary winding is nearly the same as the power passed on to the load. But the transformer must be large enough to handle the maximum power required by the load. This makes for a bulky and costly device. Also, the problem of switching large amounts of current (whether by means of a tap switch or a slider) presents serious difficulties.

B. THE SATURABLE REACTOR

We know that if a variable resistor is placed in series with a voltage source and a load, the current flowing through the load may be controlled by varying the resistance. The greater the resistance, the smaller will be the current, and vice versa. Similarly, in an a-c circuit, a variable reactor may be employed to control the current. The greater the reactance, the smaller the current will be, and vice versa. A variable reactor frequently employed for this purpose is the **saturable reactor.** To understand its operation we first must review some of our electrical theory.

As a direct current flows through a coil of wire, a magnetic field is created in its core. The strength of this magnetic field — that is, the total **magnetic flux,** designated by the Greek letter **phi** (ϕ) — depends upon the material of the core and the **ampere-turns,** which is the product of the number of turns of the coil and the current flowing through it.

If the coil has a fixed number of turns and its core is of air, the total

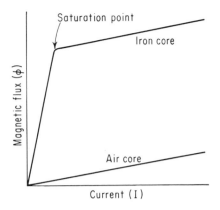

Fig. 16-3. Graph showing the difference between the magnetic curves of an air-core and iron-core coil.

flux will be directly proportional to the current, as indicated by the air-core curve of Figure 16-3. Note that the increase in flux is very small, even for a large increase in current. This is because air offers a high-reluctance path to the magnetic flux.

If the coil is wound on an iron core that furnishes a complete flux path, the flux, again, will be directly proportional to the current, as indicated by the iron-core curve of Figure 16-3. But this time, a relatively small increase in current will produce a large increase in flux up to the **saturation point.** Beyond that point the curve behaves as the air-core curve — that is, the increase in flux becomes small, even with large increases in current.

The flux rises rapidly because the iron core offers a much lower reluctance to magnetic flux than does the air core. That is, the iron can absorb much more flux than air can. However, when the saturation point is reached, the iron core has absorbed practically all the flux it can. Any increase in current beyond that point will produce a flux that must take the high-reluctance air path. This accounts for the upper portion of the iron-core curve, which, you will note, resembles the air-core curve.

So far, we have been dealing with direct current. If the direct current is replaced with an a-c source, the current-flux relationship remains the same, except that the current and flux change direction in accordance with the changes in polarity of the voltage of the source. As a result of the changing flux, a counter voltage is induced in the coil which reduces the current flowing through it.

This opposition to current flow is called **reactance** and is dependent, together with certain other factors, upon the amount of flux absorbed by the core. Thus an iron-core coil (called a **reactor**) provides a larger reactance than one with an air core.

The current set flowing by the supply voltage produces the flux. In the iron-core reactor most of the voltage is required to produce the flux absorbed by the core (up to the saturation point). Hence, relatively little voltage is left to drive current through the external load. We say the reactance of an iron-core reactor is high.

In the air-core reactor very little of the flux is absorbed by the core. Most of the supply voltage is left to drive current through the load. The reactance of an air-core reactor is relatively low.

In an iron-core reactor, if the core is saturated by any external means, the source voltage will be required to overcome the reactance offered only by the equivalent air-core reactor. As a result of this lower reactance, a larger current will flow through the load.

With all this in mind, let us return to the saturable reactor whose basic circuit is illustrated in Figure 16-4. In appearance it resembles an ordinary transformer. Its core is a closed, laminated-iron type. Upon one leg of the core a winding of many turns of relatively fine wire (the **control winding**) is wound. Upon the other leg is wound a winding with fewer turns of heavier wire (the **output winding**). The output winding is connected in series with the load and an a-c source. The control winding is connected in series with a variable resistor (R), a switch (Sw), and a d-c source.

If the switch is open, we may ignore the control winding and its associated components. The device then acts as a simple iron-core reactor. Because its reactance is fairly high, a relatively small current will flow from the a-c source through the load.

Now the switch is closed. Direct current flowing through the control winding produces a steady magnetic flux which partially fills the core. (It is obvious that the direction of the control current is immaterial,

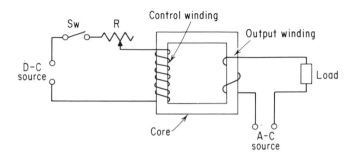

Fig. 16-4. Basic circuit of the saturable reactor.

since a current of opposite polarity merely reverses the direction of the flux, not its magnitude.) Hence less flux can be absorbed from the output winding before the core becomes saturated. As a result, the reactance of the output winding is reduced and a larger current flows through the load.

The greater the amount of flux produced by the control winding, the less will be the amount of flux required from the output winding to drive the core to saturation, the lower will be the reactance of the output winding, and more current will flow through the load. Thus, you can see, by varying the amount of flux produced by the control winding, we may control the current flowing through the load.

The amount of flux produced by the control winding depends, among other factors, upon the amount of current flowing through it from the d-c source. This current may be controlled by varying the resistance of the variable resistor. The adjustment of this resistor, then, controls the load current and, so, the power.

The saturable reactor whose circuit is illustrated in Figure 16-4 suffers from a serious drawback. It is obvious that, as alternating current flows through the output winding, a voltage will be induced in the control winding which will interfere with the voltage of the d-c source. To eliminate this interference, the saturable reactor may be constructed with a three-legged core, as shown in Figure 16-5.

The control winding is wound on the central leg of the core. The output winding is divided into two equal parts: one part is wound on each outside leg of the core, and the two are connected in a series-opposing connection. Thus the voltage induced in the control winding by one part of the output winding is canceled out by the voltage induced by the other part.

The electrical symbol for the saturable reactor, as recommended by the American Standards Association, is ⎯⎯⎯⎯⎯⎯ . The winding with the line through it represents the output winding. The letter symbol usually employed for the saturable reactor is SX.

Since the control winding contains many turns, it requires a relatively small current flowing through it to produce a flux sufficient to saturate the core. Thus a small power input to the control winding may be used to control relatively large power in the output-winding circuit. This is amplification, and hence the saturable reactor and its associated components are known as a **magnetic amplifier.**

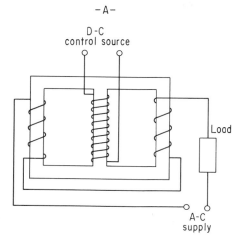

−A−

D-C
control source

Load

A-C
supply

Fig. 16-5. Saturable reactor employing a three-legged core.

−B−

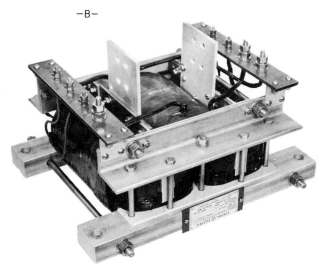

General Electric Company

As is characteristic of other amplifying devices, magnetic amplifiers are capable of controlling large amounts of power in response to small control signals. Thus the very small outputs of photo cells, thermocouples, and similar devices may be employed to govern the delivery of thousands of watts of electrical power.

As an example of how the saturable reactor is employed, let us consider the theater-lighting control problem previously discussed. (See

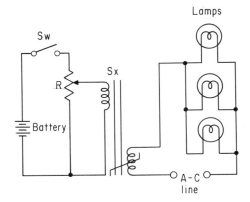

Fig. 16-6. The saturable reactor in a light-control circuit.

Figure 16-6.) The output winding of the saturable reactor is connected in series with the banks of lamps and the a-c power mains. A small battery supplies the direct current to energize the control winding. The variable resistor (R) controls the magnitude of this current and, hence, the current flowing from the a-c line through the lamps.

As the slider of R is moved up, the current flowing through the control winding increases, the reactance of the output winding decreases, and the lamps burn brighter. As the slider is moved down, the current through the control winding is decreased, the reactance of the output winding increases, and the lamps become dimmer. The battery and the variable resistor that controls the lamps may be located at a considerable distance from the rest of the equipment, and the slider may be positioned by means of a small lever or knob.

Because the power consumed in the control winding of the saturable reactor is a small fraction of the power in the lamps, the variable resistor in the d-c circuit may be much smaller than the rheostat that would be required if the control circuit illustrated in Figure 16-1A were employed. Hence there is less wasted power and much less heat to be dissipated. Also, the saturable reactor is more rugged and less costly than a variable transformer of comparable size.

C. THE AMPLIDYNE

In the d-c generator an armature, wound with many turns of wire on an iron core, is rotated within a magnetic field by some mechanical means, such as an electric motor. As the turns of wire cut across this magnetic field, a voltage is induced in them. This voltage is alternating in nature, but by using a commutator (such as described in Chapter 13, subdivision B) we may obtain a d-c output.

The magnetic field is produced by the flow of direct current through a field coil of many turns. This direct current may be from some external source, or it may be a portion of the direct current generated by the rotating armature.

In a sense, the generator acts as an amplifier, since it requires only a small amount of power in the field coil to control a great deal more power in the output of the generator. The extra power, of course, is obtained from the mechanical power used to rotate the armature. Note that in this respect the generator resembles the magnetic amplifier, where a small power input controls a large power output, the excess power being furnished by the power mains.

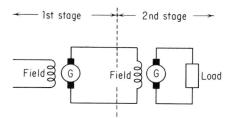

Fig. 16-7. Two generators connected as a two-stage amplifier.

We may construct a two-stage amplifier by connecting two generators, as illustrated in Figure 16-7, with the armature output of the first generator applied to the field winding of the second. (Note that the symbol ⌇ represents the field winding and the symbol -ⓖ- represents the armature, commutator, and brushes.) Thus, for example, assume that two watts applied to the field winding of the first generator produces an output of 100 watts from its armature, a gain of 50. When this output is applied to the field winding of the second generator (which we assume has a similar gain), its armature output is 5000 watts, which is delivered to the load. The power gain of this two-stage amplifier, then, is 2500.

It is possible to combine such a two-stage amplifier into a single machine. To understand how this may be accomplished, let us turn again to the ordinary d-c generator whose circuit is shown in Figure 16-8A. Assume the field winding (which we now will call the **control winding**) requires 10 watts of electrical power from a 10-volt source to produce the field magnetic flux. Thus one ampere will flow through the winding. The field magnetic flux and its direction are indicated by $\phi_F \rightarrow$.

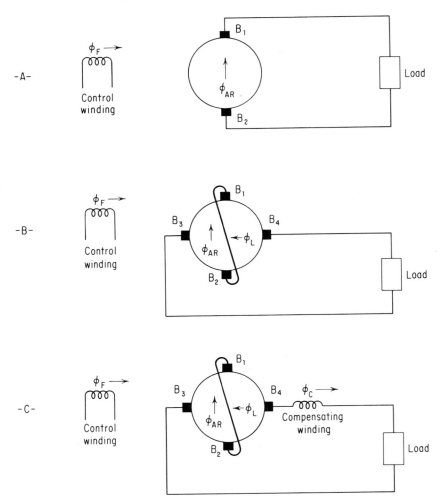

Fig. 16-8. A. Circuit of the generator as an amplifier.

B. Circuit of the amplidyne.

C. Circuit of the amplidyne employing a compensating winding.

Further assume that the resistance of the armature winding is 0.1 ohm and that it delivers 1000 watts at 100 volts to its load. Thus 10 amperes flows through the armature winding and the load. The generator's brushes are designated B_1 and B_2.

As the output current flows through the armature winding, it sets up its own magnetic flux, called the **armature reactance flux.** The strength of this flux is approximately the same as that of the field flux but lies at right angles to it. The armature reactance flux and its direction are indicated by $\phi_{AR}{}^{\uparrow}$.

Suppose, now, we short-circuit the output of the generator by connecting a jumper across brushes B_1 and B_2, as indicated in Figure 16-8B. Since the voltage is 100 volts and the resistance of the armature winding is 0.1 ohm, 1000 amperes would attempt to flow through the winding ($I = E/R$). This is a 100-fold increase in the current normally flowing in the armature.

Such a large current would, undoubtedly, burn up the armature winding. To avoid this eventuality, we may reduce the field flux 100-fold by reducing the current in the control winding from one to 0.01 ampere. Thus, with only 0.01 ampere in the control winding we obtain the original 10-ampere flow through the armature winding.

As the turns of the armature cut through the armature reactance flux, a voltage is generated which appears at brushes B_3 and B_4. Note that the plane of these brushes is at right angles to the plane of brushes B_1 and B_2.

A complication now becomes apparent. As the output current flows through the armature and the load, still another magnetic flux is created ($\leftarrow\phi_L$) whose direction is opposed to the direction of the field flux and thus tends to neutralize it. This would, of course, interfere with the control action.

To overcome this undesirable effect, a **compensating winding** is wound over the same poles as is the control winding and is connected in series between the armature winding and the load (see Figure 16-8C). This winding is in such direction that its magnetic flux ($\phi_c\rightarrow$) aids the field flux, and thus the neutralizing effect of ϕ_L is overcome.

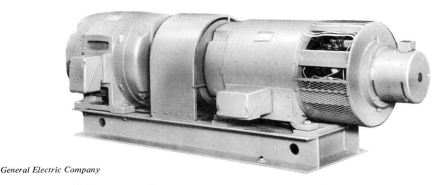

General Electric Company

Fig. 16-9. Amplidyne-motor unit.

Now let us see what we have accomplished. With only 0.1 watt (0.01 ampere × 10 volts) applied to the control winding we may control an output of 1000 watts (10 amperes × 100 volts). This is a power gain of 10,000. Thus a slight variation at the input will produce a similar, but enormously amplified, variation at the output. A rheostat, in series with the control winding, may be employed to control the power in that winding and, so, the power output of the generator. Because the current flowing in the control winding is small, this rheostat, too, may be small.

The device we have discussed is called an **amplidyne** (Figure 16-9). It is used to supply large quantities of controlled electric power in response to weak signals. The electrical symbol for the amplidyne (which is also known as a **rotary amplifier**) with its control and compensating windings is .

QUESTIONS

1. Explain how a rheostat may be used to control the electric power in a circuit.

2. Explain how a variable transformer may be used to control the electric power in an a-c circuit.

3. Describe and explain the theory of operation of the saturable reactor.

4. What are the advantages of using a saturable reactor to control the power in a circuit as compared to the use of a rheostat or variable transformer?

5. Describe and explain the theory of operation of the amplidyne.

PRACTICAL APPLICATIONS OF ELECTRICITY

VI

APPLICATIONS
DEPENDING UPON
THE THERMAL EFFECT

17

In this section we shall discuss the applications of electric current at home and in industry — applications that make possible our present-day civilization. Obviously, we cannot consider all such applications here, but we shall choose examples of those that are most frequently encountered and that represent an entire class.

To facilitate our discussion, we shall group these applications according to the electrical effects they employ. The effects of electricity, you will recall, are **thermal, luminous, chemical,** and **magnetic.** For some of these applications more than one effect may be employed; however, for the purpose of grouping, we shall consider the effect that is most pertinent to the particular application. In this chapter we shall consider applications that depend primarily upon the heating effect of the electric current.

As you know, a current flowing through a conductor encounters resistance and, as a result, heats the conductor. Thus electrical energy is

345

converted to heat energy. The common unit of electrical energy is the **wattsecond** or **joule** (Chapter 4, subdivision C, 5). The common unit of heat energy is the **calorie.** The calorie is the amount of heat necessary to raise the temperature of a gram of water 1°C. In 1840 James Joule, an English physicist, found that one wattsecond of electrical energy is equivalent to 0.24 calorie of heat energy.

There is another unit of heat energy frequently used — the **British thermal unit** (abbreviated **Btu**). This unit is the amount of heat required to raise the temperature of one pound of water 1°F. One British thermal unit is equal to 252 calories and is the equivalent of 1050 wattseconds of electrical energy.

EXAMPLE.   How many calories of heat are produced by a current of 10 amperes flowing through a resistor of 5 ohms for 5 minutes? How many British thermal units?

Electrical energy (wattseconds)

$$= (\text{current})^2 \times \text{resistance} \times \text{time (seconds)}$$
$$= (10)^2 \times 5 \times (5 \times 60) = 150{,}000 \text{ wattseconds,}$$

Heat (calories) $= 0.24 \times$ wattseconds

ANSWER. $= 0.24 \times 150{,}000 = 36{,}000$ calories.

British thermal units $= \dfrac{\text{calories}}{252}$

ANSWER. $= \dfrac{36{,}000}{252} = 142.8$ Btu.

A.   DEVICES EMPLOYING HEATING ELEMENTS

Under certain circumstances the heating effect of the electric current may be considered an unavoidable evil that wastes power. Nevertheless, for some devices, such as the electric stove, flatiron, toaster, waffle iron, heating pad, and many others, it is highly desirable. The basic principle of all these devices is the same. Current is passed through a high-resistance **heating element,** thus producing the required heat.

The heating element must be a conductor that has sufficient resistance so that a reasonable length may be employed to change large amounts of electrical energy to heat energy. At the same time it must have a melting point high enough to withstand the heat produced. Further, it must be able to withstand the oxidizing effect of the air at high temperatures.

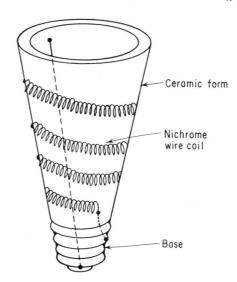

Fig. 17-1. Heating element of electric
room heater.

There are a number of substances used for this purpose. One that is
commonly employed is an alloy of nickel and chromium, called **nichrome.**
This alloy has a resistance more than fifty times that of copper. The heating element may be either nichrome wire or ribbon, wound on some insulating material that is able to withstand the heat. Where a long length
of wire is required to produce the necessary resistance, it may be made
into a long coil and this coil then wound on the insulating form. (See
Figure 17-1.)

The heating element illustrated is of a type used in certain kinds of
electric room heaters. The nichrome coil is wound in a spiral on a ceramic
form. The entire heating unit screws into a socket mounted in the center
of a metallic parabolic reflector. The reflector is used to concentrate the
heat rays and direct them forward.

Another device using a heating element is the **electric soldering iron**
illustrated in Figure 17-2. The heating element is wound with nichrome
wire on a ceramic spool. It is inserted into a metal tube, or barrel, that is
set in a wooden handle. The heat of the element is transferred to a copper
tip and the soldering is done with the hot tip. The wires connecting the
element to the power line usually are asbestos-covered, since the ordinary
cotton-and-rubber insulation cannot withstand the heat well.

The **immersion heater** is merely a variation of the soldering iron. A
similar heating element is completely sealed in a metal tube. The heat

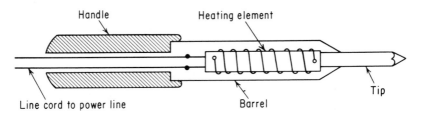

Fig. 17-2. Electric soldering iron.

from the element makes the tube hot, and when the whole is immersed into a liquid, the liquid is heated.

The **electric flatiron** is another typical household device utilizing the heating effect of the electric current. Its heating element is illustrated in Figure 17-3. Nichrome ribbon is wound on a form made of mica or some other insulating material that is able to withstand the heat produced. When the heating element is connected to the house mains, the nichrome ribbon becomes red-hot, heating the iron.

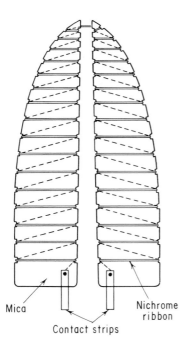

Fig. 17-3. Heating element of electric flatiron.

EXAMPLE. An electric flatiron operating from 120-volt house mains is rated at 720 watts. What is the resistance of the heating element?

$$P \text{ (watts)} = E \text{ (volts)} \times I \text{ (amperes)},$$

$$720 = 120 \times I,$$

$$I = \frac{720}{120} = 6 \text{ amperes;}$$

$$R \text{ (ohms)} = \frac{E \text{ (volts)}}{I \text{ (amperes)}},$$

ANSWER.

$$R = \frac{120}{6} = 20 \text{ ohms.}$$

The electric flatiron can be adjusted to produce varying degrees of heat for ironing different types of materials. This is accomplished by means of a **thermostat** that automatically breaks the circuit and shuts off the iron when the proper temperature is reached. If the iron becomes too cool, the thermostat closes the circuit and the iron heats up again.

The thermostat operates on the principle that different metals have different rates of expansion on heating and contraction on cooling. For example, brass expands and contracts more than iron. If a strip of brass is securely fastened on top of a similar strip of iron and the whole heated, the brass strip will expand more than the iron one, forcing the bimetallic strip to curve downward. If the bimetallic strip now is cooled, the brass, contracting more than the iron, will cause the strip to straighten.

In the flatiron the bimetallic strip is mounted so that it curves downward when heated. (See Figure 17-4.) When the proper heat has been at-

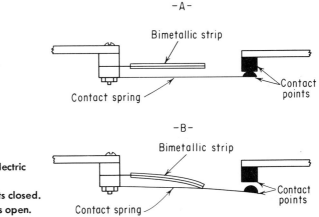

Fig. 17-4. Thermostat of electric flatiron.
A. Contact points closed.
B. Contact points open.

tained, the strip will have bent sufficiently to separate a set of contact points that are in series with the heating element of the iron. Current ceases flowing through the iron. As the iron cools, the bimetallic strip resumes it horizontal position, the contact points touch once more, and current again flows through the iron. In this way the temperature is kept fairly constant.

The thermostat may be adjusted by means of a screw that presses down on the bimetallic strip from above. The screw is operated by a knob on the iron. As the screw forces the bimetallic strip closer to the contact spring, less heat is required for the strip to bend sufficiently to separate the contact points. Consequently, the iron is turned off at a lower temperature. The thermostat is used, of course, as a heat regulator for a great many devices other than the flatiron.

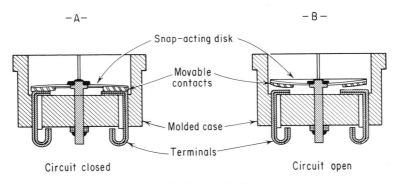

−A−   −B−

Snap−acting disk

Movable
contacts

Molded case

Terminals

Circuit closed   Circuit open

Fig. 17-5. A. Cross-sectional view of the bimetallic-disk switch closed.
B. Bimetallic-disk switch open.

An interesting variation of the thermostat shown in Figure 17-4 is the **bimetallic-disk switch** illustrated in Figure 17-5. The bimetallic element consists of a springy disk. Normally, the curvature of this disk is such that the movable contacts fastened to it touch a set of fixed contacts, thus completing the circuit. As the disk is heated, the difference between the expansions of the two metals of which it is composed causes the disk to curve in the opposite direction, thus separating the contacts and opening the circuit.

Because of the springiness of the disk, the action is not gradual. The disk snaps from one curvature to the other when its temperature reaches a certain predetermined level. When the temperature falls to another predetermined level, the disk snaps back to its original curvature.

Such switches frequently are mounted inside the housing of electric motors to prevent overheating. The switch is connected in series with the motor and the power line. Normally, the switch is closed. But should the temperature within the motor housing rise to a dangerous level, the switch opens the line circuit, thus stopping the motor. When the temperature drops to a safe level, the disk snaps back to its original curvature, and the motor may resume its rotation. Some switches contain a reset button that can be used manually to reverse the curvature of the disk from its "off" position to its "on" position.

The thermostat can be combined with a heating element to make a **flasher,** a sort of time switch that automatically turns some device, such as an electric lamp, on and off at regular intervals. (See Figure 17-6.) The bimetallic strip is mounted so that it curves upwards when heated. An insulating material is placed over a portion of the strip and over this the nichrome wire is wound to form a heating element.

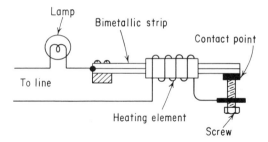

Fig. 17-6. Electric flasher.

A contact point is attached to the bottom of the free end of the bimetallic strip and a screw is so mounted that it normally makes electrical contact with this point. When the flasher is placed in the circuit, current flows from the line, through the heating element to the screw, to the contact point, through the bimetallic strip, through the lamp (or any other device that is to be turned alternately on and off), and back to the line. Because the circuit is complete, the lamp lights.

However, the heater element is heated by this flow of current. This causes the bimetallic strip to become warm. After a short period of time, determined by the physical and electrical characteristics of the flasher, the bimetallic strip becomes warm enough to curve upwards. This causes the contact between the contact point and the screw to be broken. The circuit is opened and the lamp goes out.

Since there now is no current flowing through the heating element,

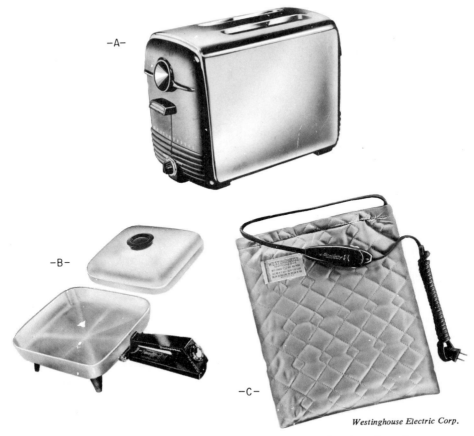

Fig. 17-7. Several of many electrical appliances using a heating element.

    A. Electric toaster.

    B. Electric fry pan.

    C. Electric warming pad.

the bimetallic strip starts to cool. When it has cooled sufficiently, the strip straightens, and the contact point touches the screw again. The circuit is completed, current flows, and the lamp lights again. Then the whole cycle is repeated.

The **electric toaster** contains two sets of heating elements somewhat similar to the type used in the flatiron, mounted vertically, with a space between them into which a slice of bread may be inserted. Some toasters also have a thermostat that controls the degree of toasting and a clockwork mechanism that pops the bread out when it has been heated sufficiently. The **electric stove** is equipped with a somewhat similar heating element, generally arranged in the form of a circle or disk upon which a pot may be placed.

B. FUSES AND CIRCUIT BREAKERS

Consider the electrical wiring in the walls of your house. Suppose that some electrical device that is connected to this wiring becomes defective and develops a **short circuit** (that is, the resistance of the device drops to approximately zero). Under such circumstances the current that will flow through the wiring will become very large. The heat produced by this current may make the wires red-hot and the house may be set on fire.

To avoid this catastrophe, we need some device that automatically will break the wiring circuit when too much current flows through it and before the wires can become hot. This device is the **fuse.** It consists of a strip of metal that will melt at a comparatively low temperature, placed in series with the wiring circuit. Generally, the fuse is made of zinc or of an alloy of tin and lead. Because its resistance is higher than that of copper, it heats more quickly than the wires. Because of its low melting point, it will melt before the wires become too hot.

One form of fuse (Figure 17-8A) is made of porcelain or some other insulating material, and resembles the base of an ordinary incandescent lamp. The strip of fuse metal is inside this porcelain cup and connects between the bottom contact and the shell contact. This fuse, which is called a **plug fuse,** screws into an ordinary lamp receptacle and is connected in series with the house wiring circuit at the point where this cir-

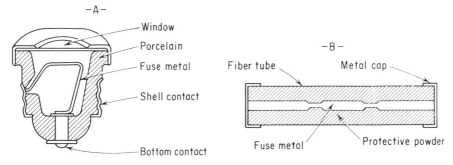

Fig. 17-8. A. Plug fuse.
B. Cartridge fuse.

cuit connects to the outside mains (generally at the watthour meter). A window of glass or mica prevents the spraying of hot metal when the fuse melts, or "blows," and permits visual observation of its condition.

Another type of fuse is the **cartridge fuse** (Figure 17-8B). The fuse metal is contained in a fiber tube. Contact is made through metal caps at each end of the tube. A special receptacle is required for this fuse. Usually the tube is filled with a protective powder that helps to break the circuit quickly, preventing current from flowing in an arc between the unmelted portions of the fuse metal when the fuse is blown. Like the plug fuse, this type, too, is placed in series with the house wiring circuit.

Fuses are rated by the number of amperes of current they will pass without melting. Thus, for example, a 15-ampere fuse will permit 15 amperes of current to flow through it. Should the current rise a little above 15 amperes, the fuse will carry the overload for a short time without blowing. But if a large overload occurs, the fuse quickly melts before the house wires become hot. It is a safe rule to use fuses rated no higher than 15 amperes for ordinary home use, except in certain special installations where extra-heavy wires are employed.

In addition to protecting house wires, fuses may be used in series with any electrical device to protect it against a too-high current. In each instance the fuse must be rated to blow when the current becomes higher than that which the device being protected can safely carry.

There are times when a momentary overload may blow a fuse needlessly. For example, when a motor is started, it draws a heavier current than when it is running steadily. Thus, a certain type of motor operating on a line that is fused for 15 amperes may draw 25 amperes at start. Within a few seconds it reaches the normal running speed and the current drain drops to below 15 amperes. The momentary overload cannot harm the house wires or the motor. But the fuse may have blown.

To avoid this needless blowing of fuses, a special **slow-blowing fuse** has been developed. It consists of a fuse strip and a **thermal cutout** in series. The thermal cutout is a device similar to the flasher previously described. For the motor under discussion, the fuse strip is rated to blow on currents above 25 amperes. The thermal cutout is designed to open the circuit on currents above 15 amperes.

Should a dangerous overload occur and the current go above 25 amperes, the fuse strip will blow at once, opening the circuit. However, the normal starting current of the motor would not be high enough to melt the strip. As for the cutout, the 25-ampere starting current would be great enough to cause it to open. However, the cutout cannot operate quickly. It takes a certain length of time for its element to heat up. Before it has a chance to do so, the motor will have reached its normal running speed and the current will have dropped to below 15 amperes. Hence the cutout remains closed. Should the running-speed current of the motor rise above 15 amperes because of some defect and this excessive current be maintained for more than a few seconds, the thermal cutout will open the circuit.

Although fuses are essential to open the circuit on overload and thus prevent fires and protect valuable appliances or instruments, there are certain drawbacks. Once blown, the fuse must be replaced. Hence there are the cost and the nuisance of replacement. It is for these reasons that the **circuit breaker** frequently is used instead of a fuse.

There are several different types of circuit breakers. However, we shall describe here one that operates upon the thermal principle. Look at Figure 17-9. A latch at the end of a bimetallic strip holds fixed and movable contact points in contact with each other, despite a spring that would separate them. Since the circuit breaker is in series with the rest of the circuit, as long as the contact points are together, the circuit is closed.

Should an overload occur, the excess current flowing through the bimetallic strip would heat it beyond normal. As a result, the end of the strip would curve upwards. This would release the latch and the spring would pull the contact points apart, thus breaking the circuit. Once the current stopped flowing, the strip would cool and become straight again. However, the contact points would remain separated. After the cause of

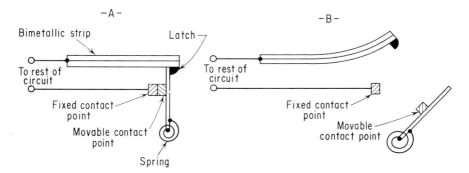

Fig. 17-9. Thermal circuit breaker.
A. Circuit closed.
B. Circuit open.

the overload had been determined and corrected, the circuit breaker would be relatched (generally by pushing a button that brings the contact points together) and it would be ready for operation once more.

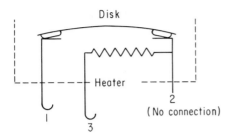

Disk

Heater

2
(No connection)

1

3

Fig. 17-10. Thermal circuit breaker using a bimetallic-disk switch.

There is a variation of the thermal circuit breaker that employs a bimetallic disk. (See Figure 17-10.) Here, a heater is inserted so that the line current flows through it and the switch, which normally is closed. As the current rises to a dangerous level, the heat produced by the heater becomes great enough to cause the disk to change its curvature, thus opening the switch and stopping the flow of current. The circuit breaker may be reset by means of a manually operated button that changes the curvature of the disk, thus closing the switch again.

C. THE INFRARED LAMP

You are aware that the incandescent-filament lamp (which will be discussed in greater detail in the next chapter) emits heat as well as light. The heat energy given off by the lamp is known as **infrared, or radiant-heat, rays.** These rays are similar to the ordinary visible light rays, except that they are invisible to the human eye.

The heat of the infrared rays is used in a number of industrial processes. For example, when the paint has been sprayed on automobile bodies, these bodies are moved slowly through large chambers, which are lined with banks of infrared lamps. These lamps resemble the ordinary incandescent-filament type but are designed to produce a larger percentage of infrared rays. The automobile bodies emerge from these chambers with the paint dry and hard, without the long waiting period normally required to dry the paint. The infrared lamp also is used at home to supply heat for the relief of pain due to neuritis and arthritis.

Fig. 17-11. Infrared lamps drying paint on automobile bodies. Floor conveyor lines bring the car through the drying oven in about seven minutes.

D. WELDING

There are many occasions when we wish to join two pieces of metal together. There are a number of ways of doing this. We may bolt or rivet them together. Certain metals may be soldered together. Another method is to **weld** them by heating the metals at their junction until they melt and fuse together.

Welding may be accomplished by several methods. One is through the use of an oxyacetylene blowtorch, which produces heat by burning a mixture of oxygen and acetylene gases. There also are two types of electrical welding. One type makes use of the resistance of the metal itself to produce the heat required for the weld. The other utilizes the heat of the electric arc.

Resistance welding generally is used for joining metal sheets. The two sheets are placed so that the edge of one overlaps the edge of the other. (See Figure 17-12A.) Two heavy copper electrodes are placed firmly at either side of the sheets at the point where the joint is to be made. An extremely heavy current flows between the electrodes, and the higher re-

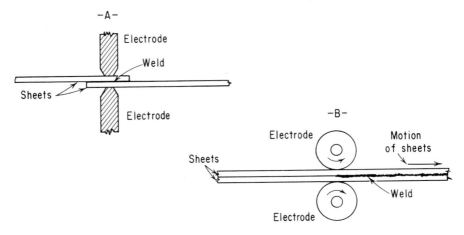

Fig. 17-12. Electrical resistance welding.
A. Spot welding.
B. Seam welding.

sistance of the sheet metal produces enough heat to melt the metal at the junction of the two sheets. At the same time, a heavy pressure is applied between the electrodes, forcing the sheets together and forming the weld. Then the electrodes are moved to another spot along the seam where another weld is made. We call this **spot welding.**

If we wish to make a continuous weld along the seam, we may use electrodes in the form of rollers (Figure 17-12B). As the sheet metal is moved between these rollers, a continuous weld is formed. We call this **seam welding.** The electrodes often must be water-cooled to keep them from overheating.

Currents up to 100,000 amperes may be employed in resistance welding, depending upon the nature of the materials to be welded. Pressures applied to the electrodes at the instant of weld may reach 30,000 pounds. The voltages required, though, are small — from one to two volts. A step-down transformer usually is employed to produce the required current. Sometimes a large capacitor is charged up from a direct-current source and then permitted to discharge instantaneously through the electrodes.

In **arc welding,** the metal work to be joined forms one electrode. (See Figure 17-13.) A conductive rod forms the other electrode. This rod is touched to the work, completing the circuit. Then the rod is withdrawn slightly, forming an intense electric spark, or **arc.** The arc heats the edges

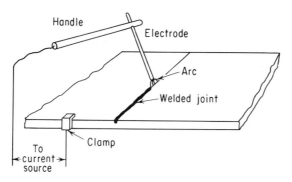

Fig. 17-13. Electric arc welding.

of the metals until they melt and flow together. The rod is moved slowly along the seam and the arc follows.

Sometimes a carbon or tungsten rod is employed as the electrode. More frequently, however, a rod of the same material as the work is used and, as this rod becomes hot, it melts and contributes some of its metal to help make the joint. Frequently, a second similar rod, called a **filler rod,** is inserted into the arc to add its metal to the joint as it melts. The rods generally are coated with various chemicals, which aid the welding by preventing oxidation of the material of the joint.

Less current is required for arc welding than for resistance welding. Generally about 20 amperes are employed, although at times as much as 1000 amperes may be required. The voltage, however, must be relatively high in order to form the arc — from 15 to 50 volts. Current may be supplied from a generator, from storage batteries, or from a step-down transformer.

Not all materials may be arc-welded satisfactorily. The intense heat produced may damage the structure of certain steel alloys. Aluminum, too, presents a difficulty because it oxidizes so readily, especially at high temperatures. The oxide coating prevents the formation of a satisfactory weld. This problem was solved by enclosing the work in an envelope of hydrogen or ammonia gas. The gas envelope keeps the air away from the aluminum, preventing the formation of the oxide coating.

The welding of magnesium presented a unique problem. This metal burns fiercely at the temperatures obtained by the arc. The problem was solved by enclosing the work in an envelope of helium gas, which keeps out the air and prevents the magnesium from burning.

Fig. 17-14. Arc welding.

Resistance welding is faster than arc welding and the lower heat it produces does not heat up the material as much. Arc welding, however, can be used for joining heavier pieces of metal, such as building beams and ship plates. The welder must wear special goggles to protect his eyes from the ultraviolet rays and the intense light produced by the electric arc.

E. ELECTRIC OVENS AND FURNACES

The ordinary household **electric oven** consists of a metal box with flat, nichrome-wire heating elements along the inner sides of the top and bottom. An adjustable thermostat usually is incorporated to keep the heat in the oven constant at any desired temperature. The **electric kiln** used for baking ceramic objects is constructed in similar fashion, except that the box usually is made of some heat-resistant material such as fire brick.

On the other hand, the electric **resistance furnace,** which is used to melt metallic materials, does not employ this kind of a heating element. Instead, a high current is passed through the material to be melted and the resistance of the material to this current produces the required heat. Look at Figure 17-15. The furnace is a box made of some heat-resisting material, such as fire brick. The material to be melted is placed in a heavy, graphite crucible, which is set in the furnace. Since graphite is a conductor of electricity, the crucible acts as one of the electrodes. The circuit is completed by insertion of another electrode, usually a heavy carbon rod, through the top of the furnace to make contact with the material in the crucible. Furnaces of this type may reach a temperature of 2200°C.

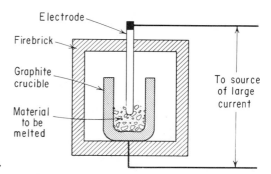

Fig. 17-15. Electric resistance furnace.

A similar device is the electric **arc furnace,** which employs the heat of an electric arc rather than the heat produced by resistance. This furnace resembles the resistance furnace, except that after the carbon electrode touches the material in the crucible, it is withdrawn slightly to form an electric arc. This arc produces a heat that may reach 3500°C. It is used in the manufacture of steel and for the heating or melting of other metals. It may be used to heat materials other than metal, provided such materials are conductors of electricity. For example, carborundum usually is made in an arc furnace.

F. INDUCTION AND DIELECTRIC HEATING

The most recent advances in electrical heating are in the area of **induction** and **dielectric heating.** Basically, such heating is produced by

American Iron and Steel Institute

Fig. 17-16. Tapping a 70-ton electric arc furnace used in manufacturing steel.

means of alternating currents of extremely high frequencies — often millions of cycles per second.

In the case of induction heating, the object to be heated must be a conductor of electricity. This object is placed within the magnetic field of a coil carrying alternating current. The magnetic field around the coil is built up and collapses in step with the alternations of the current. As the magnetic field sweeps across the object, a voltage is induced that sets up a flow of current within it. This current is called an **eddy current**. Heat in the object is produced by the resistance to the flow of the eddy current.

If the object to be heated is magnetic, there is an additional source of heat. The magnetic field around the coil magnetizes the object, a process which involves the arrangement of its molecules. Each time the magnetic field reverses, which occurs once during each a-c cycle, the molecules are rearranged. As a result of a sort of molecular friction called **hysteresis loss,** this rearrangement produces heat.

Of course, any alternating current flowing through the coil will induce a current in a conductive object within its magnetic field. But the induced voltage, which sets the induced current flowing, is a function of the frequency of the alternating current. The higher the frequency, the greater will be the induced voltage. Hence, high frequencies are employed.

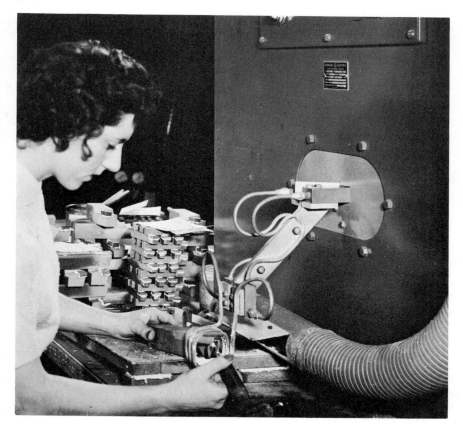

General Electric Company

Fig. 17-17. Brazing a carbide tip on a tool with a five-kilowatt high-frequency induction heater.

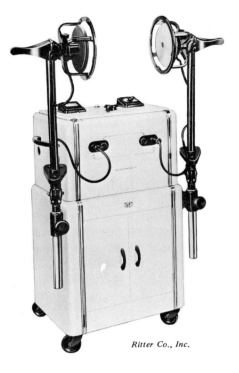

Ritter Co., Inc.

Fig. 17-18. Diathermy machine.

For purposes of melting, soldering, brazing, etc., of metals, induction heating performs as well and better than conventional gas and electric heating. Relatively low-frequency currents (up to about 10 kHz) are employed. Because the heat-producing currents flow inside the work, heating is rapid. Also, because magnetic fields, not heat, are being applied, the air surrounding the work remains cool.

Further, induction heating can perform tasks that are difficult, if not impossible, by other methods. For example, as the glass envelope of the electron tube is being evacuated, it is necessary to heat the electrodes in order that any gas contained in the pores of the metal be driven out. This is easily accomplished by induction heating, which heats the metal electrodes while the glass envelope remains cool.

The advantage of induction heating is expressly marked in the process of surface hardening. An example is the case of a gear wheel whose surface is to be hardened to reduce wear. This is accomplished by heating and quenching in water or oil. The heated steel becomes hard and brittle. It is desirable, however, that the body of the gear wheel remain relatively soft so that it may retain its toughness. This means that a skilled operator must heat the surface just enough so that the body of the wheel remains relatively cool.

In induction heating, the induced currents tend to flow near the surface of the work. Although this tendency is not marked at the lower frequencies, it increases as the frequency of the induced current is increased. By a proper choice of frequency the induced current and, hence, the heat produced by it, may be made to flow through the top few hundredths of an inch of the work. Currents with frequencies from 100 to 500 kHz usually are employed for this process.

Another use for induction heating is in diathermy, where heat is applied to the human body for healing purposes. The high-frequency current flows through a coil surrounding the body or affected member. Since the body is a conductor (though a rather poor one), heat is produced deep within it, though the skin remains cool.

When the substance to be heated is a nonconductor, or a very poor conductor, another method is employed. This method is called **capacitive**, or **dielectric, heating**. Basically, it consists of placing the substance to be heated between a set of metal plates that are connected to a source of high-frequency voltage. Thus the substance becomes the dielectric of a capacitor consisting of itself and the metal plates.

As you have learned, when an alternating voltage is applied to the plates of a capacitor, the orbits of the electrons around the nuclei of the dielectric are distorted, first in one direction and then in the other, in step with the alternations of the voltage. (See Chapter 10, subdivision C.) If the frequency of the applied voltage is high, these changes in orbit, too, have a high frequency. As a result of a sort of internal friction, considerable heat is produced in the dielectric.

The frequencies used for dielectric heating generally are considerably higher than those employed for induction heating. Frequencies up to about 40 MHz are fairly common. For some applications the frequencies are even higher.

Two characteristics of dielectric heating are quite advantageous. Unlike customary heating, where the heat travels from the outside toward the center of the object being heated, the heat is generated throughout the entire object. Hence, heating is more rapid and uniform. Also, since the dielectric properties of different substances vary, the heating effect upon these substances, too, will vary.

These qualities are shown to good advantage in the manufacture of plywood, which is made by gluing several sheets of wood together. Previously, it was made by placing a layer of glue between the sheets of wood and inserting the sandwich thus formed between steam-heated plates of a press in order to melt and set the glue. Not only did it take days for the heat of the steam to penetrate the wood, but even so, the glue would tend to harden unevenly. For this reason, plywood generally was limited to a thickness of about one inch.

Today, the wood-glue sandwich, acting as the dielectric of a capacitor, is placed between the two metal plates of the press. The plates are connected to the high-power source of high-frequency alternating voltage and a job that used to take days is performed in hours. Because of the difference in their dielectric properties, the glue is quickly melted whereas the wood remains comparatively cool. And because the heat is uniformly distributed through the layers of glue, it melts and hardens evenly. As a result, plywood today can be manufactured with much greater thickness and in a great variety of shapes.

Another widespread use of dielectric heating is in the "electronic sewing" of flexible plastic materials. Such plastics (vinyl, nylon, Orlon, saran, and so on) become soft when heated. Where two pieces are to be joined, the ends are overlapped and placed between the two metal plates that press them together. As the high-frequency voltage is applied to the plates, the ends of the plastic quickly soften and fuse together. When cool, the plastic returns to its normal state and the seam is made.

Dielectric heating is also used for the heat-sterilization of foods that are packed in nonmetallic containers. By the use of suitable frequencies, the food may be heated while the container remains comparatively cool. Frequencies as high as 150 MHz are employed for this purpose. Because the food is raised to its proper temperature very quickly, its flavor and nutritional values are retained.

We have listed here but a few of the rapidly-growing list of applications for high-frequency heating.

Rotating mechanical generators may be used to produce currents with frequencies up to about 10 or 15 kHz. Such generators are not practical for higher frequencies.

For frequencies between about 50 and 200 kHz, a **spark-gap converter** may be employed. Its basic circuit is illustrated in Figure 17-19. The line voltage is stepped up to a high value by transformer T and is applied to the *L-C* circuit composed of capacitor C and a tapped inductor L. This circuit acts to store up the electrical energy and, at a certain point, discharges through the spark gap in the form of an electric spark. The taps on L are adjusted for optimum results.

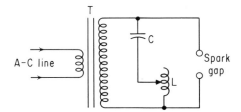

Fig. 17-19. Basic circuit of the spark-gap converter.

This spark is not a one-shot affair. Actually, each discharge produces many rapid back-and-forth surges across the gap in a manner similar to the oscillations of a pendulum set in motion. And, as in the case of the pendulum, the surges are greater at first, rapidly dying away. At the next discharge the process is repeated. Thus the 60-Hz line voltage is converted to a high-frequency alternating voltage across the spark gap. The frequency is determined by the values of C and L.

Where higher frequencies are required, electronic oscillators are employed.

QUESTIONS

1. The heating element of an electric stove operating on a 240-volt line has a resistance of 48 ohms. How many calories of heat are produced by this heating element in one-half hour? How many British thermal units?

[Ans. 518,400 calories, 2,057.1 Btu.]

2. A 600-watt heating element of an electric stove that operates on a 120-volt line was accidentally inserted into a stove operating on a 240-volt line. Explain the result.

3. An electric stove has two types of heating elements. One produces a higher heat, the other a lower heat. Which element has the higher resistance? Explain.

4. Explain the structure of a **thermostat** and show how it can be used to regulate the heat of an electric iron.

5. Explain what is meant by a **short circuit** and why it is dangerous.

6. A 120-volt line in a house is protected by a 15-ampere fuse. Attached to this line is a television receiver rated at 330 watts, a light fixture using a three-way lamp rated at 100-200-300 watts, and other light fixtures employing six 100-watt lamps and four 60-watt lamps. It is desired to add to this line another light fixture containing one 100-watt lamp and two 60-watt lamps. Can this be done safely? Explain. [*Ans.* Yes.]

7. Explain the action of the **thermal circuit breaker.**

8. Explain the **resistance-welding** process.

9. Explain the **arc-welding** process.

10. Explain the structure and operation of the **resistance furnace.**

11. Explain the structure and operation of the **arc furnace.**

12. Explain the principle of **inductive heating.** What are its advantages over ordinary heating methods?

13. Explain the principle of **capacitive heating.** What are its advantages over ordinary heating methods?

APPLICATIONS
DEPENDING UPON
THE LUMINOUS EFFECT

18

A. INCANDESCENT-FILAMENT LAMP

In Chapter 6, subdivision B, we saw how Edison heated a carbon filament by passing a current through it until it became incandescent. The temperature of the carbon filament at that point was approximately 1350°C. Because the melting point of carbon is about 3500°C., there was no danger that the filament would melt.

Although the carbon-filament lamp was used for many years, it nevertheless had serious defects. True, the filament did not burn up or melt. But some of the carbon evaporated from the hot filament. When this carbon vapor came in contact with the cooler inner surface of the glass bulb, it condensed. In this way an opaque layer of carbon was gradually deposited on the inner surface of the bulb, cutting down the amount of light that could pass through. Moreover, the carbon filament was quite brittle and easily broken by vibration. A search was therefore

369

conducted for a filament that would not evaporate so easily, and one that was not so brittle as carbon.

The search centered on a metal called **tungsten** which melts at about 3300°C. and does not evaporate so readily as does carbon. However, tungsten was too brittle and could not be drawn out into a thin filament.

In 1910 William D. Coolidge, an American scientist, discovered a process by means of which tungsten could be made extremely ductile and thus be drawn into fine filaments. These filaments are used in lamps today instead of carbon. Because the tungsten filament does not evaporate so readily, it can be heated to a greater temperature than a carbon filament. Tungsten filaments are heated to about 2100°C. The result is a more intense and whiter light than that produced by carbon-filament lamps.

It is necessary to explain why the tungsten filament at 2100°C. produces a more intense and whiter light than does the carbon filament at 1350°. If we heat an iron bar to about 525°C., it will become red-hot and emit a red light. If we raise the temperature to about 1000°, the light will be yellow. At 1200° the bar glows white-hot and white light is emitted. Thus, as the temperature is increased from 525° to 1200°, the color of the emitted light shifts from red to white. After that, increasing the temperature causes the emitted light to be whiter and more intense.

Were it not for the fact that evaporation of the filament would be increased, the tungsten filament could be heated still more and thus produce a more intense light. To reduce the evaporation and permit a greater heat, Irving Langmuir, another American scientist, introduced a certain amount of inert gas, such as nitrogen or argon, into the bulb. This gas is introduced at a pressure of about 5 pounds per square inch. When the filament is heated, the pressure of the hot gas rises to about 15 pounds per square inch. Increasing the pressure reduces the rate of evaporation. Because the gas is inert, the filament does not burn up.

The presence of the gas, however, tends to cool the filament by conducting some of its heat to the glass bulb by means of convection currents. To compensate for this undesirable effect, the filament is constructed as a tight coil of fine wire and thus can be concentrated into a small space. This results in greater heat and more brilliant light. (See Figure 18-1.) This is the type of lamp most commonly used today. The coiled tungsten filament is supported by two wires that are sealed into the base of the glass bulb. These wires carry the current to the filament. One wire is attached to the metal shell of the base. The other goes to the bottom contact. When the lamp is screwed into a socket, the shell and

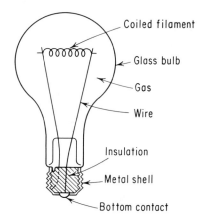

Coiled filament

Glass bulb

Gas

Wire

Insulation

Metal shell

Bottom contact

Fig. 18-1. Modern type of incandescent-filament lamp.

the bottom contact are connected to the source of electric current. The inner surface of the glass bulb is generally frosted to diffuse the light in order to reduce glare. The filament of such a lamp operates at about 3000°C., producing a white light of great intensity.

We may measure the light intensity of any lamp by comparing it with some standard. Such a standard of light intensity is the **candlepower.** This is the light intensity of a burning standard candle, which is made to certain chemical and physical specifications and which burns at a specified rate. If a lamp produces a light that is, say, 40 times as intense as the light of the standard candle, the lamp is said to produce **40 candlepower** of light. Some of our larger searchlights may produce a light intensity of millions of candlepower.

Westinghouse Electric Corp.

Fig. 18-2. Some special types of incandescent-filament lamps.

You may have noticed that our ordinary electric lamps are rated in **watts** instead of candlepower. Since we pay for the electricity used to operate these lamps by the number of watts per hour that they consume, such a rating is more practical. The carbon lamp produces about ⅓ candlepower per watt; the non-gas-filled tungsten lamp produces about 1 candlepower per watt; and the gas-filled lamp produces about 1⅓ candlepower per watt.

B. ARC LAMP

We have learned that the passage of an electric current may heat a gas to incandescence. This is the principle underlying the **carbon-arc light** (Chapter 6, subdivision B). The temperature of the carbon arc may reach as high as 3500°C., producing a brilliant white light. In addition to this visible light, the carbon arc produces large quantities of infrared and ultraviolet rays.

Although capable of producing an intense white light, the carbon arc lamp was abandoned in favor of the more practical incandescent-filament lamp for ordinary lighting purposes. One reason was that the carbon-arc lamp required more manipulation — the tips had to be brought together and separated to start the arc. Another reason was that, as the carbon rods burned away, the arc became longer and longer, until finally the gap between the tips was too great for the current to bridge, and the lamp would go out. Although mechanical devices were invented to feed the carbon rods automatically, they merely introduced further complications. The carbon rods required frequent replacement. The flame of the burning carbon vapor produced a flickering light and had to be protected against gusts of wind. Moreover, the ultraviolet rays it produced were harmful to the eyes. (Fortunately, ordinary glass can stop these ultraviolet rays.) For all these reasons, the carbon-arc lamp is used today only for special purposes, such as in motion-picture projectors and for certain photographic work where the strong ultraviolet rays are desired.

A new arc lamp has been developed that uses **zirconium** metal instead of carbon. The heat of the arc causes the metal to form a small pool, and the passage of current heats the molten metal to about 3600°C. The result is a dazzling white light that is about 20 times as bright as that produced by a comparable tungsten-filament lamp.

C. VAPOR LAMP

We have seen how light is produced by heating a solid, a liquid, or a gas to incandescence. What causes this light was, for many years, one of nature's deepest secrets. But with the development of the electron theory of the structure of matter we began to get some of the answers.

All atoms, you will recall (Chapter 1, subdivision B, 1), contain a central nucleus around which revolve electrons in various orbits or shells. The electrons of the outer shell are most loosely held and, in certain types of atoms, may be detached quite easily. The electrons of the inner shells, however, being closer to the nucleus, are held more firmly. Now should a certain amount of energy, such as heat energy, be applied to the atom, one or more electrons contained in an inner shell may be forced to leave that shell and jump to an outer one. We say that the atom is **excited.**

The displaced electron has acquired extra energy. However, the attraction of the nucleus soon pulls the displaced electron back to its normal position. As the electron falls back, it loses its extra energy. It is this extra energy that is given off in the form of light.

There are other methods besides the use of heat to excite atoms, and modern lighting techniques are relying more and more on these other methods to produce illumination. Let us turn again to the electron theory. The normal atom is neutral — that is, there are as many electrons in its shells as there are protons in its nucleus. Since the negative charge on the electron is equal and opposite to the positive charge on the proton, these charges neutralize each other. Should a neutral atom lose an electron, it becomes a positive ion. Should it gain an electron, it becomes a negative ion.

We know that opposite charges attract each other. If we were to place a positively charged electrode to one side of a neutral atom and a negatively charged electrode at the other side, nothing would happen, since the neutral atom is attracted to neither the positive nor the negative electrode. However, if the charges on the electrodes were to be made sufficiently strong, an electron from the outer shell of the atom would be attracted to the positive electrode. This loss of an electron would convert the atom into a positive ion, which would be attracted to the negative electrode.

The speed of the electron to the positive electrode is very great, as fast as 1000 or more miles per second. Thus, although the weight of the electron is extremely small, should it strike another atom in its flight, the force of the impact might knock an electron free from that atom. This new free electron then would rush toward the positive electrode. The remainder of the atom, having become a positive ion, would be attracted to the negative electrode. This process is cumulative, and soon there would be a stream of electrons speeding toward the positive electrode and a stream of positive ions rushing to the negative one.

The ion moves more slowly than does the electron. But the ion's mass is so much greater than the mass of the electron that the force of the impact between it and a normal atom may also knock an electron loose. Thus the movement of the ions, too, adds to the number of free electrons and ions produced. This process is known as **ionization.**

What happens when a positive ion and an electron collide? Since they are oppositely charged, they attract each other. The ion combines with the electron and becomes a normal atom once more, until another collision ionizes it again.

Should a free electron succeed in reaching the positive electrode, it is drawn off into the external electrical system. Should a positive ion reach the negative electrode, it receives an electron there and becomes a normal atom again. The whole process is then repeated.

But what has all this to do with light? Well, when a positive ion combines with an electron to become a normal atom, light energy may be emitted. Again, the impact of collision between a normal atom with a speeding electron or ion may cause the atom to lose an electron (ionization), or the impact may cause one of the atom's electrons to jump from an inner shell to an outer one (excitation). In the latter case, you will recall that the attraction of the nucleus quickly pulls the displaced electron back to its original position. As the electron falls back to its normal shell, light energy is emitted.

It is this principle upon which the familiar "neon lamp" operates (see Figure 18-3). A glass tube is bent into the desired shape. A metal plate, or electrode, is sealed into each end. The air is evacuated from the tube, and in its place neon gas at low pressure (about $\frac{1}{40}$ atmospheric pressure) is introduced. An electrical source capable of placing a charge up to 15,000 volts is connected to the electrodes. The neon gas near the electrodes becomes ionized. This ionization almost instantly extends throughout the tube as a result of collisions between the electrons and ions and the normal gas atoms. Light is produced as the excited atoms return to their normal states, and as ions recombine with electrons to become normal atoms once more.

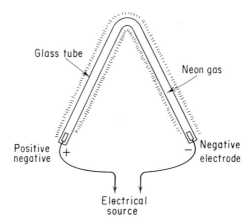

Fig. 18-3. Neon lamp.

If the gas in the tube were not at reduced pressure, there would be too many atoms present. The electrons and ions could not travel very far before colliding with normal atoms. The longer the electrons or ions travel without collision, the greater is their acceleration, the greater is their speed, and, hence, the greater the force of impact on collision. If the collisions take place too soon, the force of impact may not be great enough to ionize the normal gas atoms.

The electrical source is generally a step-up transformer which changes the 120-volt, 60-Hz alternating current obtained from the house mains in most localities to the 15,000 volts required to operate the neon lamp. This source reverses the charges on the electrodes 120 times per second. As a result, the electrons and ions change their directions of flow at this rate. This change, however, does not affect the light produced.

Other gases besides neon may be used in these lamps. Each gas produces light of a characteristic color. Neon produces a reddish light. Helium produces a pinkish light; argon a bluish-white light; mercury vapor a greenish-blue light. The gases may be mixed to produce combinations of colors, and colored glass tubes may be used to produce additional color effects.

The mercury-vapor lamp is of particular interest. In addition to visible light, it also emits powerful ultraviolet rays. Normally, these rays cannot penetrate the glass tube. But if this tube be made of fused quartz or of certain recently developed types of glass, the ultraviolet rays can pass through readily.

Ultraviolet rays tan the skin and are believed to be beneficial in moderate quantities. However, when the body is exposed to the mercury-vapor lamp, glass goggles should be worn to protect the eyes from the ultraviolet rays. Ultraviolet rays can destroy bacteria, and mercury-vapor lamps are used for this purpose in air-conditioning units, refrigerators, food counters, and other similar installations. They are also used in the treatment of certain skin diseases.

Ultraviolet light is also used to produce Vitamin D, the "sunshine" vitamin, in certain foods. Thus, milk, for example, has its Vitamin-D content increased by exposure to mercury vapor lamps.

Photographic film is particularly sensitive to ultraviolet rays. For this reason, mercury-vapor lamps are used for lighting purposes in photographic studios. However, because such lamps are weak producers of red light, neon lamps usually are used in conjunction with them to compensate for this defect.

Lamps such as we have described here are called **vapor lamps** and have a number of advantages over the incandescent-filament types that we have previously discussed. First of all, lamps that operate by ionization produce a "cool" light. Whereas the incandescent lamps operate at temperatures as high as 3000°C., the heat produced by the glowing gas in the neon-light type of lamp may be only 140°.

Second, whereas filament lamps can produce only a yellowish-white light, vapor lamps can produce light of a great number of colors. In this way, they lend themselves readily to many decorative and novel effects.

Another advantage of the vapor lamp is the fact that illumination is produced over a large area, instead of in a concentrated point as in the case of the incandescent filament. Consequently, glare and shadows are reduced.

Finally, a greater proportion of the electrical power is converted to light energy in the vapor lamp than is converted in the incandescent lamp, which uses a major part of the power to heat the filament. This means that more light per watt is obtained from the vapor lamp, or, in other words, more candlepower per dollar is produced.

The vapor lamps we have been describing, however, suffer from a serious disadvantage. Whereas the incandescent-filament lamps may operate directly from the 120-volt house mains available in most parts of the country, the vapor lamps require a high voltage across the electrodes, often as high as 15,000 volts. This requires a special device, such as the transformer, to convert the 120-volt current to the required voltage. It would be very convenient if the vapor lamp could operate at lower voltages.

One lamp that can operate at a lower voltage is the "night light" illustrated in Figure 18-4. Two electrodes, consisting of halves of a split metal disk separated by a small gap, are sealed in a small glass bulb. The air is evacuated and neon or argon gas is introduced at reduced pressure. The lamp terminates in a screw base which fits an ordinary lamp socket. The 120-volt mains are connected to this socket and, through it, to the electrodes.

When the current is turned on, the gas near the electrodes becomes ionized. The electrons stream to the positive electrode, and the positive ions travel toward the negative electrode. When these ions reach the negative electrode, they each receive an electron and become normal atoms again. In the process, light is produced. In this way the negative electrode is covered by a glow of light.

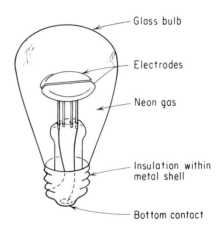

Glass bulb

Electrodes

Neon gas

Insulation within metal shell

Bottom contact

Fig. 18-4. "Night light."

Since the type of electricity commonly used in the United States reverses the charge on the electrodes 120 times per second, the glow of light will change from electrode to electrode at this rate. The eye cannot detect this change, and both electrodes will appear to glow.

Lamps of this type use very little current, and the light produced is quite feeble. They are used where little illumination is needed, as in a child's bedroom at night. Hence the name "night light."

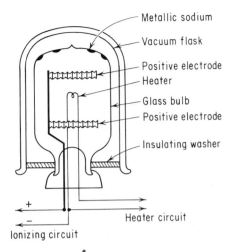

Metallic sodium

Vacuum flask

Positive electrode
Heater

Glass bulb
Positive electrode

Insulating washer

+
−
Heater circuit
Ionizing circuit

Fig. 18-5. Sodium-vapor lamp.

Another vapor lamp that is widely used for highway lighting is the **sodium-vapor lamp,** one type of which is illustrated in Figure 18-5. A specially shaped set of electrodes and a coil of tungsten wire **(heater)** are sealed in a glass bulb. The air within the bulb is evacuated and a small amount of neon gas under reduced pressure and some metallic sodium are introduced. The entire bulb then is inserted into a vacuum flask, somewhat similar to a Thermos bottle. This flask, however, is not silvered and will transmit light.

Electric current is passed through the tungsten coil in the heater circuit. The flow of current heats the coil, exciting the tungsten atoms sufficiently to cause a large number of electrons to fly off.

By means of a second circuit (the **ionizing circuit**) a negative charge is placed on the coil and a positive charge is placed on the positive electrodes. The electrons emitted by the hot coil speed toward these positive electrodes. In passage, the electrons collide with the atoms of neon gas, causing them to become ionized.

The heat produced by this ionization soon causes the metallic sodium to vaporize, and the bulb quickly fills with the sodium vapor. As a result of collisions between the speeding electrons and ions and the sodium-vapor atoms, the latter, too, become ionized and excited. As these sodium-vapor ions collect electrons and become normal atoms once more, or as the excited sodium-vapor atoms return to their normal states, a characteristic yellow light of high intensity is produced.

The vacuum flask that surrounds the bulb is used to prevent the sodium vapor from cooling and condensing. Both the heater and the ionizing circuit use low-voltage current, eliminating the need for devices such as the high-voltage transformer.

The advantage of the sodium-vapor lamp over the ordinary incandescent-filament lamp is evident from the fact that the former may produce 5 candlepower per watt, compared with 1.3 candlepower per watt for the filament lamp.

Somewhat similar is the **mercury-vapor lamp** illustrated in Figure 18-6. Here the inner tube contains a few drops of mercury and a little argon gas. The starting voltage is applied between the starting probe and the upper electrode and causes the argon gas between them to ionize. Very little current flows from the probe to the electrode at any time because of the high resistance in series with the probe.

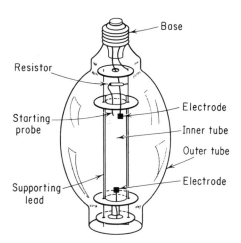

Fig. 18-6. The mercury-vapor lamp.

The flow of current through the gas produces sufficient heat to vaporize the mercury, forming mercury vapor which fills the inner tube. The ionizing voltage, applied between the upper and lower electrodes, causes the mercury vapor to ionize, thus producing an intense greenish-blue light.

D. FLUORESCENT LAMP

Perhaps the greatest advance in recent years in the science of illumination is the development of the **fluorescent lamp,** which was introduced about 1938. This lamp resembles the mercury-vapor lamp previously discussed, except that the inside of the glass tube is coated with certain chemicals, called **phosphors,** that glow (**fluoresce**) when struck by the ultraviolet rays produced by the mercury vapor.

Different phosphors glow with different colors. Cadmium borate glows with a pinkish light; zinc silicate produces a green light; calcium tungstate a blue light; and magnesium tungstate a bluish-white light. By combining these and other phosphors, light of other colors may be produced.

The fluorescent lamp is illustrated in Figure 18-7. A tungsten filament is sealed into each end of a long glass tube whose inner surface is coated with a suitable phosphor. The air in the tube is pumped out and, in its place, a little argon gas and a few drops of mercury are introduced.

First, current is passed through the filaments and the heat produced vaporizes the mercury, filling the tube with mercury vapor. Then a relatively high voltage is placed across the tube, using the filament in each end as an electrode. The argon gas ionizes first and then the mercury vapor, producing ultraviolet rays. These rays strike the phosphor coating, causing it to glow and produce visible light. When the lamp starts to glow, the filament circuit is opened. The heat produced by the ionization of the gases is sufficient to keep the mercury vaporized. Note that the filaments are not required to produce any light; they act as heating coils and as electrodes for the high-voltage ionizing circuit.

Note the **guards** that are placed over the heating coils. As the speeding electrons and ions strike, the force of impact may be great enough to damage the coils. Therefore, guards of heavy wire are placed in front of the coils to absorb the shock.

From the description above, you can see that the fluorescent lamp requires two circuits. One is the **filament** circuit, which remains closed only long enough for the filaments to heat and vaporize the mercury. Then this circuit is broken. The other is the **high-voltage** circuit, which

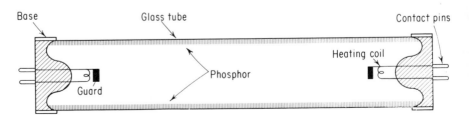

Fig. 18-7. Fluorescent lamp.

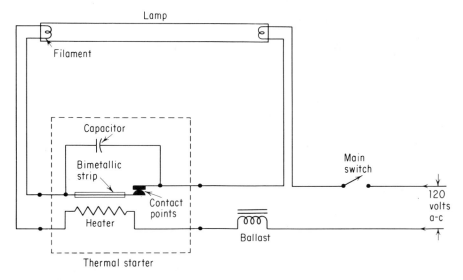

Fig. 18-8. Fluorescent-lamp circuit using thermal-switch starter.

places a high voltage across the electrodes to ionize the mercury vapor in the tube.

The filament circuit is controlled by a **thermal switch,** illustrated in Figure 18-8. The switch contains a resistor that heats when the main switch is closed and current flows through it. The current that flows through the resistor also flows through the filaments of the lamp, causing them, too, to become heated. The heat from the resistor causes a nearby bimetallic strip to bend, separating a pair of contact points that normally touch each other. As these contact points separate, they open the filament circuit.

Note the capacitor that is connected across the contact points. Its function is to reduce sparking across these points as they are separated. Were the capacitor not there, the electrical pressure would cause current to arc across the slight gap formed as the contact points start to separate. This arcing would ionize the air and form a path over which the current would continue to flow, even after the contact points were a considerable distance apart. Because of the capacitor, however, the electrical pressure is used in charging the capacitor rather than in forming the arc, and sparking is reduced.

The thermal switch and its capacitor generally are enclosed in a small, cylindrical metal can. This unit is called the **starter.** Most fluores-

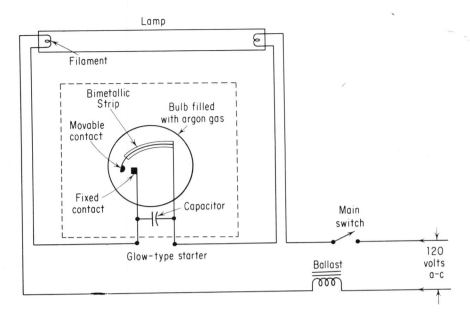

Fig. 18-9. Fluorescent-lamp circuit using glow-type starter.

cent lamps today employ the **glow-type** starter illustrated in Figure 18-9. Here, the movable and fixed contact points are enclosed in a small glass bulb containing argon gas. The movable contact point is attached to a bimetallic strip and, normally, the two points are separated. When the main switch is closed, the 120-volt current ionizes the gas in the bulb and, as a result, a small current flows between the two contact points. The flow of this current produces heat, which causes the bimetallic strip to bend until the points touch each other. When this occurs, the flow of current through the gas ceases, but now current flows through the closed contact points and through the filaments of the lamp.

Because the current has ceased flowing through the gas, the bimetallic strip cools off and moves to separate the contact points once more, thereby opening the filament circuit. However, enough residual heat remains to keep the points closed long enough for the filaments to vaporize the mercury in the lamp.

Note that the fluorescent lamp is connected in shunt across the starter. Once the lamp is started and current flows through it, the voltage across the starter drops to too low a value to cause the argon gas in its bulb to glow. Hence the contact points remain separated and the filament circuit of the lamp remains open.

Now let us consider the high-voltage circuit. Note the **ballast** in series with the lamp and starter. The ballast consists of many turns of fine wire on an iron core. Because of its self-induction, a high counter electromotive force is generated at the instant that the contact points of the starter separate and break the circuit (see Chapter 10, subdivision B, 2). This high voltage is applied across the filament electrodes of the lamp, ionizing the mercury vapor. Once this ionization starts, current will continue to flow through the lamp, even though the voltage across the electrodes is sharply reduced.

After the ionization has started, the ballast performs a second function. You will recall that this ionization tends to grow and, unless it is checked, the flow of current through the lamp may become great enough to wreck it. But the ballast acts as a safety device. After it has set up the high-voltage surge that starts the ionization of the vapor, the ballast acts as a reactor that keeps the current flowing through the lamp within safe limits.

You will note that the fluorescent lamp is, essentially, an alternating-current device. It is possible, however, to use such a lamp on direct current. The high-voltage surge set up at the moment the contact points are separated will be produced by either type of current. However, the protective action of the ballast is absent since there is no counter electromotive force generated when a steady direct current flows through it. For this reason a resistor must be connected in series to perform the current-limiting action. Because of the power loss in this resistor, fluorescent lamps do not operate as well on direct current as they do on alternating current.

There are a number of reasons why the fluorescent lamp is rapidly replacing the incandescent type. First, the fluorescent lamp gives more light per watt. For example, the white-light fluorescent lamp produces about 2.5 candlepower per watt, and so it is cheaper to operate than the incandescent-filament type, which produces about one candlepower per watt. Then, the light of the fluorescent lamp is "softer" and produces less glare, because the illumination comes from a long line of light rather than from a concentrated filament. Again, the number of colors that can be produced and the various shapes in which the lamp can be formed make the fluorescent lamp more suitable for decorative purposes than the incandescent lamp. Further, the light from the so-called "daylight" fluorescent lamp more nearly approaches natural daylight. The incandescent lamp operates at about 3000°C. and produces a yellowish-white light. The fluorescent lamp produces a whiter light and its temperature is less than 50°. To produce light of a color similar to that produced by the fluorescent "daylight" lamp, the incandescent filament would have to be heated to over 6000°C.! Finally, the life of the fluorescent lamp is several times longer than that of the incandescent type. This is because the former does not have a filament that operates continuously at high temperature.

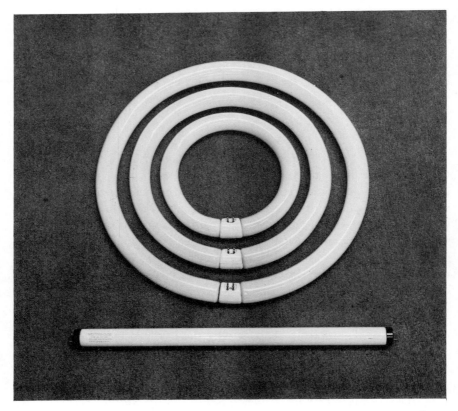

Westinghouse Electric Corp.

Fig. 18-10. Fluorescent lamps are made in various shapes and sizes.

Fluorescent lamps may be obtained in sizes ranging from six inches to over four feet in length. In addition, they are made in various forms such as straight tubes, circles, and semicircles. Thus they lend themselves to various decorative designs.

E. ELECTROLUMINESCENCE

Electroluminescence — the property of certain phosphors to glow when subjected to an alternating voltage — gives promise of revolutionizing the science of indoor lighting. No longer would such lighting be restricted to the use of bulbs or tubes. Instead, a room could be lighted by the glow from flat panels set in the ceiling, the wallpaper, or even hanging drapes. Windows could transmit sunlight in daytime and glow with a soft light at night. Further, by the turn of a knob, the brightness of the light or its color could be changed at will.

Electroluminescence was first discovered by Vladimirovich Lossev, a Russian physicist, in 1923 when he found that a crystal of silicon carbide would glow when placed between two electrodes connected to a source of direct voltage. In 1936 a French scientist, George Destriau, found that alternating voltage could cause certain phosphors to glow. Since that time there has been intensive research to convert this phenomenon to practical use.

Basically, a sort of capacitor is made of two electrically conductive plates separated by a thin dielectric consisting of a translucent plastic in which are embedded certain phosphors. Alternating voltage is applied to the two plates. As the voltage reaches the peak of its alternation, the electric field set up between the plates causes the electrons of the phosphors to become excited. As the voltage drops to zero, these electrons return to their normal states, releasing the exciting energy as light of a color characteristic to those phosphors. At the peak of the next alternation the electrons become excited once more. This process continues with alternate excitations and returns to normal (with the production of light) in step with the alternations of the voltage.

At least one of the conductive plates must be translucent to permit the light to be visible. These plates may be made of electrically conductive glass, or they may be made of a translucent, conductive plastic which could be sprayed on fabric or other material.

The brightness of the glow depends upon the voltage and frequency of the exciting current. The greater the voltage and the higher the frequency, the greater will be the light produced. At the present time, voltages of about 350 volts and frequencies of about 3 kHz are commonly employed, although there is considerable research in the use of higher and lower voltages and frequencies.

Different phosphors produce light of different colors. Also, different phosphors respond best to voltages of different frequencies. Thus, by embedding several types of phosphors in the plastic dielectric and varying the frequency of the applied voltage, we may vary the color of the light.

Certain problems must be solved before electroluminescent lighting becomes commonplace. For one, the house mains generally furnish alternating current at 120 volts and 60 Hz. Either a way must be found to employ such voltage and frequency, or else a cheap method must be devised to convert the house current to the higher voltages and frequencies required.

Further, although it gives promise of high efficiency, electroluminescence is still considerably less efficient than fluorescent lighting. Since the brightness is proportional to the voltage and frequency of the exciting current, it would appear that we could increase its efficiency merely by raising the voltage, the frequency, or both. However, too high a voltage introduces the risk of shock, dielectric breakdown, and so on. Also, too high a frequency would interfere with the time required for the electrons of the phosphors to become excited and return to their normal states.

F. ILLUMINATION

If you wish to see an object — for example, the page of this book — it is not enough that the lamp be bright. Enough of its light must fall on the page to make the page bright enough to be read. When we talk of **illumination,** we are concerned with the amount of light falling upon each unit area of the illuminated object.

The illumination of an object depends upon how close it is to the source of illumination. We know that when we want to see better, we go "nearer to the light." The illumination of an object at a distance of one foot from a light source of one candlepower is **one foot-candle,** which is the unit of illumination. If, instead of a standard candle, we use, say, a 100-candlepower lamp, the illumination on the object at a distance of one foot would be 100 foot-candles.

However, if we held the object 2 feet away from the 100-candlepower lamp, the illumination would not be 50 foot-candles, as you might expect, but rather, 25 foot-candles. You will see this more clearly if you examine Figure 18-11.

Assume that the source of light is a 100-candlepower lamp. A one-foot-square screen, one foot away from the source, receives a certain amount of light from the lamp. If, now, this screen were removed, since light rays travel in straight lines, the same amount of light would cover 4 square feet of a second screen 2 feet removed from the source. Similar-

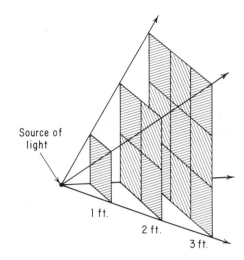

Source of
light

1 ft.

2 ft.

3 ft.

Fig. 18-11. Diagram illustrating the inverse-square law of illumination.

ly, the same amount of light would cover 9 square feet of a screen 3 feet removed from the source.

The illumination of the first screen is 100 foot-candles. At the second screen (2 feet from the source) the same amount of light is spread over 4 square feet. Hence, the light falling upon each unit area of the second screen, its **illumination,** is only one-fourth that of the first. That is, the illumination is 25 foot-candles. The same amount of light, falling upon the third screen (3 feet from the source) is spread over 9 square feet and the illumination now is only one-ninth, or 11⅑ foot-candles.

Note that at twice the distance from the source, the illumination is one-quarter as great. And at three times the distance from the source the illumination is only one-ninth as great. Thus the illumination varies inversely as the square of the distance from the source. This is known as the **inverse-square law of illumination.**

Thus, for proper illumination, we need not only a source of proper intensity, but we also need to have the object close enough to the source so that adequate illumination falls upon it. Or, if the object is farther from the source of light, this source must be made more intense to furnish the proper illumination.

Illumination engineers have set certain standards of illumination for various types of activities. Libraries and classrooms should have an illumination of at least 15 foot-candles at the desks. Manufacturing plants need from 15 to 50 foot-candles at the machines, depending upon the type of work done. These are minimum standards, and greater illumination is desirable.

There is another factor in illumination. If the source of light is too concentrated, the reflection from a bright surface, such as the page of this book, may produce a glare which will strain the eye and cause fatigue. Hence, it is better if the light comes from a large surface rather than from a small point. That is why the long fluorescent lamp is superior to the small incandescent-filament lamp. In addition, modern homes use indirect lighting, whereby the light is directed against the white ceiling and then is reflected from this large surface to the rest of the room.

Weston Instruments Div., Daystrom, Inc.

Fig. 18-12. Light meter.

There are a number of methods used for measuring illumination. The simplest method, perhaps, is to use a photronic cell such as described in Chapter 14, subdivision D, 2, and a microammeter. The combination of cell and meter form a **light meter.** This light meter is placed where we wish to measure the illumination and the light is permitted to fall upon the cell. The greater the intensity of illumination, the greater will be the voltage generated by the cell, and more current will flow through the meter. This meter is calibrated directly in foot-candles.

QUESTIONS

1. In an incandescent-filament lamp, what is the relationship between the light produced and the temperature of the filament?

2. What is the function of the inert gas introduced into the bulb of the incandescent lamp?

3. In terms of the electron theory, explain the operation of the neon-type vapor lamp. How are different colored lights obtained?

4. Give three uses for the **mercury-vapor lamp.**

5. In the **night light** illustrated in Figure 18-4, will one or both plates "light up" if the lamp is used on (*a*) alternating current; (*b*) direct current? Explain.

6. Explain the operation of the fluorescent lamp in terms of the electron theory. How are different colored lights obtained?

7. Explain the function and operation of the **glow-type starter** in the fluorescent-lamp circuit.

8. Explain the functions of the **ballast** in the fluorescent-lamp circuit.

9. Draw the circuit of a fluorescent lamp using a glow-type starter. Label all parts.

10. What are the advantages of the fluorescent lamp over the incandescent-filament type? What are its disadvantages?

11. Explain the theory of electroluminescence.

12. An engineer, measuring the illumination of a room, found that at a distance of 1 foot from a light source, the illumination was 1000 foot-candles. What would be the illumination 5 feet from the source?

[*Ans.* 40 foot-candles.

APPLICATIONS
DEPENDING UPON
THE CHEMICAL EFFECT

19

The chemical applications of the electric current are based upon the movement of ions through a liquid toward charged electrodes (see Chapter 5, subdivision B). For example, let us examine a commercial method of obtaining chlorine gas.

We start with ordinary table salt **(sodium chloride),** which is a compound consisting of one atom of sodium and one atom of chlorine. When the salt is dissolved in water, it ionizes, breaking up into positive sodium ions and negative chlorine ions. Two electrodes are placed into the solution and these electrodes are connected to a source of direct current. The positive sodium ions are attracted to the negative electrode and the negative chlorine ions move toward the positive electrode.

When a positive sodium ion reaches the negative electrode, it receives an electron and becomes a neutral, or uncharged, sodium atom. Similarly, when a negative chlorine ion reaches the positive electrode, it surrenders its negative charge to the electrode and becomes a neutral chlorine atom. The neutral sodium atom reacts with the water of the solution to form sodium hydroxide, but the neutral chlorine atom bubbles off and is collected.

If we wish to recover the sodium as well, we cannot use a water solution of sodium chloride since the sodium atom formed at the negative electrode would react with the water. Instead, we use molten salt. The molten salt ionizes and, since there is no water, pure sodium is collected at the negative electrode.

A similar method is used to obtain metallic magnesium. Pure magnesium chloride, obtained from sea water, is melted and a direct current is passed through it. The pure magnesium is collected at the negative electrode.

This process of passing electric current through a chemical compound and breaking it down is called **electrolysis.** Note that direct current must be used to attract ions of only one kind to each electrode. The electrodes generally are made of carbon or platinum so that they may resist any chemical action.

Electrolysis is used to break down water, which is a compound of hydrogen and oxygen, into its two component gases. The difficulty here, however, is that pure water is not an electrolyte and will not conduct electricity. An indirect method must be employed, therefore.

A little sulfuric acid is added to the water. Each molecule of this acid contains two hydrogen atoms combined with a group of atoms consisting of one sulfur atom and four oxygen atoms. This group of atoms is called the **sulfate** portion (whose chemical symbol is SO_4). As each molecule of sulfuric acid dissolves in the water, it breaks down into two positive hydrogen ions (H^+) and one sulfate ion bearing two negative charges (SO_4^{--}).

The hydrogen ions are attracted to the negative electrode, where each ion obtains an electron and becomes a neutral hydrogen atom. Hydrogen is a gas and it is collected as it bubbles off around the negative electrode. The negative sulfate ion is attracted to the positive electrode, where it becomes neutral by surrendering its two excess electrons. It immediately reacts with the water around it, forming sulfuric acid as it combines with the hydrogen of the water. The oxygen remaining when the hydrogen of the water combines with the sulfate portion is released as a gas, which is collected as it bubbles off around the positive electrode.

The newly formed sulfuric acid ionizes once again and the entire process is repeated. The sulfuric acid in the solution remains constant, but the water is used up. Since each molecule of water contains two atoms of hydrogen and one atom of oxygen, twice as much hydrogen gas is produced.

Electrolysis is also used to extract metallic aluminum from its ore. Vast quantities of aluminum exist in the crust of the earth, generally in the form of an oxide called **bauxite.** Yet, it was so difficult to extract the metal from its ore that for many years aluminum remained a precious metal. Today, thanks to a process invented in 1886 by an American, Charles Martin Hall, aluminum costs but a few cents a pound.

The key to the problem was in finding a way to change the bauxite into a liquid so that it could ionize. It does not dissolve in water. Nor can it be melted readily, since its melting point is about 2000°C. Hall discovered that bauxite will ionize in molten **cryolite,** a mineral composed of sodium, aluminum, and fluorine, which melts at about 1000°C.

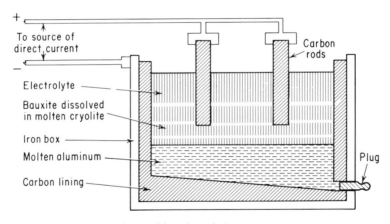

Fig. 19-1. How aluminum is obtained by electrolysis.

The bauxite is dissolved in molten cryolite and the solution is placed in a carbon-lined iron box, which acts as the negative electrode. (See Figure 19-1.) The positive electrode consists of several carbon rods that dip down into the electrolyte. The negative oxygen ions are attracted to the positive carbon rods, where they surrender their negative charges and become neutral oxygen atoms. The positive aluminum ions migrate to the negative electrode, where they receive electrons and become neutral aluminum atoms. The molten aluminum collects at the bottom of the box. From time to time it is drained or siphoned off.

Fig. 19-2. Interior view of an aluminum smelting works, showing a row of electrolytic cells in operation.

A variation of the electrolysis process, called **electroplating,** is used to deposit a metal coat on the surface of some object. For example, suppose we wish to deposit a copper coat on a piece of iron. A compound of the metal to be deposited — in this case, copper sulfate — is dissolved in water. As it dissolves, it forms positive copper ions and negative sulfate ions. If a direct current is passed through the solution, the copper ions will be attracted to the negative electrode and the sulfate ions to the positive one.

The object to be plated is made the negative electrode. As the positive copper ions reach this electrode, they obtain electrons and become neutral copper atoms. These atoms adhere to the electrode and the object thus becomes coated with a fine, uniformly distributed copper plate.

The negative sulfate ions migrate to the positive electrode, which may be a carbon rod. There the ions become neutralized as they yield up their excess electrons, and they immediately react with the hydrogen of the water to form sulfuric acid, releasing oxygen in the process. Hence the electrolyte gradually changes from copper sulfate to sulfuric acid as the copper is deposited on the negative electrode.

In practice, the positive electrode usually is made of a bar of pure metal of the type being deposited — in this example, copper. Now, as the sulfate ion is neutralized, it reacts with the copper instead of the water, forming fresh copper sulfate. Thus the copper sulfate of the electrolyte is replenished and the process may continue until the entire copper electrode is "eaten away."

Since the object to be plated forms the negative electrode, it must be a conductor of electricity. However, nonconductors may be plated, too, if they first are covered with a coating of powdered graphite. This graphite adheres to the object and, since it is a conductor, the metal plate will be deposited on its surface.

International Silver Company

Fig. 19-3. Electroplating spoons. The spoons move around the tank, keeping the solution agitated and assuring an even distribution of silver during plating. The length of time the pieces remain in the tank and the amount of electrical current used determines the thickness of the silver deposited on spoons.

Fig. 19-4. Electrotype.

Almost any metal can be used for plating. Nickel, cadmium, and chromium frequently are deposited for protective purposes. Silver and gold, too, are used for plating, but they generally are employed for decorative and ornamental purposes rather than for protecting the surface of the object from the air.

Only direct current may be used. This current usually is supplied by storage batteries or direct-current generators. A great deal of skill is necessary to form a satisfactory plate. The current and temperature of the electrolyte must be carefully controlled if a smooth, durable plate is to be obtained.

Electroplating enters into the manufacture of books. As you know, printing is accomplished by the use of type. The metal employed for type generally is an alloy of lead, antimony, and tin. Although a large number of impressions can be taken from such type, it is relatively soft. If several hundred thousand copies are to be printed, type metal would soon wear out. Accordingly, we must print a book like this one from a harder metal.

The ordinary type is arranged to form a page. Then a special kind of wax is poured over it. When this wax hardens and is peeled off, it bears an impression of that page. The wax then is coated with graphite to make it a conductor and becomes the negative electrode of a copper electroplating setup. When the copper plate becomes about as thick as an ordinary visiting card, it is peeled off the wax. We thus have an impression of the page in copper, which is a good deal harder than type metal. The copper sheet is backed by a layer about an inch thick of a special metal alloy to give it stiffness. The page is printed from this copper sheet.

This process is called **electrotyping.** A similar process is used for making phonograph records. The sound to be recorded is converted into vibrations of a sharp needle as it passes over the surface of a wax disk.

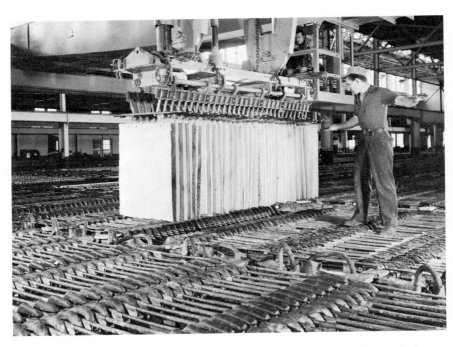

The Anaconda Company

Fig. 19-5. Lifting a load of cathode (negative) copper electrodes from electrolytic refining tanks.

This wax disk, which bears the impression of the vibrations of the needle, then is coated with graphite and electroplated. The metallic coating, which is stripped off the wax disk, bears these impressions, too. This plate, which is called the **master copy,** then is suitably backed to give it stiffness; duplicate records are made by pressing against it disks of a special composition, such as shellac or plastic. The impressions on the master copy are transferred to the duplicate records somewhat in the fashion that the impressions of the type are transferred to paper.

The electroplating process may be used for purposes other than coating some object. For example, copper obtained directly from its ore may contain certain impurities. First it is dissolved in sulfuric acid, forming copper sulfate. This copper sulfate solution will, of course, still contain these impurities.

This solution is the electrolyte. A thin sheet of pure copper forms the negative electrode. The positive electrode consists of a slab of the raw, impure copper. As a direct current passes through the solution, pure copper will be deposited on the negative electrode, which will grow in size. The copper of the positive electrode will react with the sulfate ion, forming copper sulfate and thus replenishing the solution. The impurities sink to the bottom of the tank as a sort of mud or sludge. This process is called **electrolytic refining.**

QUESTIONS

1. In terms of the electron theory, explain the electrolysis of water.
2. Describe and explain how metallic aluminum is extracted from its ore.
3. In terms of the electron theory, explain the electroplating process.
4. Explain how electrotypes are made.
5. In terms of the electron theory, explain the electrolytic refining of copper.

APPLICATIONS
DEPENDING UPON
THE MAGNETIC EFFECT

20

As current flows through a coil of wire, a magnetic field is built up around the coil. The strength of this field is a function of the number of turns of the coil and the strength of the current flowing through those turns. The magnetic field can be employed in several ways.

It can be used to attract magnetic material. This is the principle behind the lifting magnet, the relay, and many other devices. Or the magnetic field can be used to magnetize magnetic material, as in the tape recorder. A third use is to induce an electromotive force in a conductor that cuts through the field. This is the principle behind the generator, transformer, and so on. A fourth use of the magnetic field is to react with another magnetic field and so produce a mechanical motion, as in the electric motor (which will be discussed in the next chapter).

A. THE MAGNETIC FIELD CAN ATTRACT MAGNETIC MATERIAL

As previously stated, as current flows through the turns of a coil of wire, the coil becomes a magnet (see Chapter 6, subdivision D). If the coil is supplied with a core of magnetic material, its magnetic field is strengthened. The combination of coil and its magnetic core is called an **electromagnet.** The fact that the electromagnet is a temporary magnet — that is, it is surrounded by a magnetic field only as long as current flows through its turns — is a distinct advantage. It allows us to turn the magnetic field on or off, merely by the flick of a switch. Consequently, the electromagnet is employed in a great many devices.

Where large quantities of magnetic materials, such as scrap iron, iron sheets, iron castings, are to be moved, the **lifting magnet** (illustrated in Figure 20-1) generally is employed. This is a large electromagnet made up of a coil of wire and a soft-iron core. This core not only passes through the center of the coil but also surrounds the outside. Thus, when a direct current is sent through the coil, the effect is that of a number of U-magnets arranged in a circle with all their north (or south) poles at the center. Thus the magnetic force is concentrated, and the coil is protected from outside damage as well.

Since the strength of the electromagnet depends upon the number of turns of the coil and the current flowing through them, as well as the material and shape of the core, electromagnets that must lift tons of material at a time are constructed with many turns of wire heavy enough to carry safely the large currents required. Such lifting magnets generally are equipped with portable d-c generators that furnish the required current. Of course, the electromagnet does not lift the load; it merely holds it fast while a motor-driven crane lifts both magnet and load.

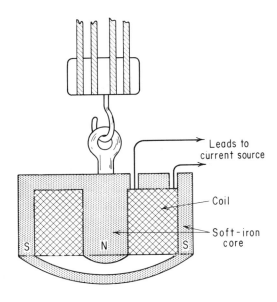

Fig. 20-1. Cross-sectional view of lifting magnet.

If alternating current is passed through an electromagnet, the polarity of the magnet will, of course, change periodically in step with the alternations of the current. Also, once each alternation the current drops to zero. However, in most applications it makes no difference whether it is a north or south pole that is doing the attracting. If a 60-Hz current is used, there are 120 alternations each second and, hence, the interruptions of the magnetic field are of such short duration as to be negligible for most purposes. Accordingly, it is possible to operate electromagnets on alternating currents.

Another use for the electromagnet is in the **electric bell** illustrated in Figure 20-2. The electromagnet consists of two coils of wire on a U-shaped soft-iron core. These coils are wound in opposite directions so that, as current flows through them, a north and south pole appear next to each other, thus concentrating the magnetic field.

The soft-iron armature is mounted on a flat spring and so placed that, normally, the two contact points touch each other. If a source of direct

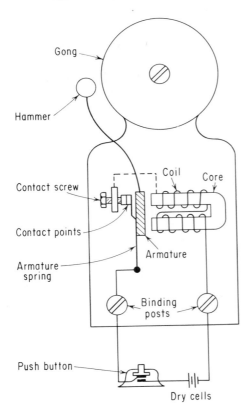

Fig. 20-2. Electric bell.

current is connected to the binding posts, current flows through the coils of the electromagnet, through the two contact points, through the armature spring, and back to the source. As it does so, however, the electromagnet is activated, attracting the armature. As the armature is pulled toward the electromagnet, the contact points are separated, breaking the circuit. This causes the electromagnet to release the armature and its spring forces it back to its normal position. The contact points touch and the circuit is completed again. The entire cycle repeats itself, the armature vibrates back and forth, and the hammer attached to this armature repeatedly strikes the gong.

The source may be several dry cells which are connected to the binding posts through a push button, a sort of switch. Note that the direction of current flow makes no difference so far as the operation of the bell is concerned. Accordingly, the bell will operate on alternating current as well as on direct current. For a-c operation we generally use a **bell-ringing transformer,** which steps the 120-volt line down to about 6 volts.

The wire coil of the electromagnet, without its core of magnetic material, is called a **helix,** or **solenoid.** If this helix is provided with a movable soft-iron core (called the **plunger**) and current flows through the turns of the coil, the magnetic field tends to pull or suck the plunger into the center of the coil. Accordingly, the coil together with its movable core is called a **sucking coil** or, more frequently, a **solenoid** (although, strictly speaking, only the coil is the solenoid).

The solenoid (see Figure 20-3) is used in a great many applications to

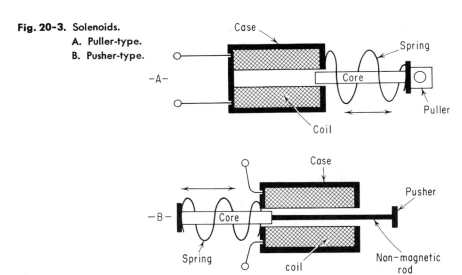

Fig. 20-3. Solenoids.
   A. Puller-type.
   B. Pusher-type.

provide a mechanical push or pull to operate some device such as a valve (as, for example, in the washing machine). In the **puller-type** solenoid (Figure 20-3A), as the coil is energized by the current flowing through its turns, its magnetic field sucks the plunger into the coil. When the current is cut off, the coil is de-energized and a spring pulls the plunger out of the coil. The device that is to be actuated is attached to the plunger and moves with it. Thus it is pulled as the coil is energized and restored to its original position by the spring as the coil is de-energized.

The solenoid may also be employed as a **pusher,** as illustrated in Figure 20-3B. As the coil is energized, the plunger is sucked in and the nonmagnetic rod attached to it pushes the object to be moved. When the coil is de-energized, the spring forces the plunger, and the rod, to their original positions. Such pushers frequently are used to remove objects from a moving belt upon a signal to the solenoid.

Either a direct or alternating current may be used to energize the solenoid, since either type will produce the magnetic field around the coil. (See Figure 20-4.) There is one precaution, however. The core of the electromagnet finds itself in the magnetic field of the coil. If a steady direct current flows through the coil, no current will be induced in the core since both the core and field are stationary. But if an alternating current flows through the coil, the changing magnetic field will cause a current to be induced in the core. This is called the **eddy current.**

The eddy current is undesirable on two counts. The flow of current through the core represents a power loss, which must come from the source. Also, the flow of current may cause the core to get quite hot. To reduce the eddy current, the core is not built solid but is made up of many thin slices, called **laminations.** Each lamination is insulated from

−A− −B−

Automatic Switch Co. *Guardian Electric Mfg. Co.*

Fig. 20-4. A. D-C solenoid.
B. A-C solenoid.

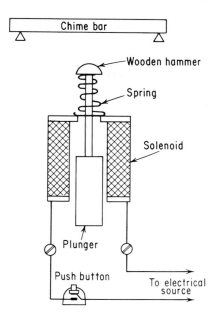

Fig. 20-5. Electric door chime.

its neighbor by a coat of varnish or some similar material. This offers considerable resistance and, as a result, the eddy currents are cut down. In solenoids operating on alternating current, too, the plunger is built up of laminations.

Keep in mind that the coil of an electromagnet will offer a greater opposition to the flow of alternating current than to the flow of direct current. The ohmic resistance of the coil is present for both types of current, but with alternating current there is also the inductive reactance. Hence, if a coil designed to operate on alternating current at a certain voltage should be connected to a source of direct current at the same voltage, the flow of current might be great enough to burn out the winding.

The solenoid can be used as a door chime, as shown in Figure 20-5. The coil and its plunger are mounted vertically so that the plunger can slide down by gravity. The spring prevents it from falling through. As the push button is pressed, the circuit is closed and current flows through the solenoid. The plunger is drawn up into the hollow core of the coil and the wooden hammer strikes the chime bar, producing a musical note. When the push button is released, the circuit is opened and the plunger drops back.

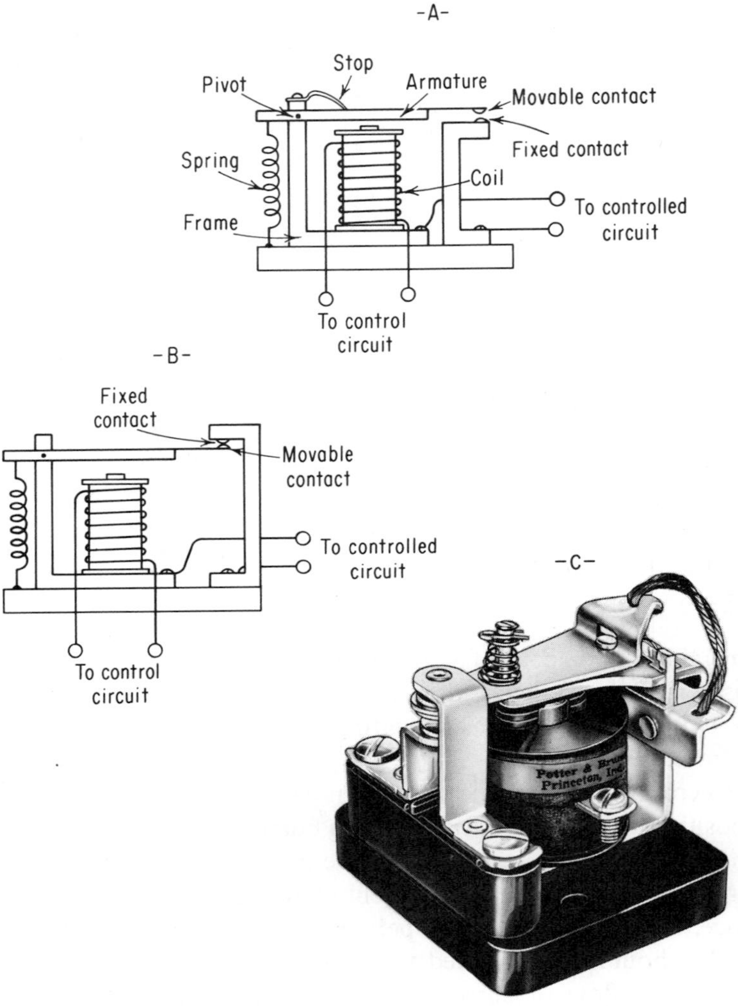

Potter & Brumfield, Inc.

Fig. 20-6. A. Normally-open (NO) electromagnetic relay.
B. Normally-closed (NC) electromagnetic relay.
C. Commercial electromagnetic relay.

One of the most common uses for the electromagnet is to open or close a switch as the coil is energized by an electric current or de-energized by removal of this current. The electromagnet together with the switch it operates is called an **electromagnetic relay.** (See Figure 20-6.)

In Figure 20-6A a **normally-open** (NO) relay is illustrated. It contains a coil of wire wound upon a soft-iron core. Normally, with the coil unenergized, the spring pulls the **armature** away from the coil. Thus the **movable contact** mounted on the armature is separated from the **fixed contact**. The switch is open.

When the coil is energized by current flowing through it, the magnetic field so created attracts the armature to the coil. The two contacts touch and the switch is closed.

A **normally-closed** (NC) relay is illustrated in Figure 20-6B. Here, with the coil unenergized, the spring brings the two contacts together and the switch is closed. As current energizes the coil, the armature is attracted and the contacts separate. The switch is open.

The basic normally-open relay circuit is illustrated in Figure 20-7. The symbol for the relay coil and its core is ⊏↯⊐. The contacts are symbolized by ↦↓. The straight line indicates the movable contact; the arrow the fixed one. When the relay is of the normally-open type, the contacts appear as ↦↓ or ⊣⊢. The contacts of the normally-closed relay appear as ↦↓ or ⊣⧸⊢. The letter symbol for the relay is K.

The **control circuit** consists of the relay coil, a light-power source (such as a battery), and a switch. The **controlled circuit** consists of the relay's contacts, the load, and a heavy-power source (such as the power line). When the switch is closed, the normally-open relay is energized and the armature is attracted, closing the contacts. As a result, power flows from the heavy-power source to the load. When the switch is

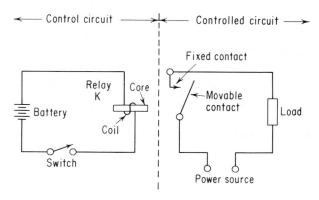

Fig. 20-7. Basic relay circuit.

opened, the contacts are forced apart, opening the controlled circuit. The normally-closed relay circuit functions in the same way except that the action of the contacts is reversed — that is, they are closed when the relay is unenergized and opened when the relay is energized.

By winding the coil of the electromagnet with many turns of fine wire, we can make the relay very sensitive. That is, it will operate with very little current in the control circuit. Since it is of the sensitive type, all the contacts of the relay must be light and so cannot be used to switch a large current. As a result, a large current cannot flow in the controlled circuit.

Where we desire to carry larger currents in the controlled circuit, the contacts of the relay must be heavier. As a result, we must employ a more rugged relay, one whose coil is wound with fewer turns of heavier wire. This, of course, reduces the sensitivity of the relay, and a larger current is required for the control circuit.

It is possible to construct relays that can perform several different functions simultaneously. In the previous illustrations we have indicated the switch portion as a single-pole, single-throw type, which can only open or close a single controlled circuit. In Figure 20-8 you see the circuit of a relay with a single-pole, double-throw switch, which can simultaneously open one controlled circuit and close another controlled circuit.

Here, with the coil unenergized, the armature is not attracted to the coil and the movable contact (2) touches one fixed contact (3), thus closing the controlled circuit #2. As the coil is energized, the movable contact is separated from contact 3 (thus opening controlled circuit #2) and touches another fixed contact (1), thus closing controlled circuit #1. Of course, more movable and fixed contacts, all operated by the relay, may be employed. In this way more controlled circuits may be opened or closed simultaneously as the relay coil is energized or de-energized.

The relay coil may be energized by either direct or alternating current. Where direct current is employed, we have no special problem. Alternating current may be employed, too, since the polarity of the field does not affect the attraction of the armature. However, the rapid alternations of

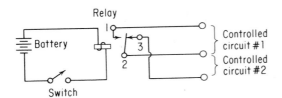

Fig. 20-8. Relay employing a single-pole, double-throw switch.

the magnetic field cause the armature to vibrate, or "chatter." Since the contacts are controlled by the armature, the controlled circuit, too, will be affected.

One method for remedying this fault is to rectify the alternating current before applying it to the relay coil. Small semiconductor diodes generally are employed for this purpose. Another method is to connect a fairly large capacitor across the coil. Then, as the coil is energized, the capacitor is charged simultaneously. In between half-cycles, as the current goes from one alternation to the next, the capacitor discharges through the coil, thus keeping it energized and preventing chatter.

Relays energized by alternating current also have their frames laminated to reduce eddy-current losses. Further, a metal ring is fastened around the top of the core of the coil. This ring acts as a single turn, and the magnetic field produced by the current induced in it, acting in opposition to the magnetic field produced by the coil, keeps the armature attracted when the latter field is reduced as the current goes from one alternation to the next. This, too, prevents chattering.

The relay has several distinct advantages over the mechanical switch. First, a relatively small current in the control circuit may be used to control a relatively heavy current in the controlled circuit. Secondly, since the current in the control circuit is small, we may employ thin wires for that circuit. Also, the power source and switch of this circuit may be at a considerable distance from the relay. The heavy current in the controlled circuit, on the other hand, requires heavier wires, and these wires must not be too long if excessive power loss is to be avoided.

A variation of the relay may be used to protect electrical circuits against overloads. For example, look at the illustration of an **overload circuit breaker** shown in Figure 20-9. Current flows from the line, through the contact points, through the movable arm, through the coil of the electromagnet, through the circuit being protected, and back to the line.

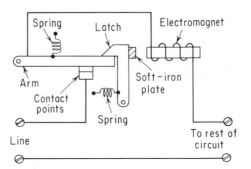

Fig. 20-9. Overload circuit breaker.

As a normal amount of current flows through the electromagnet, its attraction for the soft-iron plate attached to the latch is overcome by the pull of the spring and, hence, the latch remains in place. However, should the current rise to a dangerous value, the pull of the electromagnet is increased and the soft-iron plate and latch are pulled over to the right. This releases the arm, whose spring pulls it upward. The contact points are separated and the circuit is broken, thus protecting the rest of the electrical circuit. This circuit remains open until the cause of current rise is corrected and the circuit breaker is relatched.

Some of the most interesting applications of the magnetic effect of the electric current are in the field of communications. Man constantly seeks to extend both the distance over which he is able to transmit intelligence and the speed with which it is sent. Yet, up to the nineteenth century, messages sent by sight or sound signals were limited to a few miles. Longer distances could be covered only through the use of the mail or by personal messengers.

With the invention of the electromagnet, however, a new field was opened. An electromagnet can be operated from any distance, provided we have wires long enough and an electromotive force strong enough to send the current through the wires. If we connect the battery and the switch that opens and closes the circuit to one end of the wires, and the electromagnet to the other end (Figure 20-10), a current will flow through the electromagnet as the switch is closed. As a result, a soft-iron armature is attracted to the electromagnet.

When the switch is opened, the circuit is broken and the electromagnet loses its attractive force. A spring pulls the armature back. A U-shaped metal stop is provided against which the armature strikes. Thus, as the switch is closed, the armature is pulled down until it strikes against the bottom arm of the stop. When the switch is opened, the

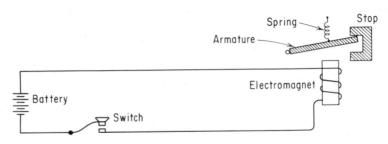

Fig. 20-10. Simple one-way telegraph circuit.

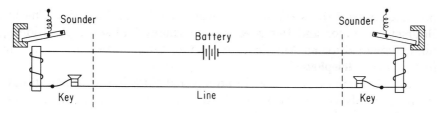

Fig. 20-11. Simple two-way telegraph circuit.

spring pulls the armature up and it strikes the upper arm of the stop. As the armature strikes the stop, it produces a metallic click. So, if the switch is closed and opened, we hear two clicks — once when the armature strikes the bottom arm, and again when it strikes the upper arm.

The interval of time elapsing between the two clicks at the electromagnet end of the line is determined by the length of time the switch at the other end is kept closed. If the switch is closed for a short period, the interval between clicks is short. If it is closed for a long period, the interval between clicks, too, will be long.

Here, then, is a method for transmitting intelligence over wires. All we must do is arrange a code composed of short intervals between clicks, called dots (·), and long intervals between clicks, called dashes (—). For example, the letter **A** may be sent as **dot dash** (·—); the letter **B** as **dash dot dot dot** (— · · ·); and so forth. Because the current flows through the wires at a speed of thousands of miles per second, messages can be sent almost instantaneously as far as we can string our wires and send current through them.

This is the principle of the **telegraph.** Although scientists all over the world had been working on the problem, its invention generally is credited to an American, Samuel F. B. Morse, in 1837.

The telegraph circuit, in simplified form, is illustrated in Figure 20-11. Each end of the circuit contains a switch (called a **key**) and an electromagnet-armature-stop combination (called a **sounder**). Note that the keys, electromagnets, and battery are connected in series.

Normally, both keys are kept closed. When the operator at one end of the line wishes to send a message, he opens his key, thus breaking the circuit. When he closes his key, both sounders operate. In this way, the message is sent. After he is through, he closes his key again.

No sooner had man learned how to send intelligence over the long wires of the telegraph circuit, when the next step was suggested. Was it possible to send the spoken word over such wires? Men like Charles Boursel of France and Philip Reis of Germany had some good ideas, but it remained for an American, Alexander Graham Bell, to invent the first practical **telephone** in 1876.

Speech originates with the vibrating vocal cords in the human throat. As these cords vibrate back and forth, they alternately compress and expand the air in front of them. These alternate compressions (**condensations**) and expansions (**rarefactions**) travel through the air in the form of a sound wave.

There are two factors that determine the intelligence carried by the sound wave. The tone or pitch of the sound is determined by the **frequency** of the sound wave — that is, by the number of condensations and rarefactions per second. The intensity or loudness of the sound is determined by the **amplitude** of the sound wave.

At the transmitting end of the telephone, the sound wave enters a **microphone** where the variations of the sound wave produce corresponding fluctuations in an electric current. One type of microphone commonly used in the telephone consists of a large number of small carbon granules loosely packed in a cup between two carbon plates (Figure 20-12). The back plate is firmly held in place, but the front one can move back and forth. This movable carbon plate is fastened to a thin, flexible diaphragm that is held firmly at its rim.

Suppose, as in Figure 20-12A, that a rarefaction of the sound wave approaches this diaphragm. Because of the low pressure produced by

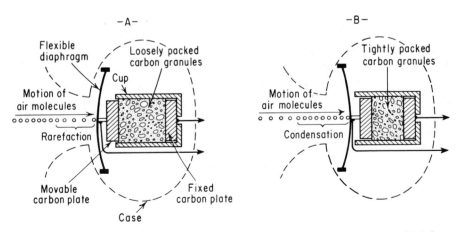

Fig. 20-12. Carbon-granule microphone, showing the effects of rarefactions and condensations of the sound wave.

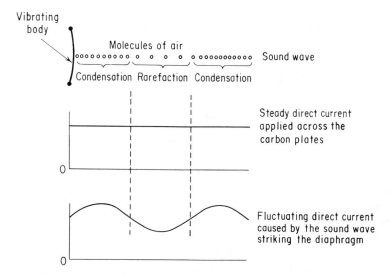

Fig. 20-13. Graphs showing how the sound wave modulates the electric current.

this rarefaction, the diaphragm bulges out, carrying with it the movable carbon plate. As a result, the carbon granules are more loosely packed than before. When a condensation approaches the flexible diaphragm (Figure 20-12B), the latter is pushed in and the carbon granules become more tightly packed.

The electrical resistance between the two carbon plates through the carbon granules depends upon how tightly the granules are packed. If they are packed loosely, the resistance is greater. If they are packed tightly, the resistance is smaller. Thus the granules form a variable resistor whose resistance depends upon whether a rarefaction or a condensation of the sound wave approaches the diaphragm.

Now, suppose that we apply a steady direct current across the two carbon plates. The current will remain steady so long as the carbon granules are not disturbed. But if a rarefaction approaches the diaphragm, the resistance of the granules increases. As a result, the current drops. As a condensation approaches the diaphragm, the resistance decreases and the current rises. You can see this relationship in the graphs of Figure 20-13.

Here, then, is a method for superimposing intelligence upon an electric current. If the sound wave produced by the human voice strikes the diaphragm, its condensations and rarefactions will be impressed

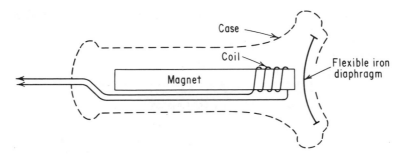

Fig. 20-14. Telephone receiver.

upon the electric current. We call this process of superimposing intelligence on the electric current **modulation,** and say that **the sound wave modulates the electric current.**

The modulated current, carrying the intelligence superimposed on it, can be sent along wires to a distant point. The next problem to be solved is how to remove the intelligence from the current. This is the function of the **telephone receiver.**

In the telephone receiver a permanent magnet is so placed that it attracts a thin, flexible, iron diaphragm that is held along its rim. (See Figure 20-14.) A coil is placed over one end of the magnet and it is so wound that, as current flows through this coil, a magnetic field is set up opposed to that of the permanent magnet. The diaphragm, then, is attracted with a force that is the resultant of the magnetic field of the permanent magnet and the opposing field of the coil. The current that flows through this coil comes from the microphone.

If a steady current flows through the coil, the resultant pull on the diaphragm, too, will be steady. However, if a fluctuating current flows through the coil, its magnetic field will fluctuate with the variations in current. Thus, the resultant pull on the diaphragm will fluctuate in step and, accordingly, the diaphragm will vibrate back and forth. The vibrating diaphragm will set in motion the air next to it and a sound wave will be produced. Since the vibrations of the diaphragm are in step with the variations of the original sound wave, the sound wave it produces will be identical with the original.

B. THE MAGNETIC FIELD CAN MAGNETIZE MAGNETIC MATERIAL

The fact that the magnetic field created as a current flows through a coil of wire can be used to magnetize magnetic material is the principle

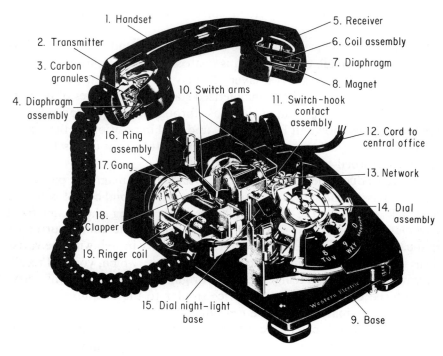

1. Handset
2. Transmitter
3. Carbon granules
4. Diaphragm assembly
5. Receiver
6. Coil assembly
7. Diaphragm
8. Magnet
10. Switch arms
11. Switch-hook contact assembly
16. Ring assembly
17. Gong
12. Cord to central office
13. Network
18. Clapper
14. Dial assembly
19. Ringer coil
15. Dial night-light base
9. Base

Fig. 20-15. Cutaway view of modern telephone showing transmitter, receiver, dialing, and ringing apparatus.

behind a number of devices. For example, the intelligence carried by the sound wave can be stored in a magnetic wire or tape for future reproduction by means of the **magnetic recorder** invented by Valdemar Poulsen, a Danish scientist. In this device the sound wave strikes a microphone, producing a fluctuating current, as in the telephone. This fluctuating current is made to flow through the coil of an electromagnet (the **recording coil**), thus producing a magnetic field that fluctuates in step with the variations of the current.

An unmagnetized wire made of some magnetic material passes through this magnetic field at a uniform rate. As it does so, the wire becomes magnetized to a degree depending upon the strength of the magnetic field at that particular instant. Since the magnetic field is a fluctuating one, various portions of the wire are magnetized to a greater or lesser degree, depending upon the strength of the field at the instant each portion is passing through. Thus the intelligence contained in the sound wave is transferred into a corresponding magnetic pattern in the wire.

In modern practice, a magnetic tape is used instead of the wire. This tape usually is made of paper and is coated with some plastic material. Imbedded in this plastic is a layer of finely divided magnetic material, such as iron powder. In its unmagnetized state, the molecules of iron are arranged helter-skelter. But as they pass through the magnetic field, they are lined up, as shown in Figure 3-5B, and held in position by the plastic coating. Once lined up, the particles remain that way unless they are further disturbed by a magnetic field.

To reproduce the sound, the tape bearing the magnetic pattern is made to pass at the same uniform rate through a second coil of wire (the **playback coil**). As the magnetic fields surrounding the magnetized particles of the tape cut across the conductors of the coil, a current is induced in that coil. This induced current will vary in step with the variations in strength of the magnetic fields of the tape, which, you will recall, vary in step with the intelligence of the original sound. Passing this varying induced current through a device similar to the telephone receiver causes the original sound to be reproduced.

With reasonable care, a recording can be reproduced many thousands of times. If we wish to erase the recording from the tape, all we need do is to pass the tape through the magnetic field produced by a high-frequency alternating current flowing through another coil (the **erase coil**). As a result, the magnetic molecules will become arranged in a helter-skelter pattern once more. The tape then is ready to receive a new recording.

The magnetic tape may be used for other purposes than the recording of speech or music. In the data-processing and computer systems so widely employed today, magnetic tapes are used to store tremendous quantities of data or information. The data first is translated by machine into a **binary code,** a code which has only two significant states such as ONE and ZERO, or ON and OFF, or PULSE and NO PULSE.

As the magnetic tape passes through the machine, the coded information is transferred to it as a series of magnetized and unmagnetized spots. If, as the tape moves beneath the recording coil, the code indicates the ONE state, a pulse of current flows through the coil. The magnetic field thus created forms a magnetic spot on the tape at that point. If the ZERO state is indicated, there is no current flow through the recording coil, and that spot on the tape remains unmagnetized. In this way the data is recorded upon the tape as a series of magnetized and unmagnetized spots.

Webster Electric Co.

Fig. 20-16. Magnetic tape recorder.

To recover the coded data, the tape is passed beneath a playback coil, just as in the case of the recorder. The electric pulses coming from the playback coil then are sent to a device that translates them back to human language and rapidly prints out the data. In this way data can be stored and retrieved at an extremely rapid rate. Millions of bits of data can be stored in a single reel of tape.

C. THE MAGNETIC FIELD CAN INDUCE AN EMF IN A CONDUCTOR

We have already seen how the magnetic field produced by current flowing through a coil can induce an electromotive force in a conductor that cuts across the field. [See the transformer (Chapter 11) and the mechanical generator (Chapter 13).] The same principle is employed by a number of other devices.

The transformer is an alternating-current device, since it utilizes a changing magnetic field to induce the voltages in the windings. Suppose the primary winding were connected to a steady direct-current source. At the moment the primary circuit is completed, current flows, a magnetic field expands around the coils, and a voltage is induced in the secondary winding. Then, as the current in the primary circuit reaches its steady value, the magnetic field, too, becomes steady and the induced voltage in the secondary winding drops to zero. When the primary circuit is broken, the magnetic field collapses and, again, an induced voltage

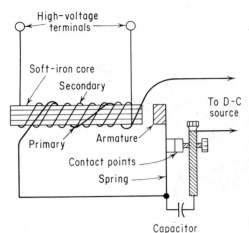

Fig. 20-17. Induction coil.

appears momentarily in the secondary winding, though now in the opposite direction.

Thus, you see, a voltage is induced in the secondary winding every time the primary circuit is completed or broken. If we periodically interrupt the direct current flowing in the primary circuit, an alternating voltage will be induced in the secondary winding.

This is how the **induction coil** (Figure 20-17) operates. The core consists of a bundle of soft-iron wires. A primary winding of few turns and a secondary winding of many turns are wound over this core. An **interrupter,** somewhat similar to the mechanism of the electric bell, is connected in series in the primary circuit to interrupt periodically the flow of direct current through that circuit.

A soft-iron armature is mounted on a flat spring and is placed near the end of the core. A contact point, also mounted on the flat spring, normally touches another fixed contact point. The primary current then flows from the source (which may consist of several dry cells connected in series), through the contact points, through the primary winding, and back to the source.

As it does so, the primary winding and core become an electromagnet, attracting the armature and spring. This causes the contact points to separate, opening the primary circuit. The electromagnet loses its magnetism and the spring pulls back the armature, causing the contact points to touch once more. The primary circuit is completed again and the entire cycle is repeated.

Each time the contact points separate, the self-induction of the primary winding will cause a spark to arc across the points. This, in time, will pit and burn these contact points. A capacitor is connected across the points to cut down the arcing. Now, as the contact points are separated, the electrical energy between them is utilized to charge up the capacitor, rather than to produce a spark.

Before the use of alternating current became so widespread, the induction coil was employed to step up a relatively low-voltage direct current to the high alternating voltage needed to operate devices such as X-ray tubes. It also was employed in early radio transmitters to produce sparks across its high-voltage terminals. Hence the induction coil is sometimes called a **spark coil.**

Today, we use a variation of the induction coil in the automobile to produce the high-voltage spark needed to explode the gasoline-and-air mixture in the cylinders of the engine. Look at Figure 20-18. The primary and secondary windings and the soft-iron core are housed in a plastic insulating case. The interrupter in the primary circuit is a portion of the distributor. It consists of a rotating cam that, as it turns, periodically

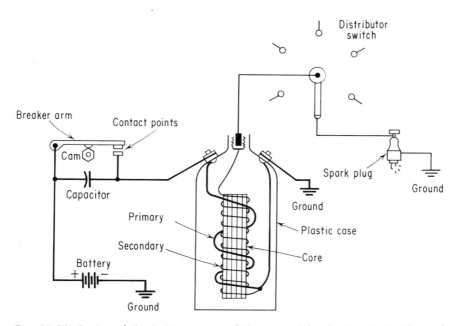

Fig. 20-18. Portion of the ignition system of the automobile, showing the ignition coil. (Low-voltage circuit appears in heavy lines.)

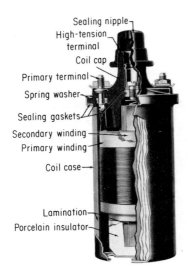

Sealing nipple
High-tension terminal
Coil cap
Primary terminal
Spring washer
Sealing gaskets
Secondary winding
Primary winding
Coil case
Lamination
Porcelain insulator

General Motors Corp.

Fig. 20-19. Cross-sectional view of an automobile ignition coil.

raises and lowers the breaker arm that causes the contact points to separate and touch. The voltage source is the storage battery of the automobile. The primary circuit is completed through the metal frame of the

automobile, as indicated by the symbol $\equiv$ (ground).

Each time the contact points touch and separate, a high voltage is induced in the secondary winding. However, the shape of the cam is such that the separation of the points is much quicker than the touching. Since the induced voltage is a function of the speed with which the points separate or touch, a much higher voltage is induced when the points separate. It is this higher voltage that is sufficient to fire the mixture in the cylinders.

The high-voltage secondary circuit consists of the secondary winding, the distributor switch, and the spark plug. This circuit, too, is completed through the frame of the automobile. The distributor switch consists of a rotating arm and a series of contact points, each connected to one of the spark plugs. It is so constructed that it is rotated by the engine in step with the cam. Switch and cam are so adjusted that, as the contact points are separated and a high voltage is induced in the secondary, the rotating arm is touching one of its contact points and a high-voltage surge is sent to its respective spark plug. A spark is produced and the mixture in the corresponding cylinder is fired.

QUESTIONS

1. Explain the operation of the electric bell.

2. Explain why an electromagnet is able to operate on either direct or alternating current. What changes should be made if an electromagnet designed to operate on direct current is to operate on alternating current? Explain.

3. Explain the operation of the magnetic sucking coil. Give an example of how it is used.

4. Explain the operation of (*a*) the normally-open electromagnetic relay; (*b*) the normally-closed electromagnetic relay.

5. Discuss three methods for preventing "chattering" in the relay.

6. Draw the basic relay circuit.

7. Explain the operation of the magnetic circuit breaker.

8. Draw a simple circuit of a two-way telegraph system. Label the important parts and explain the functions of each.

9. Explain how sound waves can modulate an electric current.

10. Explain the action of a carbon-granule microphone.

11. Explain the action of the telephone receiver.

12. Explain how sound can be recorded on magnetic tape. How can this tape be used to reproduce the sound?

13. Explain the operation of the ignition coil in the automobile ignition system.

ELECTRIC MOTORS

21

As previously stated, when an electric current flows through a coil of wire a magnetic field is set up around the coil. This magnetic field can react with another magnetic field to produce a mechanical motion. We have seen an example of motion so produced when we discussed the moving-coil meter movement (see Chapter 7, subdivision C).

Assume a conductor is placed in the magnetic field existing between the poles of a magnet, as in Figure 21-1A. Since we are looking at a cross section of this conductor, it appears as a circle. The magnetic field of the magnet is indicated by the dotted lines going from the north pole to the south pole.

Now assume that a current is sent through the conductor in such a direction that the current is flowing into the page, as indicated in Figure 21-1B. (The **x** appearing in the center of the conductor represents the tail feather of an arrow that indicates the direction of current flow.) If we apply the left-hand rule illustrated in Figure 6-4, we find

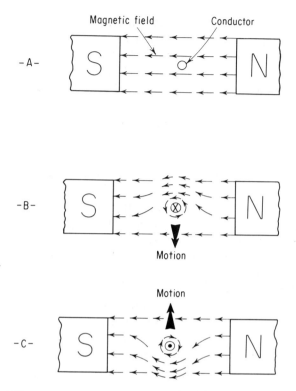

Fig. 21-1. Why a current-carrying conductor in a magnetic field is caused to move. The direction of movement is indicated.

that the conductor is surrounded by a counterclockwise magnetic field. (This rule, you may remember, states that if the conductor be grasped by the left hand with the thumb outstretched and pointing in the direction of current flow, the curved fingers will point in the direction of the magnetic field around the conductor.)

Note that the magnetic field around the conductor aids the magnet's field above the conductor, but opposes the field below it. The effect, then, is to distort the magnet's field, as indicated. Since the magnetic lines of force act somewhat as stretched rubber bands, the effect of the distorted field is to push the conductor downward.

If, as in Figure 21-1C, we reverse the current flow through the conductor (as shown by the dot in the center of the conductor, which represents the tip of an arrow indicating the direction of current flow), the magnetic field around the conductor aids the magnet's field below the conductor, but opposes the field above it. The effect, then, is to distort the magnet's field, as indicated. As a result, the conductor is pushed upward.

We can formulate a **motor rule** to find the direction of motion imparted to a current-carrying conductor in a magnetic field. Extend the thumb, forefinger, and middle finger of the right hand so that they are at right angles to each other. Let the forefinger point in the direction of the magnetic field (from north to south pole). Let the middle finger point in the direction of current flow in the conductor. The thumb then points in the direction of motion imparted to the conductor.

The force acting upon a current-carrying conductor in a magnetic field depends upon the strength of the field, the length of the conductor in that field, and the amount of current flowing through the conductor. The greater any of these components becomes, the greater is the force exerted on the conductor.

Suppose we were to mount a single loop of wire between the poles of a magnet so that it was free to rotate on its horizontal axis. Now let us send a current through this loop as indicated in Figure 21-2. (The two sides of the loop are seen in cross section and therefore appear as the two circles between the poles of the magnet. The current flows in one direction in one side and in the other direction in the other side.) One side of the loop is moved down and the other up. As a result, the loop tends to rotate in a counterclockwise direction around its axis. The rotating effect produced by the reaction between the magnetic field of the magnet and the magnetic fields around the conductors of the loop is known as **torque.**

A. DIRECT-CURRENT MOTORS

If, as in Figure 21-2, a direct current is passed through the loop, the reaction between the magnetic fields tends to push the left-hand side of the loop down in a counterclockwise direction. However, when the conductor reaches the bottom of its sweep, it encounters an equal and opposite push. Hence it becomes stationary. Simultaneously, the right-hand side of the loop becomes stationary at the top of its sweep. The result, then, is that the loop remains fixed in its vertical position. Since there is no rotating effect and, hence, no torque, we call this the **zero-torque** position of the loop.

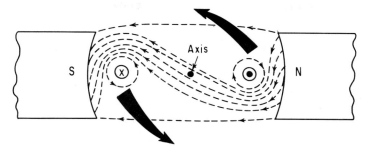

Fig. 21-2. How the magnetic field is distorted to give the loop a rotating motion.

However, if we had some way of reversing the direction of current flow in the loop every time it reached the zero-torque position, the loop would continue to rotate counterclockwise. This is accomplished by the **commutator,** which we discussed in the Chapter 13, subdivision B.

This, then, is how the direct-current motor operates. The loop, or **armature,** rotates in a magnetic field. This **field** may be set up by permanent magnets or, more commonly, by electromagnets. The **commutator** periodically reverses the direction of current flow through the armature.

A simplified version of such a motor is shown in Figure 21-3. The soft-iron armature core upon which the armature loop is wound is

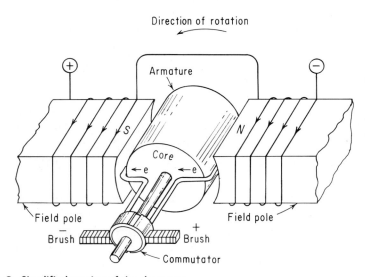

Fig. 21-3. Simplified version of the d-c motor.

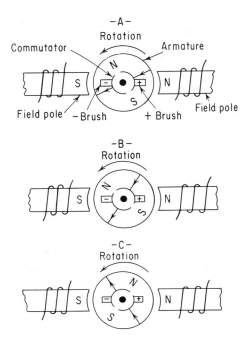

Fig. 21-4. D-C motor rotates because of the attraction between unlike magnetic poles and the repulsion between like poles.

used to concentrate the magnetic field. Note the resemblance between the d-c motor and the d-c generator. As a matter of fact, in 1873 Zénobe T. Gramme, a Belgian engineer, had two identical d-c generators standing side by side. One of these generators was being rotated by a steam engine. When the current output of this generator was accidentally fed into the second generator, the latter started to rotate as a motor.

There is another way of looking at the motor. Look at Figure 21-4A. Note that, as current flows as indicated through the armature winding, the iron core becomes a magnet with its north pole at the top and its south pole at the bottom. Since unlike poles attract, the north pole of the core is attracted to the south field pole and the south pole of the core to the north field pole. As a result of this attraction, the armature rotates in a counterclockwise direction until the opposing poles line up (Figure 21-4B).

However, as the armature rotates, so does the commutator. Thus, as the opposing poles come in line, the commutator causes the current to reverse its direction through the armature winding. This, in turn, causes a reversal of the polarity of the core (Figure 21-4C). The south field pole finds itself facing the south pole of the core and the north field pole the north pole of the core. Since like poles repel, the armature is forced to rotate in a counterclockwise direction. As the polarity of the core is reversed every half-revolution, the armature continues to rotate.

If, as in Figure 21-4, the armature contains a single winding, the rotary motion will be jerky since the attraction between opposite poles becomes greater as these poles approach each other. Accordingly, if, instead of one, a number of windings be employed, each with its set of two commutator bars, the rotary action becomes smoother. This produces a multicoil armature similar to the type employed in the d-c generator (Chapter 13, subdivision B).

Look at Figure 21-5. Note that, as current flows as indicated through the armature windings, the iron core becomes a magnet with its north pole at the top and its south pole at the bottom. Thus the armature's magnetic field is at right angles to the field between the field poles. Since unlike poles attract, the core rotates counterclockwise. However, so does the commutator which controls the current flow to the windings. As a result of its action, the magnetic field of the core is always maintained at right angles to the field between the field poles. The armature thus continues to rotate as unlike poles seek to line up.

Modern motors have their windings placed in slots along the surface of the armature core in a manner similar to that employed by the drum type armature used in d-c generators (see Figure 13-12). The slot winding presents a number of advantages over a surface-wound armature. For one, since the wires lie in slots where there is very little of the field flux, very little force is exerted on them. Hence there is less danger of damage to their insulation. The main force is exerted on the teeth-like

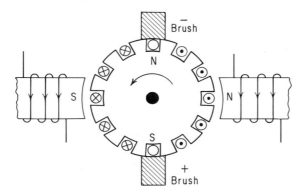

Fig. 21-5. D-C motor using a drum-type multicoil armature.

portions of the iron core between the wires where most of the flux is found. Secondly, the air space between the armature and the field poles can be made much smaller, thus reducing the reluctance of the magnetic path through the motor.

Like the d-c generator, practical d-c motors may have two, four, or more field poles. For each set of poles there is a set of brushes, and all brushes of the same polarity are connected together.

The motor does indeed resemble a generator. In fact, it is a generator. Current is sent into the windings from the line, causing the armature to rotate. But as the armature rotates, its windings cut across the magnetic field set up by the field poles. As a result, an induced voltage is generated in the armature windings just as in the generator. However, by Lenz's law, this induced voltage opposes the line voltage. Hence it is called a **counter electromotive force** (cemf).

We may see the effect of this counter electromotive force if we connect an ammeter in the line to the motor. As the line switch is closed, the ammeter indicates that a large current is flowing to the motor. Since the armature is at rest, no counter emf is generated. Hence the full line voltage is applied and the current is determined mainly by the resistance of the armature, which generally is quite low.

As the motor starts to speed up, a counter emf is generated in the armature winding. Since this counter voltage opposes the line voltage, the **net** voltage applied to the motor is reduced. Since the armature resistance remains constant and the applied voltage is reduced, the line current starts to fall. When the motor reaches full speed, the counter emf reaches its maximum and, hence, the net applied voltage its minimum. Accordingly, the line current, too, reaches its minimum value.

From our discussion of induced voltage (Chapter 8, subdivision A) we learned that the induced voltage depends upon the strength of the magnetic field, the number of conductors cutting that field, and the speed at which they cut it. Accordingly, in the motor,

Counter emf = strength of field × speed of rotation × K,

where K represents the number of conductors and certain other factors that are constant for any given motor. From this we can say that the counter emf is directly proportional to the speed of rotation and to the strength of the magnetic field between the field poles. Keep in mind, however, that the line voltage is always somewhat larger than the counter emf since a certain amount of current must always be made to flow into the armature if the motor is to rotate.

From the previous equation we may evolve the following:

$$\text{Speed of rotation} = \frac{\text{counter emf}}{\text{strength of field} \times K}.$$

Thus the speed of rotation is **inversely** proportional to the strength of the field. Increasing the field strength decreases the speed.

Since the motor acts as a generator, it also suffers from **armature reaction.** This, you will recall (see Chapter 13, subdivision B), is the result of the interaction between the magnetic field around the armature and that produced by the field coils. The effect is to shift the neutral running plane of the generator slightly forward in the direction of

General Electric Company

Fig. 21-6. A. D-C motor, showing commutator and brushes.
  B. Stator, showing interpoles between regular field poles.

rotation (see Figure 13-14). In the motor, on the other hand, armature reaction shifts the neutral running plane slightly backwards. If the brushes are placed in this neutral running plane, sparking will be reduced.

However, since the load on a motor rarely is constant, the armature current and, hence, the armature reaction, varies widely. This requires constant shifting of the brushes if excessive sparking is to be avoided. To avoid this nuisance, **commutating poles,** or **interpoles,** are employed as in the d-c generator (see Figure 13-15). Unlike the generator, however, in the motor these commutating poles are wound so that their polarities are the same as those of the field poles they **follow** in the direction of rotation.

1. Types of Direct-current Motors

When looking at the characteristics of a motor, we seek to answer two main questions. How does the torque of the motor behave under variations of load? How does the speed behave under such variations? The behavior of the motor determines whether a particular motor is suitable for a particular job.

To aid us in finding an answer to these questions, keep in mind the following relationships.

> 1. The torque is directly proportional to the current flowing through the armature and to the strength of the magnetic field in which the armature rotates.
>
> 2. The speed is inversely proportional to the strength of the magnetic field.
>
> 3. The counter emf is directly proportional to the speed and the strength of the magnetic field.
>
> 4. In any motor, an increase in load tends to slow it down. A decrease in load tends to speed it up.

Like d-c generators, d-c motors can be placed into three general classes, depending upon the manner in which their field windings are connected into the circuit. These are the **series motors, shunt motors,** and **compound motors.**

a. The series motor

In this motor the field coil is connected in series with the armature winding, as illustrated in Figure 21-7. Because they are in series, the armature current also flows through the field coil. The coil, therefore,

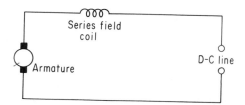

Fig. 21-7. Circuit of series motor.

is made of heavy wire to safely carry the heavy armature current, and it has relatively few turns so as not to introduce too much resistance. Since the current flowing through it is high, the coil, in spite of its few turns, can set up a powerful magnetic field.

When a series motor is started, because there is no counter emf, a large current flows through the armature and field windings. This produces a large starting torque. As the motor speeds up, however, the counter emf developed cuts down the armature and field current. As a result, the torque drops somewhat.

Should the load be increased while the motor is running, the tendency is to slow down the motor. Since the counter emf depends upon the speed of rotation, it, too, drops. As a result, the armature and field current increases and the torque rises. Should the load be lessened, the motor tends to speed up, the counter emf increases, the armature and field current decreases, and the torque drops.

Should the load be removed from the series motor while it is running, a dangerous situation arises. The removal of the load causes the motor to speed up. The counter emf rises sharply and the armature and field current drops to a very low value. Because of the small current flowing through the field coil, the strength of the field drops sharply. But we have seen that the speed of a motor is inversely proportional to the strength of the field. Hence, as the field strength drops, the motor speeds up. This process is cumulative and may cause the armature to rotate at speeds high enough to make it fly apart by centrifugal force.

It is for this reason that the series motor must always be connected directly to the load. Belts and other connecting devices that may slip or break are not used. Sometimes special centrifugal switches or other overspeed devices are attached to the armature of the motor to disconnect it from the line should the load be removed accidentally and the speed rise above safe limits. This danger is not so great for very small series motors where friction and other losses may slow the motor down before it can damage itself.

The high starting torque of the series motor makes it valuable in applications where the inertia of a heavy load must be overcome. Such applications include streetcars, electric locomotives, cranes, and the like.

b. The shunt motor

In the shunt motor the field coil is connected in parallel, or shunt, with the armature winding, as illustrated in Figure 21-8. In order that as much of the line current as possible should flow through the armature, the shunt field has a fairly high resistance. Accordingly, it is wound with many turns of relatively fine wire. Although little current flows through it, the proper field strength is maintained because of the many turn s of the coil.

When a shunt motor is started, because of the lack of counter emf, a heavy line current flows through the armature, just as in the series motor. However, since the shunt field has a high resistance, comparatively little line current flows through it. Hence the starting torque of a shunt motor is less than that of a comparable series motor.

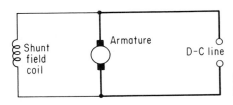

Fig. 21-8. Circuit of shunt motor. The heavy lines indicate the path of armature current.

Since comparatively little of the line current passes through the shunt coil, the current flowing through this coil varies but slightly with variations in the line current. Hence the strength of the field remains fairly constant under all conditions of load. As the load is increased, the motor tends to slow down. The counter emf is thus reduced and a greater current flows through the armature, increasing the torque and raising the speed to normal. If the load is decreased, the motor tends to speed up, the counter emf is increased, the armature current is decreased, and the torque is reduced, causing the speed to drop back to normal. Thus the speed of rotation tends to remain constant under all conditions of load.

If the load is removed from a series motor, it tends to run away with itself. In the shunt motor, however, since the field remains fairly constant, the faster the armature rotates, the greater is the counter emf developed. Hence the motor will speed up until the counter emf becomes equal to the line voltage. At that point, since the two cancel out, the armature current drops to practically zero and the motor cannot speed up any more. The speed of the motor at this point is only slightly greater than its normal running speed.

Should the field winding of a shunt motor become open, the field, and hence the counter emf, would drop to zero. As a result, the armature current and speed would rise sharply, and the motor might be damaged. As a precautionary measure, therefore, devices such as overload circuit breaker and fuses usually are connected in the line to open when the current becomes excessive.

Shunt motors have fair starting and operating torques. Their chief advantage is a fairly constant speed under varying conditions of load Hence they are used to operate machine tools, blowers, and the like, where a constant speed is desired.

c. The compound motor

In this motor two field coils are employed. One is wound with few turns of heavy wire and is in series with the armature winding. The other is wound with many turns of fine wire and is in shunt with the armature winding. (See Figure 21-9.)

Like the compound generator discussed in Chapter 13, subdivision B, 1, a, (3), the compound motor has both the series and shunt coils wound upon the same field poles. These windings may be such as to aid each other **(cumulative compound)** or they may be wound to oppose each other **(differential compound)**, as illustrated in Figure 13-25.

Cumulative compound motor. This motor acts very much like a series motor. Since the two fields aid each other, it has a very large starting torque. However, since it has a shunt field coil, it cannot run away if the load is reduced. This represents a great improvement over the series motor and, as a result, the cumulative compound motor is used where loads may vary suddenly from no-load to heavy overload.

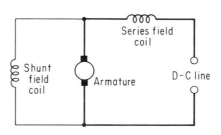

Fig. 21-9. Circuit of compound motor.

Such conditions occur, for example, in punch presses, rolling mills, and elevators.

DIFFERENTIAL COMPOUND MOTOR. This motor acts very much like a shunt motor. Since the two fields oppose each other, the resultant field is less than if only the series field were present. Hence the starting torque of this motor is considerably less than that of a comparable series motor. However, because of the opposition of the two fields, the resultant field is decreased as load is applied to the motor. Because of this decreased field, the motor is speeded up. As a result, the differential compound motor produces a very constant speed under varying conditions of load. Nevertheless, because this motor offers very little advantage over the simpler shunt motor, it is little used.

B. ALTERNATING-CURRENT MOTORS

For heavy-duty work, as in industrial plants, large motors, receiving their power from three-phase mains, generally are employed. For household purposes smaller motors, generally of fractional horsepower, receive their power from single-phase mains. Even where three-phase mains are present, a single-phase line may be obtained by connecting to two of the three lines (three-wire, three-phase system) or by connecting between the neutral and one of the lines (four-wire, three-phase system) as described in Chapter 15, subdivision B, 2.

Thus, alternating-current motors may be placed into two general categories — **polyphase motors,** those operating on polyphase current, and **single-phase motors,** those operating on single-phase current. Polyphase motors, in turn, may be subdivided into two other categories — the **induction motor** and the **synchronous motor.**

1. Polyphase Motors

a. The induction motor

The induction motor is the most commonly used a-c motor. Imagine a transformer with the primary winding stationary, but with the secondary winding wound upon a separate core that is free to rotate between a set of bearings. Further, consider the primary wound on two opposite poles of an iron ring, as shown in Figure 21-10. Here the secondary winding consists of a single closed loop wound on a spherical iron core that is pivoted between the two poles. Let us call the fixed iron

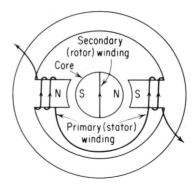

Fig. 21-10. Simplified version of induction motor.

ring, the poles, and the ring's primary winding, the **stator;** and the iron core and its secondary winding, the **rotor.**

As current flows through the primary winding (as indicated in the illustration) north and south poles appear at opposite poles of the stator and a magnetic field is produced between them. The closed loop of the secondary winding cuts across the magnetic field and, as a result, current is induced in the loop. From Lenz's law you know that the direction of current induced in the loop is such that magnetic poles opposite to those of the stator will appear in the rotor core. That is, a **south** pole will appear at the portion of rotor that is opposite to the **north** pole of the stator, and vice versa. As a result, the poles of the rotor will be attracted to the poles of the stator.

As the alternating current flows through the stator winding, its poles will alternate in step with the alternations of the current. But the poles of the rotor will change too, and will continue to be attracted to the poles of the stator. How, then, can we make the rotor rotate? If we could make the stator poles rotate, or, what amounts to the same thing, if we could make the magnetic field rotate, then the rotor would rotate with it. The problem, of course, is to obtain the rotating magnetic field.

Assume that we have two identical a-c generators, one starting up a quarter-revolution (90°) behind the other. The graph showing the current output from each generator appears in Figure 21-11. We call currents that bear this relationship to each other, **two-phase** alternating current. (You will recall that we may obtain such two-phase output from a single two-phase generator, as discussed in Chapter 13, subdivision A, 3. For the sake of simplicity, however, we shall consider the current as being obtained from two generators, as described above.)

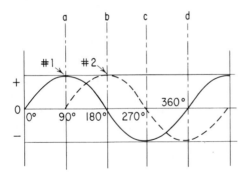

Fig. 21-11. Graph showing what is meant by a two-phase alternating current.

Now, suppose that we have four poles set in a stator ring, as in Figure 21-12. Assume that one winding, which is connected to the first generator (#1), is wound on two opposite poles, as shown. Another winding, which is connected to the second generator (#2), is wound on the other two poles.

The output from generator #1 is at its positive maximum at Point **a** on the graph in Figure 21-11. The current flowing through winding #1 produces N and S poles, as indicated in Figure 21-12A. At that instant there is no current from generator #2 and, consequently, there are no magnetic poles for winding #2.

The current from generator #1 has fallen to zero at Point **b** of the graph (90° later), and there are no magnetic poles for winding #1 (Figure 21-12B). However, the current output from generator #2 has risen to its positive maximum and winding #2 now has N and S poles as indicated.

At Point **c** of the graph current #2 has fallen to zero and current #1 is at its negative maximum. Accordingly, winding #2 has no magnetic poles (Figure 21-12C) and winding #1 has N and S poles again. However, because the current is flowing in the opposite direction now, the poles are in reverse position from those in Figure 21-12A.

At Point **d** of the graph current #1 has fallen to zero and current #2 is at its negative maximum. Winding #1 now has no magnetic poles (Figure 21-12D), and the poles of winding #2 are reversed from Figure 21-12B.

Look closely at the four illustrations in Figure 21-12. You will note that the north and south poles have, in effect, rotated in a clockwise direction. Or, what is the same thing, **the magnetic field has rotated as the currents have proceeded through their alternations.** The magnetic field has made one revolution during one cycle of the current.

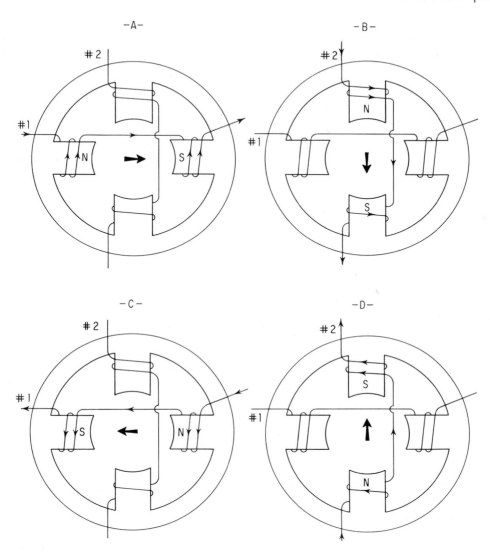

Fig. 21-12. Diagram showing how a two-phase alternating current is used to obtain a rotating magnetic field. The heavy arrows indicate the direction of the magnetic field at each instant of the cycle.

This, then, is how we obtain a rotating magnetic field. We wind the stator as illustrated in Figure 21-12 and connect its windings to a source of two-phase alternating current. As the magnetic field of the stator rotates, so does the rotor.

Note, in Figure 21-12, there are four stator poles. However, at any instant only two of these are active. Actually, when we talk about the number of poles of such a motor we mean the **number of poles per phase.** Since we are dealing here with two-phase current, the four poles represent two poles per phase and, hence, we have a **two-pole motor.**

The speed of rotation of the magnetic field depends upon the number of current reversals per minute divided by the number of poles per phase. If we are using 60-Hz alternating current, there are 120 current reversals per second or 7200 per minute. If, as in our illustration, there are two poles per phase, the speed of rotation of the field may be determined as follows:

$$\text{Speed} = \frac{\text{frequency of the current} \times \text{number of reversals per second}}{\text{number of poles per phase}}$$

$$= \frac{60 \times 120}{2} = 3600 \text{ rotations per minute.}$$

The rotor then should also rotate at 3600 revolutions per minute. But if the rotor were to turn at the same rate of rotation as the magnetic field, it would not cut across any lines of force and, accordingly, no current would be induced in it. This condition is approached when there is no load on the motor.

But as the motor is loaded, the rotor tends to lag behind the rotating field. Now it cuts across lines of force and a current is induced in it. This current produces north and south poles on the rotor and thus there is a turning effect, or torque. Therefore, the greater the load on the motor, the more the rotor lags behind the rotating field. More current is induced in the rotor and the torque of the motor is increased. In a motor running at full load, the rotor rotates at a speed of about 5 to 20 percent slower than the magnetic field, depending upon the design of the individual motor. The difference between the rotor speed and the field speed is called **slip.**

In most commercial installations **three-phase alternating current** is used. The effect is the same as if three generators were employed, operating 120° apart. Accordingly, the motor must be three-phase, the rotating field being obtained by three stator windings wound on three sets of poles, 120° apart. These windings usually are placed in slots along the inner surface of the stator frame. Thus the effect of "poles" is obtained without the use of protruding pole pieces. The result is a more compact motor with a reduced air gap between the stator and rotor.

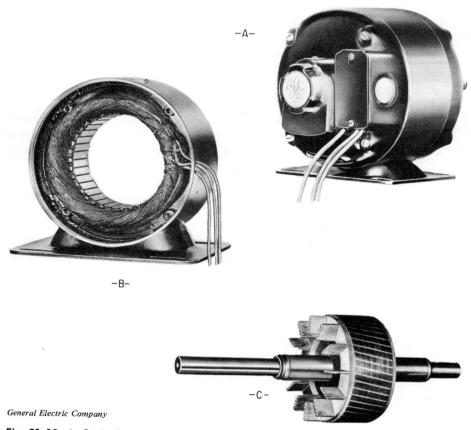

General Electric Company

Fig. 21-13. A. Squirrel-cage polyphase induction motor.
              B. Stator.
              C. Squirrel-cage rotor. Note cooling fins.

The simplest rotor employed in induction motors is the **squirrel-cage** type. Its core consists of a laminated iron cylinder. Instead of wires, copper bars are inserted into slots in the surface of the core. The ends of these bars are joined together, thus forming a series of closed loops arranged in a sort of squirrel cage; hence its name. The magnetic field set up by the stator cuts across these closed loops and large currents are induced in them. As a result of these induced currents, the rotor becomes a magnet which is rotated by the rotating magnetic field of the stator.

In the **wound rotor** a drum-type laminated iron core, very much like the one illustrated in Figure 13-12, is employed. The wire windings are placed in the slots on the surface of the core and are grouped to form a Y-connected or delta-connected three-phase winding (see Figure 13-7). The three ends of the winding are connected to three slip rings, whose sole purpose is to permit the insertion of resistance into the rotor winding for starting. When the rotor reaches its operating speed, an automatic switch cuts out the resistance and short-circuits the slip rings. The rotor then runs as a squirrel-cage type.

The characteristics of the induction motor are very much like those of the d-c shunt motor. It has a fair starting torque and a fairly constant speed with variations in load. In the wound-rotor type the insertion of resistance in the rotor winding at the start helps increase its starting torque.

b. The synchronous motor

We have seen how a polyphase alternating current can produce a rotating field in the stator of a motor. The speed of rotation of this field depends upon the number of poles and the frequency of the current. In the induction motor the current necessary to produce torque is induced in the rotor only when it rotates at a speed slower than that of the rotating magnetic field and hence cuts across lines of force. The rotor therefore must revolve more slowly than the field, especially when the motor is heavily loaded.

But if we made the rotor an electromagnet by a means other than the induced current, as, for example, by connecting it to a direct-current source, there would be no need for the rotor to slip behind the rotating field. If the rotor then were brought up to the same speed as that of the rotating field by some external means, its poles would lock in step with this rotating field, and the rotor would rotate at the same speed, regardless of its load. We say that the rotor is then **synchronized** (that is, **in time**) with the rotating field, hence the name **synchronous motor.**

The alternating-current generator, shown in simplified form in Figure 13-3, can be modified to form such a synchronous motor. The stator is rewound and connected to a source of polyphase alternating current to produce a rotating magnetic field. The rotor is made an electromagnet by connecting it to a d-c source supplied through the slip rings. An induction motor may be attached to the rotor to bring it up to the required speed for starting.

Modern synchronous motors, however, do not need external starting motors. Instead, the rotor is constructed as a combination squirrel-cage and wound type with definite poles. At the start, the direct current is not applied to the rotor. When the polyphase alternating current is applied to the stator, the motor starts running as an ordinary squirrel-cage induction motor. When the speed of the rotor reaches about 95 percent of the speed of rotation of the stator field, the direct current is applied to the wound portion of the rotor through the slip rings, thus setting up north and south poles. These poles lock in with the poles of the rotating field and, hence, the rotor revolves at the same speed as the field, that is, in synchronization.

Because the bars of the squirrel-cage portion of the rotor now rotate at the same speed as the rotating stator field, these bars do not cut any lines of force and, therefore, have no induced current in them. Hence the squirrel-cage portion of the rotor is, in effect, removed from the operation of the motor.

The speed of rotation of a synchronous motor depends upon the frequency of the current supplied to it and the number of poles per phase for which the stator is wound. We may apply the following formula:

$$\text{Speed (rpm)} = \frac{\text{frequency (Hz)} \times 120}{\text{number of poles per phase}}.$$

The greater the number of poles per phase, the lower will be the speed of rotation. Commercial motors generally are constructed with the number of their poles ranging from four to about 100.

The starting characteristics of the synchronous motor using the squirrel-cage rotor for starting are the same, of course, as those of a comparable squirrel-cage induction motor. The chief advantage of the synchronous motor is its constant speed under varying conditions of load. Its constant-speed characteristics make it suitable for operating blowers, air compressors, centrifugal pumps, and so forth, and for turning d-c generators.

Since the speed of a given synchronous motor is determined by the frequency of the alternating current applied to it, and since most power companies carefully maintain their current at a constant frequency, you can see that the synchronous motor can be used to operate an electric clock. The synchronous motor, you will recall, requires an external source of direct current for its rotor. Such a source is not available in the average home. But if the load upon the motor is very small, we may use a permanent magnet for the rotor, instead of an electromagnet, thereby eliminating the necessity for a d-c source.

Electric Machinery Mfg. Co.

Fig. 21-14. Synchronous motor, 700 horsepower, 180 rpm.

The rotor of the clock motor may consist of a steel disk mounted between two field poles. The magnetic field between the poles magnetizes the disk and, if the rotor is started rotating at a speed approximating the alternations of the magnetic poles of the stator, the poles of the rotor will lock in with the poles of the stator and the rotor will continue to rotate in step with the stator's alternations. Since synchronous motors, like induction motors, are not self-starting on a single-phase line, some electric clocks are fitted with an extension of the rotor shaft, which may be spun by hand to bring the rotor to the required speed. Most electric clocks, however, are made self-starting by use of the shaded-pole device (which will be described a little later).

The motor turns a gear wheel, which, in turn, engages a train of gears similar to that of an ordinary spring-motor clock. However, because the speed of the motor is constant, we need neither balance wheel nor escapement movement. Hence, the electric clock is much simpler than its spring-motor counterpart.

There is another use for the synchronous motor. In our discussion of power factor (Chapter 9, subdivision C, 1) you learned that if there is a phase difference between the current and voltage in a line, the true power is less than the apparent power and the power factor is less than 1. Also, if the current lags behind the voltage, the power factor is a **lagging** one. If the current leads the voltage, the power factor is **leading.** Thus, inserting an inductance in the line produces a lagging power factor. A capacitor in the line produces a leading power factor.

If we connect an induction motor to the a-c line and measure the voltage and current by suitable meters, we obtain the apparent power consumed by the motor. But if we use a wattmeter, we find that the true power is less than the apparent power. Thus the motor has introduced a component that reduces the power factor. This is called the **reactive component** and, in the case of the induction motor, it produces a lagging power factor.

The presence of this reactive component requires the power company to maintain oversize generators, transformers, and transmission lines. Accordingly, a penalty is charged the consumer who operates equipment that reduces the power factor and, the greater the reduction, the greater is the penalty. Obviously, it is to the advantage of the consumer to keep the power factor of his line as close to unity as possible.

It is here that the synchronous motor comes to the rescue. You will recall that the rotor field is excited by the use of direct current. Increasing this field strength has no effect upon the speed of the rotor, but it does change the phase angle between the voltage and current of the applied alternating current. At a certain field current, the voltage and current are in phase and the power factor is at unity. This is called the **normal** field current. Below the normal field current (the motor is said to be **underexcited**) the power factor is less than unity and lagging. Above the normal field current (the motor is **overexcited**) the power factor is less than unity and leading.

So you see that by operating an overexcited synchronous motor on a line whose power factor is lagging, we can compensate for the lag. The field current is adjusted until the leading power factor produced by the overexcited motor is about equal to the lagging power factor of the line. The power factor then approaches unity.

In some installations where a number of induction motors are employed or where the lagging power factor of the line is caused for other reasons, an overexcited synchronous motor that carries no load is used merely for the purpose of correcting the power factor. Such a motor then is called a **synchronous capacitor** since it has the same effect on the line as a capacitor would.

2. Single-phase Motors

Suppose a single-phase alternating current is applied to the stator winding of the induction motor (see the basic circuit of such a motor, as illustrated in Figure 21-10). As the current flows through the winding, the polarity of the field changes in step with the alternations of the current. But so does the polarity of the rotor. Hence, although all the magnetic polarities reverse with every alternation of the current, the poles of the rotor always face unlike poles of the stator. As a result, the rotor does not revolve.

If the rotor is now spun so that its rotational speed is brought close to that of the reversals of the stator field, the rotor will lock in with the reversals of the stator field and will continue to rotate in step with these reversals. If 60-Hz alternating current is applied, there are 3600 reversals per minute and the rotor will rotate at this speed.

As a matter of fact, such a motor is used to operate an electric clock. To start the motor, a rod attached to the rotor is spun by hand. While this can be done with a very small motor, obviously, it is not feasible for larger motors. Nor can it be done for a motor located at some inaccessible spot.

Any means that will bring the rotor of the motor close to its full running speed can be used to start the motor. Once the rotor is rotating at this speed, it will continue to rotate even though single-phase current is flowing through the stator winding. For example, if one leg of a three-phase line is opened, the remaining legs furnish a single-phase current. It has been found that if a three-phase motor is brought up to speed and one leg of the line should open, the motor will continue running on the resulting single-phase line.

There are three widely used methods for starting an induction motor operating from a single-phase a-c line: (1) the split-phase method, (2) the repulsion-start method, and (3) the shaded-pole method.

a. Split-phase induction motors

The split-phase induction motor has two stator windings displaced 90° from each other, just as in the two-phase induction motor described previously (see Figure 21-12). One of these windings, called the **main**

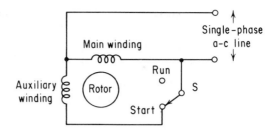

Fig. 21-15. Circuit of the inductively-split, split-phase motor.

winding, is wound with many turns of wire. The other, called the **auxiliary,** or **starting, winding,** is wound with many turns of finer wire. The rotor may be the ordinary squirrel-cage type. (See Figure 21-15.)

With the switch (S) in the START position, the single-phase a-c from the power mains is made to flow through two parallel paths, one containing the main winding and the other the auxiliary winding.

Both windings are inductive. Hence they both cause the currents flowing through them to lag behind the current of the power mains. However, since the auxiliary winding has a higher resistance than the main winding, the current through the auxiliary winding will lag less than the current through the main winding (see Chapter 11, subdivision A). Accordingly, there is a phase difference between the currents flowing through the two windings, thus setting up a rotating magnetic field which causes the rotor to revolve.

Since the auxiliary winding is wound with fine wire, it is not designed for continuous operation. Accordingly, when the rotor reaches about three-quarters of its full running speed, a centrifugal switch (S), mounted on the rotor shaft, is thrown to the RUN position, thus cutting the auxiliary winding out of the circuit. The motor continues to run as a single-phase induction motor. A motor of this type is called an **inductively split, split-phase induction motor.**

A variation of this type of motor is the **resistance-start, split-phase induction motor** whose circuit is illustrated in Figure 21-16. Here a resistor (R) is inserted in series with the auxiliary winding to increase

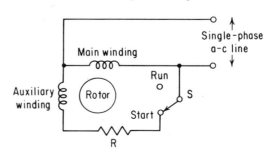

Fig. 21-16. Circuit of the resistance-start, split-phase motor.

the resistance of that path. As a result, the phase difference between the currents in the two paths is increased. As in the case of the inductively split, split-phase motor, when the motor reaches about three-quarters of its full running speed, the centrifugal switch cuts both the auxiliary winding and the resistor out of the circuit. Again, the motor continues to run as a single-phase induction motor.

Another variation is the **capacitor-start, split-phase induction motor** whose circuit is illustrated in Figure 21-17. Here a capacitor (C) is placed in series with the auxiliary winding. The purpose of this capacitor is to reduce still further the phase difference between current flowing in that branch of the circuit and the line current. (See Chapter 11, subdivision C, 1.) The phase difference between the two branches is thus further increased, thereby increasing the starting torque of the motor.

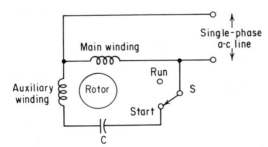

Fig. 21-17. Circuit of the capacitor-start, split-phase motor.

As in the other types of split-phase motors, when the rotor attains about three-quarters of its full running speed, the centrifugal switch (S) moves to the RUN position, thus cutting the auxiliary winding and the capacitor out of the circuit. The motor then continues to run as a single-phase induction motor.

As is true of all induction motors, the split-phase type operates at a fairly constant speed under variations in load. But the split-phase method does not produce a true two-phase relationship between currents in the two windings; it merely approximates the two-phase relationship. Accordingly, the starting torque of the split-phase motor is not as great as for a comparable two-phase squirrel-cage induction motor. In this respect the capacitor-start motor has a better starting torque than the other split-phase motors.

Split-phase motors generally are built to supply fractional horsepower, although some have been constructed somewhat larger. They are

Westinghouse Electric Corp.

Fig. 21-18. Cutaway view of a single-phase, capacitor-start induction motor. Note the centrifugal switch mounted on the rotor shaft.

used in such devices as household appliances, fans, oil burners, small power tools, and light machinery.

b. The repulsion-start motor

The **repulsion-start induction motor** is a single-phase motor whose starting principle is different from that employed by the split-phase types. Its stator resembles that of the ordinary d-c motor. Its rotor, too, resembles the armature of the d-c motor — drum-type core and winding, commutator, and brushes. However, the brushes do not connect to the line; they connect to each other. (See Figure 21-19.)

As current flows through the stator winding, a current is induced in the rotor winding. Accordingly, magnetic poles are set up on the rotor, very much as poles are set up on the rotor of the induction motor. By adjusting the positions of the brushes on the commutator, we may move the magnetic poles of the rotor to positions slightly out of line with similar poles of the stator. The repulsion between similar poles causes the rotor to turn, thus starting the motor.

When the rotor reaches about three-quarters of its full running speed, a centrifugal device mounted on the rotor shaft completely short-

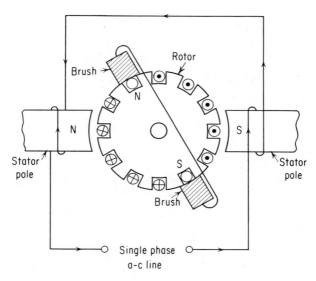

Fig. 21-19. Circuit of a repulsion-start induction motor.

circuits the commutator. The rotor then resembles the squirrel-cage type and the motor continues to run as a straight induction type. In some motors the centrifugal device also lifts the brushes from the commutator to reduce wear.

Motors of this type have fairly good starting torques. They are employed to operate devices such as refrigerators, pumps, compressors, and the like.

c. The shaded-pole motor

Perhaps the simplest method of starting the induction motor on a single-phase line is by means of the device whose principle of operation is illustrated in Figure 21-20. The rotor is the ordinary squirrel-cage type. The stator, however, is similar to the field structure of the d-c motor. A slit is cut in the face of each stator pole, making it two-pronged. A single closed loop of wire, called a **shading coil,** is placed over one of these prongs.

Now assume that the current flows through the stator winding in such a way that the pole of our illustration becomes a north pole. We have indicated the magnetic strength of this pole by means of the lines of force. Where these lines of force are numerous, the magnetic pole is strong. Where they are few, the pole is weak.

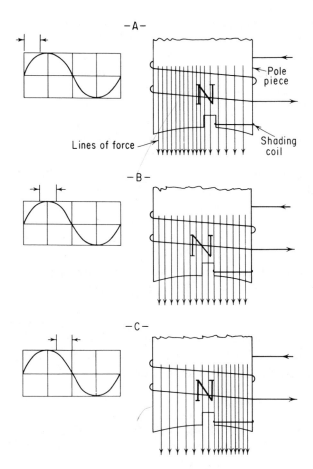

Fig. 21-20. Diagram illustrating the action of the shading coil.

Start with the portion of the alternating-current cycle indicated between the arrows in Figure 21-20A. The current is ascending rapidly and the strength of the magnetic pole consequently is increasing. A current is induced in the shading coil, and, because of this current, the coil is surrounded by a magnetic field. However, according to Lenz's law, this field will oppose the field that creates it. It tends to weaken the magnetic field of the portion of the stator pole that is enclosed by the shading coil. The effect, then, is that the strongest portion of the magnetic field appears at the left side of the stator pole.

In Figure 21-20B we advance to the portion of the a-c cycle indicated by the arrows. Now the rate of current change is slow. The induced current in the shading coil is low and so is the strength of its opposing magnetic field. Accordingly, the magnetic field is distributed more uniformly over the entire stator pole.

In Figure 21-20C we consider the next portion of the cycle, as indicated by the arrows. Now the current is decreasing rapidly, and the magnetic field around the stator pole is decreasing likewise. However, the induced current in the shading coil opposes the collapse of the original current. The effect, then, of the current flow in the shading coil is to create a strong north pole in the portion of the stator pole enclosed by the coil. The strongest portion of the magnetic field now appears at the right side of the stator pole.

If you examine the three parts of Figure 21-20, you will notice that the overall effect achieved here is as if the north pole had moved from left to right across the face of the stator pole. A similar action in the opposite direction takes place at the south pole of the stator. Hence, we have created a rotating magnetic field that can be used to start our motor.

Motors that start in this way are called **shaded-pole motors.** Because the current and force in the shading poles are small, single-phase induction motors of this type are used only where the load on the motor is light, as in small fans and similar devices.

d. The universal motor

What would happen if single-phase alternating current were to be applied to the d-c motors we have previously discussed? Theoretically, they should run, since the poles of the field coil and the armature would reverse in step as the current reverses its flow. Two south poles repel each other just as effectively as two north poles.

However, we would have difficulty with the shunt-wound motor. In this type of motor, the field coil consists of many turns of wire and the armature coil of relatively few turns. Consequently, the self-inductance of each of these coils is quite different. This produces a phase difference in the counter emf of each and, as a result, the currents flowing through these windings are retarded in different degrees. Thus, the currents flowing through them get "out of step" and the motor action soon ceases.

In the series-wound motor, however, both the field and the armature windings are in series and the same current flows through each. Thus the change of polarity in the field and the change in the armature windings are "in step." This type of motor will run on either direct or alternating current and it is called a **universal** motor.

In such a motor, all iron portions are laminated to reduce eddy currents and fewer turns are used in the windings so that their impedances will be low enough for the flow of sufficient current. The motor tends to run faster on direct current than on alternating current because the inductive reactance, which is present when alternating current is applied, reduces the line current.

One drawback is the considerable sparking that takes place between the brushes and commutator when this motor is used on alternating current. When we discussed the commutator in the section on d-c generators, you may recall learning that the brushes were set at the neutral plane. At this plane the brushes made contact with the commutator bars connected to the ends of the armature coil that, at the moment, was not cutting across the magnetic field. In effect, the brushes were short-circuiting this coil, but since there was no induced voltage in the coil at that particular moment, there was no current, and hence no sparking.

When the motor is used on alternating current, however, there are changing magnetic fields around every coil carrying a current. The short-circuited coil now finds itself in the fluctuating fields set up by the other windings of the motor. As a result, a heavy induced current flows through the short-circuited coil, producing considerable sparking at the brushes.

This is one of the reasons why the ordinary universal motor is not used for heavy-duty purposes where large currents are involved. It is widely used, however, for certain light-duty purposes such as operating vacuum cleaners, fans, and the like.

C. MOTOR CONTROL

There are a number of factors to be considered when discussing motor control. First of all, we must consider the starting and stopping of the motor. We must consider, too, certain protective devices that safeguard the motor from damage while it is operating. In some applications reversal of the direction of rotation is necessary. Finally, speed control generally is desirable.

1. D-C Motor Control

a. Series motor

When the motor is first started, the armature is at rest. Hence there is no counter emf developed and the full line voltage is applied. In very small motors the field and armature windings have fairly high resistances. The current, then, is low and thus a simple switch may be employed to start such motors.

In larger motors, however, the resistances of the armature and of the series field are quite small. A very heavy current will therefore flow through them, which may burn out the windings. Accordingly we need some protective device, such as a series resistor, to drop the voltage at the start to a safe level. After the motor has gotten up speed and a sufficient counter emf is being generated, the resistor is cut out.

There are a number of different types of starting devices. Simplest is the **manual starter box** illustrated in Figure 21-21. To start the motor, the line switch is closed and the arm of the starter is moved to contact 1. This places the entire resistance of the box in series with the armature and field windings of the motor and thus the starting current is held to a safe value. As the motor speeds up, the arm is moved progressively to the other contacts. Thus resistance is cut out gradually as the motor gains speed. When it has reached its full running speed, the arm is moved to contact 5 and now all the resistance is removed from the circuit. The holding magnet, which is in series in the line, holds the arm in place. To stop the motor, the line switch is opened, the holding magnet loses its magnetism, and the spring snaps the arm back to its OFF position.

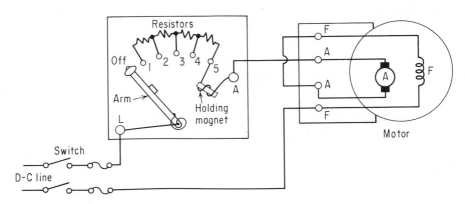

Fig. 21-21. Starter box for series motor.

The spring and holding magnet act to prevent the motor from being started without the series resistor. Otherwise, the arm might be left on contact 5 (resistance all cut out) when the line switch was opened and the motor came to rest. Then, as the switch was closed again, the full voltage would be applied to the windings of the motor, without the protection of the series resistor.

When moving the arm over contacts 1 to 4, one should be careful to hold it at each contact only long enough to permit the motor to gather up sufficient speed. Otherwise, the heavy current flowing through the resistors might damage them. On the other hand, the arm should not be moved from point to point so rapidly that a sufficient counter emf cannot be developed. Generally, a pause of about 20 seconds at each contact is sufficient.

The holding magnet, as illustrated here, is known as a **no-voltage release,** since it releases the arm if the line voltage is removed (as, for example, the line switch is opened or one of the line fuses "blows"). Some starter boxes may also be equipped with an **overload release.** This is a relay whose coil is in series with the line and whose contact points connect to the two ends of the coil of the holding magnet. The relay is **normally open;** that is, if a normal current is flowing in the line, the contact points are separated. Should the line current rise sharply (for example, as the result of a short-circuit in the motor), the contact points close, shorting out the holding magnet. The arm of the starter box flies back, opening the line.

There is another protective device that may be used with all kinds of motors. This is the **thermostatic cutout,** which is a heat-operated switch that is mounted inside the motor. (See Figure 17-5.) When the motor overheats — because of a sustained overload, for example — this switch opens the line. After the motor has cooled off sufficiently, the thermostatic switch may then be reset by pressing a push-button set in the motor frame.

A d-c motor rotates primarily because of the repulsion between the poles of the field and the similar poles of the armature. It makes no difference whether it be two south poles or two north poles that are repelling each other. Accordingly, if we reverse the polarity of **both** the field and armature, it would make no difference and the motor would continue to rotate in the same direction. In order to reverse the direction of rotation we must reverse the polarity of the field **or** the armature. Generally, the armature's polarity is reversed.

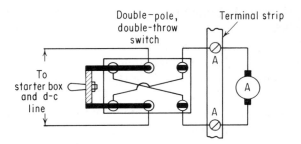

Fig. 21-22. Reversing circuit for d-c motor.

Note, in Figure 21-21, that the armature and field windings are not connected within the motor. Instead, the ends of each winding are brought out to a terminal strip at the side of the motor and connections are made there. This makes it easy to reverse the polarity of the windings. A simple double-pole, double-throw switch, wired as in Figure 21-22, may be used.

The speed of the series motor may be controlled by means of a variable resistor **(rheostat)** placed in series with it. (See Figure 21-23.) Increasing the resistance decreases the armature current and the strength of the field. Hence the torque and speed of rotation are reduced. Decreasing the resistance increases the speed. After being adjusted, the speed remains constant as long as the load, too, is kept constant. Increasing the load reduces the speed. The rheostat then must be adjusted to decrease its resistance if the speed is to be restored to normal. If the load is decreased, the motor speeds up. The rheostat then is adjusted to increase its resistance to bring the speed back. (But remember, the load must never be removed completely from a d-c series motor!)

The rheostat used for this purpose is a heavy-duty type wound

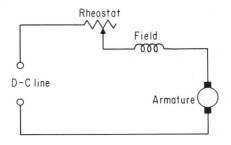

Fig. 21-23. Speed-control circuit for series motor.

with heavy resistance wire. This is because it must be able to carry the full motor current continually. In contrast, the wire need not be so heavy in the resistors of the starter box, since they carry the motor current for only short periods of time

b. Shunt motor

The starting precautions for the series motor apply equally for the shunt motor. The starter-box circuit for this type motor is shown in Figure 21-24. Note that the holding magnet is in series with the shunt field. You will recall that if this field winding should become open, the motor will speed up and may be damaged. Before this can happen, however, the holding magnet in series with the field winding loses its magnetism and releases the starter arm, opening the line, and thus stopping the motor. The holding magnet, as illustrated here, is known as a **no-field release.**

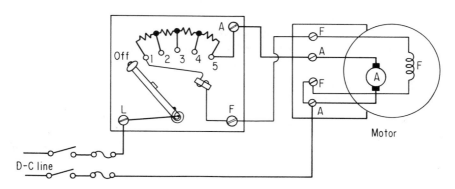

Fig. 21-24. Three-point starter box for shunt motor.

Note that when the starter arm is at the full-running position (contact 5) the entire resistance of the starter box is in series with the holding magnet and shunt field. Since the resistance of the field is high and that of the starter-box relatively low, it makes very little difference so far as field excitation is concerned. Since comparatively little current flows through the field winding (usually only about 5 percent of the total motor current), there is no danger of the resistors burning up.

The starter box illustrated in Figure 21-24 is also known as a **3-point starter box** because it contains three terminals — one to the armature winding, one to the field winding, and one to the line. Reversing the

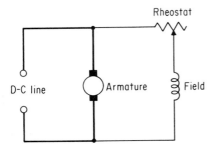

Fig. 21-25. Speed-control circuit for shunt motor.

shunt motor is the same as reversing the series motor. Either the polarity of the armature winding or the polarity of the field winding is reversed; generally, it is the armature winding. The windings of the motor are brought to a terminal strip to facilitate the connections.

The speed of the shunt motor usually is controlled by means of a rheostat placed in series with the field, as shown in Figure 21-25. Increasing the resistance of the rheostat decreases the current flowing through the field winding, thus decreasing the strength of its magnetic field. Decreasing the strength of the field, you will recall, increases the speed of the motor. If the resistance of the rheostat is decreased, the strength of the field is increased, and the speed of the motor is cut down. The rheostat used here need not be as heavy as the one used to control the speed of the series motor, since it carries only the field current which is a small portion of the full motor current.

If the speed-control rheostat is used with the 3-point starter box of

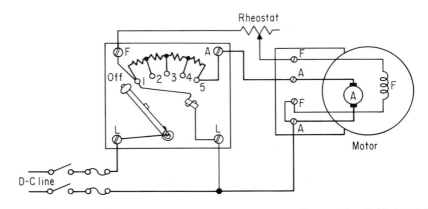

Fig. 21-26. Four-point starter box for shunt motor permitting the use of a field rheostat for speed control.

Figure 21-24, it must be connected in series with the holding magnet and the field winding. Thus, if its resistance is increased to speed up the motor, the field current may be reduced to the point where the holding magnet no longer can hold the arm. As a result, the arm flies back to its OFF position and the motor stops.

Accordingly, the **4-point** starter box illustrated in Figure 21-26 is employed. Here the holding magnet is no longer in series with the field winding, but is connected directly across the line. Thus, as long as there is sufficient line voltage, the arm will be held in place. Variations in the field current do not affect the holding magnet, which is now called a **no-voltage release.**

c. *Compound motor*

Either a 3-point or 4-point starter box may be used for the compound motor. The 3-point box is illustrated in Figure 21-27. Connections are made to the terminal strip of the motor. Note that by reversing the polarity of the series field winding it is possible to operate this motor as either a cumulative or a differential compound type.

To control the speed of the compound motor, a rheostat may be inserted in series with the shunt field, just as in the shunt motor. In such a case, the 4-point starter box should be employed. Such adjustment of the shunt field current is not too satisfactory since, as the shunt field strength is reduced, the effectiveness of the series field is increased, thus partly offsetting the change in the shunt field.

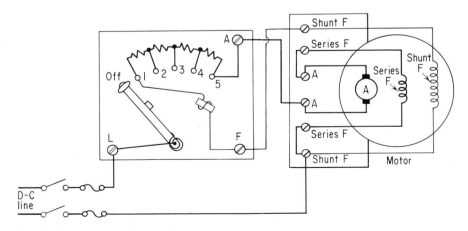

Fig. 21-27. Three-point starter for compound motor.

A rheostat may be inserted in series with the armature and series field, just as in the series motor. However, this requires a heavy duty rheostat. Some compound motors are constructed so that both series and shunt fields are used for starting. After the motor has reached its full running speed, a centrifugal device mounted on the motor shaft cuts out the series field. The motor then continues to run as a shunt motor and its speed may be controlled as such.

2. A-C Motor Control

When the three-phase squirrel-cage induction motor is at rest it is similar to a transformer whose secondary winding is short-circuited. Accordingly, when such a motor is started, a large current flows through the stator winding. Modern motors are designed to take this current surge with their stator windings specially braced to withstand the magnetic effects produced by the large starting currents. As a matter of fact, most a-c motors are started by being connected directly to the power line, except where the line is not heavy enough to pass the starting current and where certain other difficulties exist.

Where conditions do not permit full-voltage starting, some form of reduced-voltage starter is used. One method is to introduce similar resistors or inductors into each of the three-phase lines to cut down the starting voltage to the stator of the motor. After the motor reaches its full running speed, these resistors or inductors are cut out and the motor continues to run at full line voltage. Another method is to use an auto-transformer in each line that reduces the line voltage at the start. When the running speed is attained, the autotransformers are adjusted to produce the full line voltage.

Still another method is to use a switching system that changes the three-phase stator winding from a Y-connection for starting purposes to a delta-connection after the running speed is reached. In our discussion of three-phase generators (Chapter 15, subdivision B) we learned that if the armature coils are connected in Y formation, the voltage between any two of the lines is equal to 1.73 times the voltage across each armature coil. Conversely, if three windings are Y-connected across a three-phase line, the voltage across each winding is equal to the voltage between any two of the lines divided by 1.73. In other words, the voltage across the winding would be about 58 percent of the line voltage.

On the other hand, the voltage across the windings in a delta

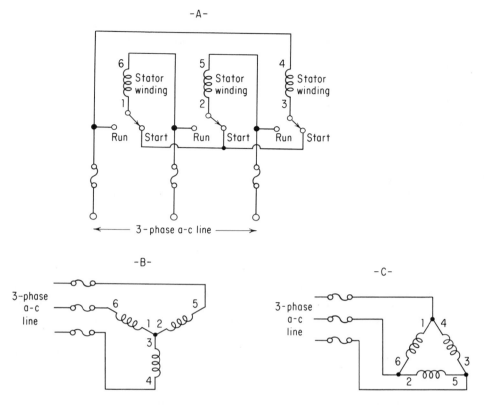

Fig. 21-28. A. Switching arrangement for changing the three-phase stator winding from a Y connection to a delta connection.
B. Y connection formed when switches are thrown to START position.
C. Delta connection when switches are thrown to RUN position.

connection is equal to the full line voltage. Hence, if the windings of the stator of the motor are connected in a Y formation at the start, only 58 percent of the line voltage is applied to the motor. When the motor reaches its running speed the stator windings are switched to a delta connection and the full line voltage now is applied.

Figure 21-28A shows the switching arrangement. When the three-pole switch is thrown to its START position, the stator windings are Y-connected (Figure 21-28B). With the switch thrown to its RUN position, they are delta-connected (Figure 21-28C).

In the wound-rotor induction motor, you will recall, the ends of the rotor windings terminate in slip rings, by means of which resistances

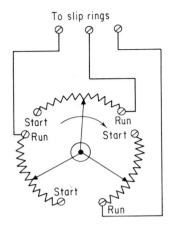

To slip rings

Fig. 21-29. Triple rheostat used for starting wound-rotor induction motor.

can be introduced for starting purposes. How this is done is illustrated in Figure 21-29. A triple rheostat is employed. The slider arms of the three sections are connected together electrically and joined so that they move in unison. One end (marked RUN) of each resistor goes to one of the three slip rings of the rotor. With the slider arms at the START position, the entire resistance of each section is inserted into the winding to which it is connected. As the motor starts rotating, the sliders are slowly moved clockwise, thus reducing the resistances. When the sliders reach the RUN position all the resistances are cut out and the slip rings are short-circuited. The motor then runs as a squirrel-cage type.

Because single-phase motors generally are small and draw relatively small currents, they are operated directly from the line. Synchronous motors that start as squirrel-cage induction motors have the same types of starting controls, of course.

It is quite simple to reverse the direction of rotation of a three-phase motor. All we need do is to interchange any two of the three line wires going to the motor. This reverses the direction of rotation of the rotor. To reverse a split-phase induction motor we must reverse the connections to either the main-field coil or the starting coil.

To reverse the repulsion-start, induction-run motor, the brushes must be shifted far enough to reverse the magnetic poles of the rotor. To reverse the shaded-pole motor, if the pole pieces of the motor can be removed, they should be rotated 180 degrees before being replaced.

Since a-c motors are essentially constant-speed devices, the disadvantage of such a motor is the lack of adequate speed control. Where variable-speed control is essential, the general practice is to change the alternating current to direct current to be used to operate d-c motors that permit such control.

Nevertheless, some speed control of the a-c motor is possible. We have seen that, for a given line frequency, the speed of the motor depends upon the number of poles. Accordingly, some motors have their stators so wound that, by suitable switching, the number of poles are halved. This doubles the speed of the motor.

To a limited extent the speed of a wound-rotor induction motor can be controlled by inserting resistances into the rotor windings. This can be done through the slip rings to which the windings are connected. Increasing these resistances decreases the speed of rotation.

Recently, another method has been developed to control the speed of an a-c motor. The speed of the motor, you will recall, varies directly as the frequency of the current. Thus, by varying the frequency of the current applied to the motor, we can vary its speed.

An electronic device, called an **adjustable-frequency inverter,** converts direct current to alternating current whose frequency may be varied between about 10 and 120 Hz. As the frequency is lowered, the motor speed is reduced; as the frequency is raised, the speed is increased.

Although this method of speed control is practical, there are some drawbacks. For one, it introduces a costly piece of apparatus. Secondly, the inverter requires a d-c source (generally obtained by rectifying the alternating current of the power mains). If such a source is available, why not use d-c motors whose speeds can be controlled more easily by simpler means?

Nevertheless, the a-c motor is simpler, more rugged, and less costly than an equivalent d-c motor. In certain applications these advantages may outweigh the disadvantages of this system.

Since speed control is not too feasible for a-c motors, variable-speed transmission systems are often employed. These generally consist of gear trains, belts with stepped pulleys, and so forth.

3. Automatic Motor Control

The various types of motor controls we have discussed are manual types. When using d-c starter boxes the operator must judge the proper time to move the arm from one contact point to the next. When starting a-c motors the operator must judge the proper time to cut the various starting devices in or out of the circuit.

Modern practice, especially where large motors are involved, tends to the use of automatic control devices. Thus, at the pushing of a button, relays and timing devices cut the necessary resistors, inductors, or whatever, into or out of the circuits at the proper times. Let us consider a simple example. You will recall that when we start a d-c motor, we cut in a resistor to limit the current flow through the armature until the counter electromotive force has had a chance to build up. We may do this manually by means of the starter box.

The illustration in Figure 21-30 shows an automatic method for doing the same thing merely by pressing the START button. This causes the line current to flow through the coil of relay #1 and, hence, contacts #1 and #2 touch. As a result, the line current flows through the series field of the motor (F), the armature (A), and the starting resistor (R). The motor starts to rotate. At the same time that contacts #1 and #2 of relay #1 close, contacts #3 and #4 close as well. This shorts out the START push button so that the motor continues to run even after the button is released.

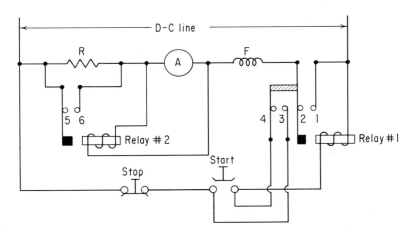

Fig. 21-30. Automatic d-c motor control circuit.

The coil of relay #2 is connected across the armature. Because its resistance is much greater than that of the armature, very little line current flows through the coil. Also, at the start of the operation the counter electromotive force developed in the armature is quite low.

Hence relay #2 remains unenergized. However, as the motor reaches its proper speed of rotation, the counter electromotive force reaches a point where the relay becomes energized. Contacts #5 and #6 close, shorting out the starting resistor, and the motor continues to run normally. To stop it, all we need do is press the STOP push button. This de-energizes relay #1, opening the line circuit.

There are many variations of such automatic controls, some of them considerably more complicated and performing other duties as well. Frequently, automatic controls are tailored to suit a particular installation and perform highly specialized tasks.

The most recent development in this field is the use of electronic control devices that change the alternating current of the line to direct current which is fed to d-c motors. Then, by controlling the voltages of the current fed to the stators and rotors of these motors, the speed of rotation is controlled. This makes for a very flexible speed-control system. For example, the motor may be set to rotate at a predetermined speed. Any variations from that speed would cause corresponding variations in the currents fed to the motor and thus it would be brought back to its original speed. Also, the motor could be speeded up or slowed down automatically to suit the specific needs of the load. Of course, these electronic devices also perform all the other functions of automatic controls (starting, protection, and so forth).

D. RATING AND EFFICIENCY OF MOTORS

Motors are rated by the amount of torque they can produce. This rating is in *horsepower*. (One horsepower = 550 foot-pounds per second. That is, one horsepower is the amount of power required to lift 550 pounds one foot per second, one pound 550 feet per second, or any combination that gives 550 foot-pounds per second.) Small motors that may be used in the home generally have an output less than one horsepower. On the other hand, the synchronous motors used to operate the pumps at the Grand-Coulee installation are rated at 65,000 horsepower each.

Horsepower output of a motor can be measured by a brake test, whereby a brake is applied to the armature shaft of a running motor and the force in pounds exerted on the brake by the shaft is measured. Horsepower input to the motor can be determined by measuring the power (in watts) consumed by the motor running with a specified load.

Then, since 746 watts are equivalent to one horsepower, we can calculate the horsepower input to the motor.

EXAMPLE. A motor operating under full load draws 20 amperes at 240 volts. What is the horsepower input to the motor?

$$P = E \times I = 240 \times 20 = 4800 \text{ watts},$$

ANSWER. Horsepower $= \dfrac{\text{watts}}{746} = \dfrac{4800}{746} = 6.4$ horsepower.

Power is always lost in a motor (as in every other type of machine). These losses fall into several categories.

Copper losses. This is the I^2R loss in the windings of the field and armature.

Hysteresis loss. This is due to the rapid magnetization and demagnetization of iron portions of the motor that find themselves in a changing magnetic field. This loss can be kept down if such iron portions (armature core, and so forth) are made of soft iron or annealed steel.

Eddy-current loss. Since these iron portions are conductors, eddy currents are induced in them by the changing magnetic field. Lamination is used to keep this loss down.

Mechanical losses. This is due to friction and includes the friction between brushes and commutator or slip rings, bearing friction, and air friction produced by rotating portions of the motor.

The power lost in the motor is transformed to heat. Accordingly, where it is feasible, the frame is left open to permit the free circulation of air. Also, fan blades usually are mounted on the rotor shaft to blow air through the motor.

Like that of any machine, the efficiency of a motor is the ratio between the power output and the power input. To obtain this efficiency in percent we may use the following formula:

$$\text{Efficiency (in percent)} = \frac{\text{power output}}{\text{power input}} \times 100.$$

EXAMPLE. A motor rated at 5 horsepower, operating under full load, was found to draw 25 amperes at 240 volts. What is its efficiency?

Power output $= 5$ horsepower $\times 746 = 3730$ watts,

Power input $= 25$ amperes $\times 240$ volts $= 6000$ watts,

ANSWER.

$$\text{Percent efficiency} = \frac{\text{power output}}{\text{power input}} = \frac{3730}{6000} \times 100 = 61.6\%.$$

The efficiency of very small motors may be less than 50 percent. On the other hand, the efficiency of large motors may be over 90 percent.

QUESTIONS

1. Explain how a current-carrying conductor in a magnetic field is made to move. Explain the right-hand motor rule for determining the direction of motion of the conductor.

2. What are the four main parts of a d-c motor? Explain the function of each part.

3. Explain the operation of the d-c motor.

4. Explain why a motor draws more current at its start than when it is running.

5. Explain how the rotational speed of the motor depends upon the strength of the magnetic field and the counter emf.

6. Explain how the torque of the motor depends upon the current flowing through the armature and the strength of the magnetic field in which the armature rotates.

7. Explain the use of commutating poles in the d-c motor. How do they differ from the commutating poles of the d-c generator?

8. Draw the schematic circuit of a d-c series motor. Label all important parts.

9. Explain the starting characteristics of the d-c series motor.

10. Explain why the load should never be removed from a running d-c series motor.

11. Draw the schematic circuit of a d-c shunt motor. Label all important parts.

12. Compare and explain the difference in starting characteristics of the d-c series and shunt motors.

13. What is the chief advantage of the shunt motor over the series motor? Give two examples where shunt motors may be used.

14. What is the chief advantage of the series motor over the shunt motor? Give two examples where series motors may be used.

15. Draw the schematic circuit of a d-c compound motor. Label all important parts.

16. Explain the advantages of the cumulative-wound compound motor. Give two examples where such a motor may be used.

17. Explain how a polyphase a-c motor obtains its rotating field.

18. Explain the basic principle of the induction motor.

19. Explain the principle of the induction motor employing a squirrel-cage rotor.

20. Upon what factors does the speed of rotation of the induction motor depend? Explain the difference between the nominal speed and operating speed of the induction motor.

21. Explain the principle of the synchronous motor. Upon what factors does its speed of rotation depend?

22. Compare the characteristics of the induction motor with those of the synchronous motor.

23. Explain how a synchronous motor may be used to correct the power factor of the a-c line.

24. Explain how the inductively split, split-phase induction motor operates.

25. Explain how the resistance-start, split-phase induction motor operates.

26. Explain how the capacitor-start, split-phase induction motor operates.

27. Explain how the repulsion-start induction motor operates.

28. Explain how the shaded-pole motor operates.

29. Explain why the universal motor is able to operate on direct or alternating current.

30. What precautions must be taken in starting large d-c motors? Explain.

31. How may the speed of a d-c shunt motor be controlled? Explain.

32. Draw the circuit of a four-point starter box for the control of a d-c shunt motor. Explain the function of each of its parts.

33. Explain how an adjustable-frequency inverter may be used to control the speed of the a-c motor.

34. List four types of power losses in a motor. Explain how these losses are reduced.

35. A motor rated at 30 horsepower, operating under full load, was found to draw 60 amperes at 600 volts. What is its efficiency? [*Ans.* 61.9 percent.]

PRACTICAL
POWER
DISTRIBUTION
SYSTEMS
VII

POWER FACTOR AND DISTRIBUTION TRANSFORMERS

22

A. POWER FACTOR

In an a-c circuit containing only resistance, the current and voltage are in phase. Hence, by measuring the voltage (E) with a voltmeter and the current (I) with an ammeter and multiplying E by I, the power, in watts (w) or kilowatts (kw) is obtained. Another method is to use a wattmeter which measures the instantaneous values of voltage and current and produces a reading of the product of these two values. This is **real** or **true power** and is capable of doing useful work.

The presence of inductance or capacitance in an a-c circuit introduces a phase difference between the voltage and current. If only inductance is present, the current is made to lag 90° behind the voltage. If only capacitance is present, the current is made to lead the voltage by 90°. (See Chapter 10, subdivisions B, 1 and C, 1.)

In both the circuit containing only inductance and the circuit containing only capacitance the true power is zero. (See Figure 9-9.) Energy alternately flows from the source to the load and back again. The only

467

power in the circuit, then, is a **reactive,** or **wattless, power** which does no useful work. This power is calculated by multiplying the voltage by the current and, because it is reactive, its unit is the **var** (volt-ampere, reactive) or **kvar** (kilovolt-ampere, reactive). Note, however, that even though this reactive power can do no useful work, it must be supplied by the source and the current flowing through the feeder lines to the load produces heat due to the I^2R loss in those lines.

Since practical circuits contain some resistance as well as inductance and capacitance, the phase difference between the voltage and current becomes some value less than 90°, depending upon the relative amounts of resistance and reactance in the circuit. (See Chapter 11, subdivisions A and B.) Hence, as shown in the graph of Figure 9-10, the power in the circuit contains two components. One is the true power due to the resistive component and the other is the reactive power due to the reactive component. The total power in the circuit, which is known as the **apparent power,** is the vectorial sum of the true power and the reactive power, and is expressed in units of **volt-amperes** (va) or **kilovolt-amperes** (kva).

Both the true power and the apparent power may be measured by the voltmeter-ammeter method. But only the true power can be measured by a wattmeter since the reactive power will not register correctly on this instrument. So if we wish to measure both the apparent power and the true power of a circuit, the voltmeter, ammeter, and wattmeter are connected as shown in Figure 22-1. The apparent power is obtained by multiplying the voltmeter (V) reading by the ammeter (A) reading. The true power is obtained from the wattmeter (W) reading. (As we shall see later, the reactive power may be found by vectorially subtracting the true power from the apparent power.)

The ratio between the true power and the apparent power is called the **power factor** of the circuit. Thus:

$$\text{power factor} = \frac{\text{true power (watts)}}{\text{apparent power (volt-amperes)}}.$$

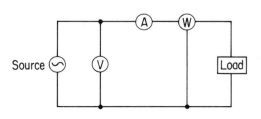

Fig. 22-1. How to measure the true and apparent power in an electrical circuit.

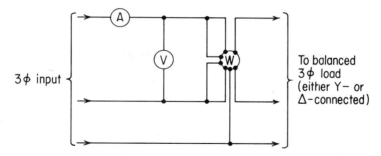

Fig. 22-2. How to measure the true and apparent power in a three-phase circuit.

In a three-phase system, the power factor of each individual phase may be obtained in a similar manner, as illustrated in Figure 22-2. (We assume here that we are dealing with a balanced system where the load is balanced so that the power in each phase is the same. In such a system the power factor of the three-phase circuit is the same as the power factor of each individual phase.)

The apparent power in a three-phase system may be obtained from the readings of the voltmeter (V) and ammeter (A). However, regardless of whether a wye or delta configuration is used, the apparent power, in volt-amperes, is equal to $1.73 \times I \times E$. The true power, in watts, is obtained from the reading of the three-phase wattmeter (W).

In circuits containing nothing but inductance or capacitance, the true power is zero. Therefore, the power factors of such circuits, too, are zero. In circuits containing nothing but resistance, the true power is equal to the apparent power. Hence the power factors of such circuits are unity, or 1.0.

In circuits containing both resistive and reactive components, both true power and reactive power are present. Hence the apparent power becomes greater than the true power and the power factor drops to some value less than 1.0. Thus the power factor of the circuit lies between zero and 1.0, depending upon the relative amounts of true and reactive power.

Although the power factor may be expressed as a decimal (e.g. 0.75), it generally is expressed as a percentage (75%). If the reactance of the circuit is inductive, the power factor is said to be **lagging.** If the reactance is capacitive, the power factor is **leading.**

The vector diagram of Figure 22-3A shows the voltage and current relationships in a circuit containing inductance and resistance. Since

−A−

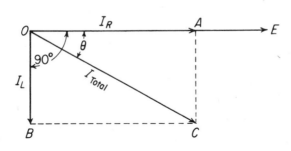

−B−

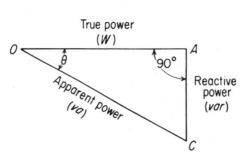

Fig. 22-3. A. Vectorial diagram showing phase relationships in a circuit containing resistive and inductive components.
B. Power triangle showing relationships between apparent power, true power, and reactive power in an a-c circuit.

industrial circuits generally employ constant-potential electrical sources, vector E is used as a reference. The resistive component of the current (I_R) is in phase with the voltage and is represented here as vector OA. The inductive component of the current (I_L) lags 90° behind the voltage and is represented here by vector OB. The resulting current (I_{Total}) is represented by vector OC. Angle θ (theta) represents the phase difference between E and I_{Total}.

Triangle OAC of this diagram can be used to illustrate the power relationships in this circuit. (See Figure 22-3B). Since OA represents the resistive component of the current, it may also be used to represent the true power (in units of watts). Side AC (which is equal and parallel to OB) represents the reactive component of the current. Hence it may also be used to represent the reactive power (in units of vars). Since OC

represents the total current, it may also be used to represent the apparent power (in units of volt-amperes). Angle θ represents the angle of lag between I_{Total} and the voltage.

Triangle OAC is a right triangle. In such a triangle the lengths of the three sides bear a definite and constant ratio to each other, depending upon the value of angle θ. These ratios are known as the trigonometric functions of the angle. The ratio between the side opposite angle θ and the hypotenuse of the triangle is known as the **sine** of angle θ (sin θ). Thus:

$$\sin \theta = \frac{AC}{OC}.$$

Similarly, the **cosine** of angle θ (cos θ) is the ratio between the side adjacent to angle θ and the hypotenuse. Thus:

$$\cos \theta = \frac{OA}{OC}.$$

In a similar manner, the **tangent** of angle θ (tan θ) is the ratio between the opposite side and the adjacent side of the angle. Thus:

$$\tan \theta = \frac{AC}{OA}.$$

(The numerical values of some of the trigonometric functions of angle θ may be found in Appendix G.)

Since the power factor is the ratio between the true power and the apparent power, the power factor, then, is equal to the cosine of angle θ. Thus, if, for example, angle θ is equal to 30°, cos θ = 0.866. Hence the power factor is 0.866, or 86.6%. Should angle θ be 45°, cos θ = 0.707 and the power factor is 70.7%.

Further, tan $\theta = AC/OA$. Since AC represents the reactive power (var) and OA represents the true power (w), if we know the true power and angle θ, we may calculate the reactive power. Thus:

Reactive power (var) = true power (w) $\times$ tan θ.

Similarly, if we know the apparent power (va) and angle θ, we may calculate the reactive power from the following formula:

reactive power (var) = apparent power (va) $\times$ sin θ.

If the reactive component of the circuit is capacitive, a similar vector diagram may be drawn except, of course, that the reactive component of the current (I_C) would be shown leading the voltage by 90°. The cosine of the angle by which the resultant current leads the voltage is the power factor of the circuit.

Most industrial circuits have lagging power factors arising from the inductive effects of devices such as induction motors and transformers. For example, circuits containing induction motors may have lagging power factors ranging from about 60% to 85%, depending upon the sizes of the motors and whether they are operating at their rated loads.

Since the reactive power, although it can do no useful work, must be supplied by the power source (such as a generator), this power source must be large enough to furnish both the reactive power and the true power which performs the useful work. Hence larger and heavier equipment and conductors are required to supply a given quantity of useful power to the load if the power factor of the circuit is low than if the power factor were closer to unity.

Thus the power company must use larger generating and distribution equipment to supply useful power to a consumer whose circuit has a low power factor. Accordingly, the company generally imposes a penalty upon such a consumer to help compensate for its greater investment.

Further, within the plant itself, current is distributed from the service entrance to the various equipments. The conductors carrying these currents must be heavy enough to safely pass, not only the true power, but the useless reactive power as well. If the power factor of the circuit is low, the ratio of reactive power to the total power is relatively large. Thus a sizable portion of the current-carrying capacity of the conductors is needed for the transport of useless current. Should the power factor of the circuit be increased, the ratio of reactive power is decreased. Thus the same conductors now can pass more useful power or, in the alternative, smaller conductors could be used to pass the same amount of useful power.

Hence it becomes both practical and economical to operate electric circuits at power factors as high as is feasible. Induction motors introduce a lagging power factor. The greater the number of poles of such a motor, the slower its speed will be. And the slower the speed, the lower is the power factor. Hence the power factor will be higher if higher-speed motors (with two or four poles) are used. If slow speeds are necessary, mechanical speed reducers may be employed.

Also, the more lightly the motor is loaded, the lower the power factor will be. Accordingly, such motors should be run at or near their full-rated loads. If necessary, the motors may be replaced by others having lower rated capacities.

There is another method for correcting a lagging power factor, using a synchronous motor. In such a motor (see Chapter 21, subdivision B, 1, b) current from a three-phase line produces a rotating field in the stator winding. The rotor winding is excited by direct current from an external source. Depending upon the amount of d-c excitation, the a-c line current can be made to lag, lead, or be in phase with the line voltage.

With a certain amount of d-c excitation (depending upon the design of the motor) the line current to the stator will be in phase with the line voltage. The motor is said to be **normally-excited** and, since the current is in phase with the voltage, the power factor of the line is unity, or 100%.

If the d-c excitation is reduced, the motor is said to be **under-excited.** Under such conditions, the line current lags behind the line voltage and the power factor of the line falls below unity. Since the current lags behind the voltage, the power factor, too, is lagging.

If the d-c excitation is raised above the normally-excited value, the motor is said to be **over-excited.** Now the line current leads the line voltage and, again, the power factor of the line falls below unity. This time, however, the power factor is leading.

Should a line have a lagging power factor due to the presence of several induction motors, one or more of the induction motors may be replaced by an over-excited synchronous motor. The leading power factor thus introduced tends to neutralize the lagging power factor, so raising the overall power factor of the line toward 100%.

Synchronous motors generally are designed so that at their rated loads they produce a power factor of 1.0 (normally-excited) or a leading power factor of 0.8 (over-excited). Even at unity power factor, such a motor will improve a lagging power factor, though not as much as will a motor with an 0.8 leading power factor. In some installations it becomes economically feasible to connect a synchronous motor which carries no load, just to improve the lagging power factor of the line. A motor so used is called a **synchronous capacitor.**

The cheapest and most widely used method for correcting a lagging power factor is by means of fixed capacitors. Since the power factor is a function of the phase angle between voltage and current of the line, any method that reduces this phase angle also increases the power factor. Most industrial circuits have lagging power factors, indicating inductive reactance. Hence, by introducing some capacitive reactance into the circuit, the total reactance may be reduced. (See Chapter 11, subdivision C, 1.) Thus the phase angle between the voltage and current is reduced, and the power factor is raised.

An example may make this matter easier to understand. Assume we have a single-phase induction motor operating on a 220-volt line. Assume, too, that an ammeter indicates that the motor is drawing 80 amperes of current. Thus the apparent power in this circuit is 220 volts multiplied by 80 amperes, or 17.6 kva.

Assume, also, that if a wattmeter is inserted into the circuit the true power is shown to be 12.3 kw. Since the power factor of the circuit is equal to the true power divided by the apparent power,

$$\text{power factor} = \frac{\text{true power (kw)}}{\text{apparent power (kva)}} = \frac{12.3}{17.6} = 0.7.$$

The power factor, then, is 70% and, since an induction motor introduces an inductive component, the power factor is a lagging type. Also, since the power factor is equal to the cosine of the phase difference between the voltage and current (angle θ), an examination of the table of trigonometric functions will show us that the angle whose cosine is equal to 0.7 is approximately 45°.

The reactive power (kvar) is equal to the true power (kw) multiplied by the tangent of angle θ. Since angle θ is 45° and the trigonometric table shows that tan θ is equal to 1.0, the reactive power is equal to the true power. Thus the reactive power, which is inductive in nature, is equal to 12.3 kvar. To neutralize the 12.3 kvar of inductive reactive power and thus bring the power factor of the circuit up to unity, 12.3 kvar of capacitive reactive power must be added.

There are several methods for calculating the size of the capacitor required to furnish this capacitive reactive power but simplest, perhaps, is the use of a table such as 22-1. Let us assume that we wish to increase the power factor of the circuit of our previous problem from its 70% value to 95%.

Using this table, with an original power factor of 70% and a desired power factor of 95%, we find a factor of 0.69. Multiplying this factor by the true power (which we had found to be 12.3 kw) we obtain 0.69 × 12.3 kw, or 8.487 kvar of capacitive reactive power required to raise the power factor to 95%.

The capacitors used for this purpose come in standard units whose name plates list, among other factors, the rated voltage and frequency at which the unit is to be operated, and the capacitive reactive power (in kilovars) it will introduce into the circuit. A standard unit, whose kvar rating is closest to the desired value, then is connected across the line. If necessary, several standard units may be connected in parallel to form a bank whose total value is equal to the sum of the kvars of the individual units.

Table 22-1. Table for calculating the size of a capacitor required for connecting a lagging power factor.

present power factor	resultant or desired power factor in percent																				
	80	81	82	83	84	85	86	87	88	89	90	91	92	93	94	95	96	97	98	99	100
50	.98	1.01	1.03	1.06	1.09	1.11	1.14	1.16	1.19	1.22	1.25	1.28	1.30	1.34	1.37	1.40	1.44	1.48	1.53	1.59	1.73
51	.94	.96	.99	1.01	1.04	1.07	1.09	1.12	1.15	1.17	1.20	1.23	1.26	1.29	1.32	1.36	1.39	1.43	1.48	1.54	1.69
52	.89	.92	.95	.97	1.00	1.02	1.05	1.08	1.10	1.13	1.16	1.19	1.21	1.25	1.28	1.31	1.35	1.39	1.44	1.50	1.64
53	.85	.88	.90	.93	.95	.98	1.01	1.03	1.06	1.09	1.12	1.14	1.17	1.20	1.24	1.27	1.31	1.35	1.40	1.46	1.60
54	.81	.83	.86	.89	.91	.94	.97	.99	1.02	1.05	1.07	1.10	1.13	1.16	1.20	1.23	1.27	1.31	1.36	1.42	1.56
55	.77	.79	.82	.85	.87	.90	.93	.95	.98	1.01	1.03	1.06	1.09	1.12	1.16	1.19	1.23	1.27	1.32	1.38	1.52
56	.73	.76	.78	.81	.83	.86	.89	.91	.97	.97	1.00	1.02	1.05	1.08	1.12	1.15	1.19	1.23	1.28	1.34	1.48
57	.69	.72	.74	.77	.80	.82	.85	.87	.90	.93	.96	.99	1.01	1.05	1.08	1.11	1.15	1.19	1.24	1.30	1.44
58	.65	.68	.71	.73	.76	.78	.81	.84	.86	.89	.92	.95	.98	1.01	1.04	1.08	1.11	1.15	1.20	1.26	1.40
59	.62	.64	.67	.70	.72	.75	.77	.80	.83	.86	.88	.91	.94	.97	1.00	1.04	1.08	1.12	1.16	1.23	1.37
60	.58	.61	.64	.66	.69	.71	.74	.77	.79	.82	.85	.88	.90	.94	.97	1.00	1.04	1.08	1.13	1.19	1.33
61	.55	.57	.60	.63	.65	.68	.71	.73	.76	.79	.81	.84	.87	.90	.94	.97	1.01	1.05	1.10	1.16	1.30
62	.51	.54	.57	.59	.62	.64	.67	.70	.72	.75	.78	.81	.84	.87	.90	.94	.98	1.01	1.06	1.12	1.26
63	.48	.51	.53	.56	.59	.61	.64	.67	.69	.72	.75	.78	.80	.84	.87	.90	.94	.98	1.03	1.09	1.23
64	.45	.48	.50	.53	.55	.58	.61	.63	.66	.69	.72	.74	.77	.80	.84	.87	.91	.95	1.00	1.06	1.20
65	.42	.44	.47	.50	.52	.55	.58	.60	.63	.66	.68	.71	.74	.77	.81	.84	.88	.92	.97	1.03	1.17
66	.39	.41	.44	.47	.49	.52	.54	.57	.60	.63	.65	.68	.71	.74	.77	.81	.85	.89	.93	1.00	1.14
67	.36	.38	.41	.44	.46	.49	.51	.54	.57	.60	.62	.65	.68	.71	.74	.78	.82	.86	.90	.97	1.11
68	.33	.35	.38	.41	.43	.46	.49	.51	.54	.57	.59	.62	.65	.68	.72	.75	.79	.83	.88	.94	1.08
69	.30	.32	.35	.38	.40	.43	.46	.48	.51	.54	.56	.59	.62	.65	.69	.72	.76	.80	.84	.91	1.05
70	.27	.30	.32	.35	.37	.40	.43	.45	.48	.51	.54	.56	.59	.62	.66	.69	.73	.77	.81	.88	1.02
71	.24	.27	.29	.32	.35	.37	.40	.42	.45	.48	.51	.54	.56	.60	.63	.66	.70	.74	.78	.85	.99
72	.21	.24	.26	.29	.32	.34	.37	.40	.42	.45	.48	.51	.53	.57	.60	.63	.67	.71	.75	.82	.96
73	.19	.21	.24	.26	.29	.32	.34	.37	.40	.42	.45	.48	.51	.54	.57	.61	.64	.68	.73	.79	.94
74	.16	.18	.21	.24	.26	.29	.32	.34	.37	.40	.42	.45	.48	.51	.55	.58	.62	.66	.70	.77	.91
75	.13	.16	.18	.21	.24	.26	.29	.31	.34	.37	.40	.43	.45	.49	.52	.55	.59	.63	.67	.74	.88
76	.10	.13	.16	.18	.21	.23	.26	.29	.31	.34	.37	.40	.43	.46	.49	.53	.56	.60	.65	.71	.85
77	.08	.10	.13	.16	.18	.21	.24	.26	.29	.32	.34	.37	.40	.43	.47	.50	.54	.58	.62	.69	.83
78	.05	.08	.10	.13	.16	.18	.21	.24	.26	.29	.32	.35	.37	.41	.44	.47	.51	.55	.59	.66	.80
79	.03	.05	.08	.10	.13	.16	.18	.21	.24	.26	.29	.32	.35	.38	.41	.45	.48	.52	.57	.63	.78
80	.00	.03	.05	.08	.10	.13	.16	.18	.21	.24	.27	.29	.32	.35	.38	.42	.46	.50	.54	.61	.75
81		.00	.03	.05	.08	.10	.13	.16	.18	.21	.24	.27	.29	.33	.36	.39	.43	.47	.51	.58	.72
82			.00	.03	.05	.08	.10	.13	.16	.19	.21	.24	.27	.30	.33	.37	.41	.45	.49	.56	.70
83				.00	.03	.05	.08	.10	.13	.16	.19	.22	.24	.28	.31	.34	.38	.42	.46	.53	.67
84					.00	.03	.05	.08	.11	.13	.16	.19	.22	.25	.28	.32	.35	.39	.44	.50	.65
85						.00	.03	.05	.08	.11	.14	.16	.19	.22	.26	.29	.33	.37	.42	.48	.62
86							.00	.03	.05	.08	.11	.14	.17	.20	.23	.26	.30	.34	.39	.45	.59
87								.00	.03	.05	.08	.11	.14	.17	.20	.24	.27	.32	.36	.42	.57
88									.00	.03	.06	.08	.11	.14	.18	.21	.25	.29	.34	.40	.54
89										.00	.03	.05	.09	.12	.15	.18	.22	.26	.31	.37	.51
90											.00	.03	.06	.09	.12	.15	.19	.23	.28	.34	.48
91												.00	.03	.06	.09	.13	.16	.21	.25	.31	.46
92													.00	.03	.06	.10	.13	.18	.22	.28	.43
93														.00	.03	.07	.10	.14	.19	.25	.39
94															.00	.03	.07	.11	.16	.22	.36
95																.00	.04	.08	.12	.19	.33
96																	.00	.04	.09	.15	.29
97																		.00	.05	.11	.25
98																			.00	.06	.20
99																				.00	.14
100																					.00

Westinghouse Electric Corporation

These capacitors generally have plates of aluminum foil, using paper impregnated with insulating oil or some other nonflammable insulating material as a dielectric. The entire unit is hermetically sealed in a steel can. (See Figure 22-4.) Since such a capacitor may retain a considerable charge, there is danger of receiving a serious shock when removing it from the circuit. Accordingly, the electrical code requires that internal resistors or inductors be included to discharge the capacitor within a stated interval of time when it is removed from the line.

In a three-phase circuit the power factor of each phase must be corrected, just as if it were a single phase. Accordingly, individual capacitors (or banks of capacitors) of proper value must be inserted into each phase. If the circuit is balanced (as we assume) each capacitor (or bank of capacitors) is of the same value. Or else we may use a three-phase capacitor consisting of three units connected in a delta or wye configuration and contained in a single case. (See Figure 22-5.) One lead from this capacitor is connected to each of the three power lines of the circuit.

There are two general methods for connecting these capacitors into

Westinghouse Electric Corp.

Fig. 22-4. Capacitor unit used for power-factor correction.

Fig. 22-5. Cutaway view of a three-phase capacitor unit used for power-factor correction.

General Electric Company

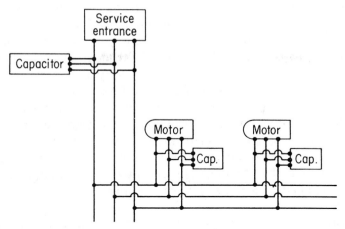

Fig. 22-6. How power-factor-correcting capacitors may be installed in an industrial plant

the circuit, as illustrated in Figure 22-6. One method is to calculate the size of the capacitor required to correct the power factor for the entire plant and connect this capacitor to the main power bus near the service entrance. Since the capacitor corrects the power factor from the point where it is inserted back to the source, there is no power-factor correction for the feeder lines within the plant from the service entrance to the individual appliances. However, this method does reduce the low-power-factor penalty imposed by the power company.

The second method is to connect a power-factor-correcting capacitor at each appliance. This method has the advantage that it also corrects the low power factor of the individual feeder lines in the plant as well. However, it generally is more costly to use a number of smaller capacitors at the appliances than one larger capacitor at the service entrance. Which method is employed depends upon the economics involved for a given installation.

B. DISTRIBUTION TRANSFORMERS

As described in Chapter 15, in this country electric power is usually generated at some central point (generally at a frequency of 60 Hz) and transmitted in three-phase, high-voltage systems to surrounding localities and industrial plants. Hydroelectric generating plants, utilizing rushing or falling water, are generally located at considerable distances from the ultimate consumers. Accordingly, the generated power is sent over long transmission lines at transmission voltages of about 345 kv, or higher, usually in a three-phase, four-wire wye configuration. At the receiving end of the lines, the voltage is stepped down by means of transformers to a value more suitable for local distribution.

Steam generating plants, using either fossil fuel (coal or oil) or nuclear fuel (uranium or plutonium) may be located closer to the consumer. Whether the power is brought in by transmission lines or generated locally, it is distributed through the nearby community as a three-phase current at distribution voltages generally ranging from about 4 kv to 13 kv. At the industrial plant or residential area, the power is supplied at a utilization voltage that is generally about 600 volts or less.

Industrial plants employing heavy-power-consuming equipment usually are supplied with three-phase voltage for this equipment and single-phase voltage for lighting and lighter equipment. Residential areas most generally are supplied with single-phase voltage at 240 or 208 volts and 120 volts. Commercial buildings are supplied with single-phase voltage, frequently at higher voltages such as 277 and 480 volts.

Power consumption grows with a community and, since power companies are understandably reluctant to scrap existing facilities to meet greater demands, new facilities are added to the old. Thus, since all communities did not necessarily start with the same system, different localities may employ different systems using different distribution voltages and configurations.

A popular distribution system employed in urban and suburban areas where population densities and power demands are large and distribution lines are relatively short is a three-phase, four-wire wye system at 4160 volts. (See Figure 22-7.) Three-phase current from some source, which may be either a generator or the secondary winding of a substation transformer, is applied to the primary winding of a three-phase distribution transformer. (As shown here, this primary winding is arranged in a delta configuration. It might just as well be a wye configuration.)

The secondary winding of the distribution transformer is arranged in a three-phase, four-wire wye configuration with a turn ratio to produce 2400 volts across each winding. Since, in such a configuration, the line voltage is equal to the winding voltage multiplied by $\sqrt{3}$ (which is approximately 1.73), the output, then, is a three-phase voltage at 4160 volts ($2400 \times \sqrt{3} = 4160$) which is distributed through the locality by

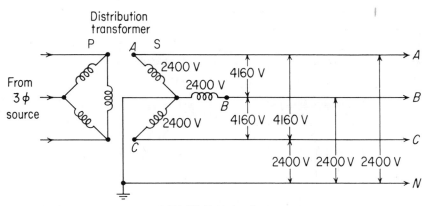

Fig. 22-7. Circuit illustrating the 2400/4160 Y distribution system.

either overhead or underground lines. Between any two of the three-phase lines (*A*, *B*, or *C*) we may obtain single-phase voltage at 4160 volts. Between the neutral line (*N*) and any one of the three-phase lines we may obtain a single-phase voltage at 2400 volts.

The system just described is known as a 2400/4160Y distribution system. Transformers whose primaries are rated at either 2400 volts or 4160 volts may be hooked into this system. In some localities a similar 4160/7200 Y distribution system is employed. Three-phase voltage at 7200 volts is distributed and transformers rated at either 4160 volts or 7200 volts may be hooked into the system.

In general, where distribution lines are long or where power demands are extremely heavy, higher distribution voltages are employed to reduce the current required for a given amount of power and thus reduce the I^2R losses in the lines. For example, in New York City, a 7620/13,200Y system is used to distribute three-phase voltage at 13,200 volts. In some localities the 7620/13,200Y system is used to serve industrial loads and the 2400/4160Y system is used to serve residential and commercial loads.

At the consumers' end of the distribution lines the distribution voltage is stepped down to the utilization voltage, generally 600 volts or lower. Industrial plants require both three-phase and single-phase voltage whereas residential and commercial establishments generally need only single-phase voltage.

There are several methods for obtaining the single-phase current at the required voltages from the higher-voltage three-phase distribution lines. One method is to connect a single-phase step-down transformer between any two of the three-phase lines or, if the line is in a four-wire

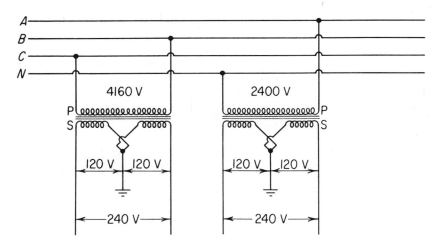

Fig. 22-8. How the distribution voltage of a three-phase, four-wire wye line may be stepped down to the single-phase utilization voltage.

wye configuration, between the neutral line and any one of the three-phase lines. (See Figure 22-8.)

The primary winding of this single-phase transformer must be matched to the voltages of the power lines. If the circuit illustrated in Figure 22-9 is used, the primary which connects between any two of the three-phase lines must be rated at 4160 volts. If this primary is connected between neutral and one of the three-phase lines, it must be rated at 2400 volts.

Fig. 22-9. Step-down system which produces three-phase voltage at 208 volts and single-phase voltage at 208 and 120 volts.

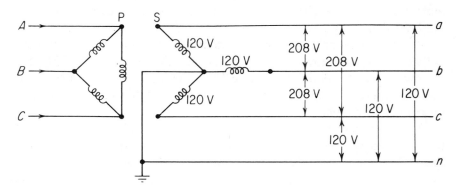

The secondary winding of this transformer is a center-tapped, 240-volt winding. Thus 240 volts are obtained from across the entire secondary winding and 120 volts from the center tap (which is grounded) to either end of the winding. (You will recognize this as the Edison three-wire, single-phase system discussed in Chapter 15, subdivision A.)

Since an industrial plant generally requires both single-phase and three-phase voltage, the method illustrated in Figure 22-9 may be employed. The step-down transformer is a three-phase type with the primary windings, in either a wye configuration or a delta configuration (as shown), connected to the distribution power lines. The secondary windings of this transformer are connected in a four-wire wye configuration with turn ratios which provide winding voltages of 120 volts. Accordingly, three-phase voltage may be obtained from lines *a*, *b*, and *c* at 208 volts ($120 \times \sqrt{3} = 208$). Single-phase voltage at 208 volts may be obtained from between any two of the three-phase lines. Single-phase voltage at 120 volts may be obtained from between the neutral line (*n*) and any one of the three-phase lines.

This four-wire wye system frequently is used in residential and commercial installations to provide the required single-phase voltage at voltages of 208 and 120 volts. In some of the more recent commercial installations step-down turn ratios are used to provide a secondary winding voltage of 277 volts. As a result, single-phase voltages at 277 and 480 volts may be obtained.

Transformers are used at substations to step down the high transmission voltages to the lower distribution voltages. They may also be used at the local generating plant for the same purpose. At the consumers' end of the distribution lines they are used to change the distribution voltages to the lower utilization voltages and, where required, to convert three-phase voltage to single-phase voltage.

Because these transformers usually are connected to resistive and reactive loads, they are rated by the apparent power (in kilovolt-amperes) they can safely handle rather than in kilowatts, which is a measure of the true, or useful, power. Transformers rated at 500 kva or less, are known as **distribution transformers.** Those rated at more than 500 kva generally are called **power transformers.**

These transformers may be either single-phase or three-phase and their winding voltages may vary to suit different needs. They are constructed in different sizes and shapes to suit their applications. The single-phase transformer (see Figure 22-10) is a step-down type whose

General Electric Company

Fig. 22-10. Single-phase distribution transformer.

primary winding is connected to the high-voltage single-phase line. It generally contains a two-winding secondary which provides the lower-voltage single-phase voltage for the load. These two windings are identical and may be connected in series or parallel, either internally or externally. If the windings are connected in parallel, they act as a secondary whose voltage is equal to that of a single winding, but with twice the current-carrying capacity. If they are connected in series, they act as a center-tapped secondary whose total voltage is twice that of a single winding.

The high-voltage primary terminals generally are located at the ends of large insulators (known as **bushings**) which insulate them from the transformer case (which is grounded). The lower-voltage secondary terminals usually have smaller insulators. Some transformers have only one high-voltage bushing for one terminal of the primary winding. The other end of this winding is connected to the grounded transformer case.

In a transformer, a voltage applied across the primary winding sets up an induced voltage across the secondary winding. If the primary and secondary windings are wound in the same direction, as is assumed in Figure 22-11A, the primary and secondary voltages are opposite in phase. That is, if the primary winding is connected to the source (as represented by -⊙-) and the secondary to a load (-ⱳ-), at a half-cycle when the left-hand side of the source is negative and the right-hand side positive, the direction of the voltage across the primary winding will be such as to make primary current flow from point 1 to point 2, as indicated by the arrow.

At the same half-cycle the induced voltage across the secondary winding will be in the opposite direction, causing the induced current to flow as indicated by the arrow. A transformer whose voltages have this relationship to each other is said to have **additive polarity.** This may be proved by joining point 1 of the primary to point 4 of the secondary. Since the directions of primary and secondary voltages are as indicated by the arrows, they should aid each other. Accordingly, a voltmeter connected between points 2 and 3 will read the sum of the primary and secondary voltages.

If the primary and secondary windings are wound in opposite directions, as is assumed in Figure 22-11B, the primary and secondary voltages will be in phase. Thus, under the same conditions as previously indicated, the directions of the voltages will be such as to make current flow as indicated by the arrows. Now the transformer is said to have

Fig. 22-11. Transformer circuits showing what is meant by
      A. additive polarity.
      B. subtractive polarity.

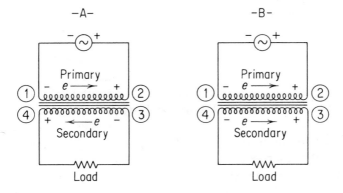

subtractive polarity. If point 1 is joined to point 4, the two voltages will oppose each other and the voltmeter between points 2 and 3 will register the difference between the two voltages.

In the individual transformer, it makes no difference whether it has additive or subtractive polarity. But whenever the transformer windings are to be connected in series, parallel, or three-phase configurations with the windings of other transformers, the polarities become important. Unless proper polarities are observed, we may find voltages bucking each other when they should be aiding, or else shorts, which can cause damage to the transformers, may be produced.

There is a conventional method for marking the terminals of distribution and power transformers so that their polarities are indicated. The high-voltage (primary) terminals are marked with the letter H. The low-voltage (secondary) terminals are marked with the letter X. Facing the high-voltage side of the transformer, the right-hand terminal is marked H_1 and the left-hand terminal is marked H_2.

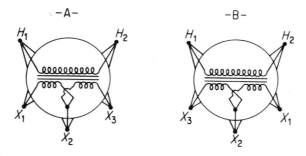

Fig. 22-12. How distribution transformer terminals are marked to indicate polarity.
A. Subtractive polarity.
B. Additive polarity.

If the windings are subtractive, the low-voltage terminal adjacent to the H_1 terminal is marked X_1 (see Figure 22-12A). The center-tap terminal is marked X_2 and the remaining terminal is marked X_3. If the windings are additive (Figure 22-12B) the low-voltage terminal adjacent to the H_1 terminal is marked X_3, the center-tap terminal X_2, and the remaining terminal X_1. (In some instances, the center-tap terminal may be marked X_0. In such a case the X_3 terminal becomes the X_2 terminal.) Where transformers are to be connected together, all the H_1 terminals are of the same polarity, as are all the X_1 terminals.

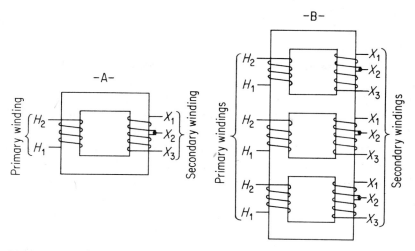

Fig. 22-13. A. Single-phase transformer, showing windings.
B. Three-phase transformer, showing windings.

In general, there are two types of three-phase transformers. One type consists of a bank of three single-phase transformers with their primary windings connected in a three-phase configuration, as are their secondaries. Thus we have three distinct transformers, each with its set of windings and core.

The other type consists of a single core upon which are wound three sets of primary and secondary windings. (See Figure 22-13B). Each set of windings with its portion of the core acts as a complete single-phase transformer. Thus the single three-phase transformer is the same as a bank of three single-phase transformers.

Both types are in use. The bank of three single-phase transformers is more flexible and, in event of failure in one phase, is easier and cheaper to repair. However, the single three-phase transformer requires less iron for its core and hence is cheaper than an equivalent bank of three single-phase transformers. Also, it is lighter and occupies less space. Further, improved design has developed the transformer to the point where failure is quite remote. Accordingly, most present-day installations employ three-phase transformers rather than banks of three single-phase transformers.

There are four basic ways in which the windings of the transformers may be connected in three-phase configurations. Since these windings may be connected in either a wye or delta configuration, we may have a

wye-wye transformer where the primary winding is connected in a wye configuration, as is the secondary winding. Or else, we may have a **delta-delta** transformer where both primary and secondary windings are connected in delta configurations. In addition, we may have a **wye-delta** transformer (primary wye and secondary delta) and **delta-wye** transformers (primary delta and secondary wye).

In addition to these four basic configurations, we may have variations of each. For example, the wye configuration may either be three-wire or four-wire. And the neutral wire may be either grounded or ungrounded. Each locality chooses the configuration best suited for its needs.

The importance of observing proper polarity becomes clear when considering the interconnections of the windings of three single-phase transformers into a three-phase bank. The connections for such a bank in a wye-wye configuration is shown in Figure 22-14. In this illustration all three transformers have additive polarity. Note that the voltages in all three primary windings are in the same direction, as are the voltages in all three secondary windings.

It would make no difference if all three transformers had subtractive polarity. The direction of the secondary voltages, although opposite to the direction shown above, would still be the same for all. Only if the polarity of one of the transformers were opposite to the polarity of the other two would one secondary voltage oppose the other two with resulting malfunction of the entire bank.

The connections for a bank of three single-phase transformers in a delta-delta configuration is illustrated in Figure 22-15. Here all three transformers have subtractive polarity. Note that the voltages in all

Fig. 22-14. Wye-wye configuration using transformers with additive polarity.

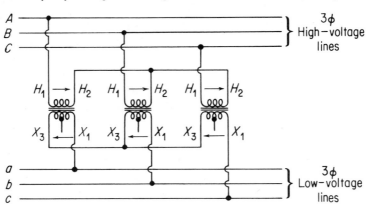

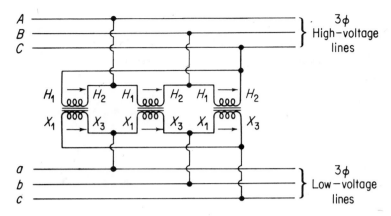

Fig. 22-15. Delta-delta configuration using transformers with subtractive polarity.

three primary windings are in the same direction, as are the voltages in all three secondary windings. Here too, it would make no difference if the transformers had additive polarities, provided all three transformers had the same polarity. If, instead of using three separate single-phase transformers, a single three-phase transformer is employed, the various windings are interconnected with the same regard for polarities as indicated above.

As current flows through the windings of a transformer, heat is produced and, unless this heat can be prevented from becoming excessive, the transformer may be damaged. When the currents involved are relatively small, air cooling may be employed. Transformers using this type of cooling are known as **dry** types.

There are two methods used for air-cooling the transformer. Those of smaller size generally employ **self-air-cooling.** Here the windings and core are placed in a metal enclosure. Cooling is accomplished by the natural convection of the heated air which carries the heat from the transformer to the sides of the enclosure. From there, the heat is radiated to the surrounding atmosphere. Sometimes, louvers or gratings in the sides of the enclosure help remove the heated air. (See Figure 22-16.) For transformers of larger size **forced-air-cooling** may be employed. Here the cooling effect is increased by means of fans or blowers that help circulate the air.

Another method for cooling consists of immersing the core and windings of the transformer into an insulating liquid contained in a

Westinghouse Electric Corp.

Fig. 22-16. Dry-type, three-phase transformer and enclosure.

metal tank. This liquid usually is a type of insulating oil. For certain special applications, a synthetic nonflammable and nonexplosive insulating liquid, known by various trade names such as Askarel, Pyranol, Inerteen, etc., is employed. Heat produced by the transformer is removed by the natural convection of the liquid which carries the heat to the sides of the tank from which it is removed by radiation and the convection of the surrounding air. Transformers using liquids for cooling are known as **wet** types.

Wet-type transformers of smaller size may employ tanks whose outer surfaces are smooth. For those of larger size the tanks usually are equipped with fins or with external tubes or radiators through which the heated liquid may circulate by natural convection. In this way the radiating area of the tank is increased and the heat can be dissipated more readily. (See Figure 22-17.) This method is known as **self-cooling.**

Where greater cooling is required, fans or blowers may be used to force a flow of air over the external radiators of the tank. Thus the cooling effect of the air blast is added to the self-cooling effect of the liquid. (See Figure 22-18.)

In some of the larger wet-type transformers cold water is circulated through copper coils placed within the tank to carry away some of the heat of the insulating liquid. Thus the cooling effect of the circulating water is added to the self-cooling effect of the insulating liquid. This

General Electric Company

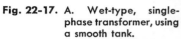

-A-

Fig. 22-17. A. Wet-type, single-phase transformer, using a smooth tank.
B. Wet-type, three-phase transformer, using radiators for cooling.

Westinghouse Electric Corp.

-B-

General Electric Company

Fig. 22-18. Wet-type, three-phase transformer using self- and forced-air cooling.

method of cooling may be used, of course, only where a continuous supply of cold water is available.

The structure of the transformer enclosure is determined by the use to which the transformer is put. Where the transformer is to be mounted outdoors on poles for the overhead distribution of power, it is encased in a **pole** enclosure equipped with special hooks or hangers for attachment to the poles. (See Figure 22-19A.)

The **platform** enclosure is used where the transformer is to stand on its own base. Where the transformer is outdoors it generally rests on a concrete pad laid on the ground. (See Figure 22-19B.) For purposes of overhead distribution the transformer may rest on an elevated platform supported by poles.

Westinghouse Electric Corp.

General Electric Company

– B –

Fig. 22-19. A. Three pole-type single-phase transformers mounted in a three-phase configuration.
B. Platform-type transformer mounted outdoors on a concrete pad.

If the transformer is installed in an underground area where there is danger of flooding, a **subway** enclosure is used. This is a watertight enclosure which will permit the transformer to be completely submerged for protracted periods without damage. Where the transformer is to operate in a damp environment, such as an underground vault, but where there is no danger of prolonged submersion, a **vault** enclosure is used. In addition to these types, there are special enclosures for special purposes.

QUESTIONS

1. Explain the difference between **true power** and **apparent power** in an a-c circuit containing a reactive component. In what units is true power measured? In what units is apparent power measured?

2. As regards to the power factor of an a-c circuit, what is the relationship between the true and the apparent power?

3. Draw a vector diagram of an a-c circuit containing inductance and resistance showing the relationship between I_R, I_L, and I_{Total} where the phase difference (θ) between the source voltage (E) and I_{Total} is 45°.

4. Draw the power triangle showing the relationship between apparent, true, and reactive power for the above circuit.

5. In a right triangle, explain what is meant by (a) the **sine** of an angle; (b) the **cosine** of an angle; (c) the **tangent** of an angle.

6. What are the disadvantages of a circuit containing a low power factor?

7. Describe and explain two methods for correcting a low lagging power factor of a circuit.

8. Draw the circuit of a 4160/7200Y distribution system. (a) What value of three-phase voltage may be obtained from this system? (b) What value of single-phase voltage? (c) Indicate three methods for obtaining this single-phase voltage.

9. Indicate two methods by which single-phase transformers may be connected to the above system to obtain single-phase voltage at a utilization value of 120-240 volts.

10. Using diagrams, explain what is meant when a transformer is said to have (a) additive polarity; (b) subtractive polarity.

11. Draw a diagram to show the terminal markings for a transformer with (a) additive polarity; (b) subtractive polarity.

12. Draw a diagram showing a wye-wye configuration using three single-phase transformers with subtractive polarity.

13. Draw a diagram showing a delta-delta configuration using three single-phase transformers with additive polarity.

14. What is meant by (a) a dry-type transformer; (b) a wet-type transformer?

APPENDIX A: GLOSSARY
OF ELECTRICAL TERMS,
SYMBOLS, AND UNITS

Term	Symbol	Unit
alternating-current (adj.)	a-c	
ampere-turn	NI	
British thermal unit	Btu	
calorie	cal	
candlepower	cp	
capacitance	C	farad (f)
		microfarad (μf)
		picofarad (pf)
capacitive reactance	X_C	ohm (Ω)
centimeter	cm	
counter electromotive force	cemf	volt (v)
cubic centimeter	cu cm or cm³	
current	I	ampere (a)
		milliampere (ma)
		microampere (μa)
dielectric constant	K	
direct-current (adj.)	d-c	
electric energy		joule or wattsecond (wsec)
		watthour (whr)
		kilowatthour (kwhr)

Term	Symbol	Unit
electric power	P	watt (w)
		kilowatt (kw)
		milliwatt (mw)
		microwatt (μw)
electromotive force (potential difference, voltage)	E or emf	volt (v)
		kilovolt (kv)
		millivolt (mv)
		microvolt (μv)
electron	e	
flux density (magnetic)	B	gauss
foot-candle	ft-c	
foot-pound	ft-lb	
frequency	f	hertz (Hz)
		kilohertz (kHz)
		megahertz (MHz)
frequency, resonant	f_r	hertz
horsepower	hp	
impedance	Z	ohm (Ω)
inductance	L	henry (h)
		millihenry (mh)
		microhenry (μh)
inductive reactance	X_L	ohm (Ω)
kilovolt-ampere	kva	
magnetic flux	ϕ	maxwell
magnetizing force	H	oersted
magnetomotive force	$\mathcal{F}$	gilbert
microsecond	μs	
mutual inductance	M	henry (h)
permeability (magnetic)	μ	
phase	ϕ	
phase angle	θ	degree
power factor	pf	
reactance	X	ohm (Ω)
reluctance (magnetic)	$\mathcal{R}$	
resistance	R	ohm (Ω)
		kilohm (kΩ)
		megohm (MΩ)
revolutions per minute	rpm	
revolutions per second	rps	
root mean square	rms	
square centimeter	sq cm or cm^2	

APPENDIX B: GLOSSARY OF ELECTRICAL FORMULAS USED IN THIS BOOK

B

Values for alternating current (or voltage) with sinusoidal waveform

Average current (or voltage) = 0.637 × peak current (or voltage)

Effective (rms) current (or voltage) = 0.707 × peak current (or voltage)

Effective (rms) current (or voltage) = 1.11 × average current (or voltage)

Peak-to-peak current (or voltage) = 2 × peak current (or voltage)

Peak current (or voltage) = 1.57 × average current (or voltage)

Peak current (or voltage) = 1.41 × effective (rms) current (or voltage)

Ohm's law for d-c circuits

$$I = \frac{E}{R}; \qquad E = I \times R; \qquad R = \frac{E}{I}$$

Ohm's law for a-c circuits

$$I = \frac{E}{Z}; \qquad E = I \times Z; \qquad Z = \frac{E}{I}$$

I = current in *amperes*
E = electromotive force in *volts*
Z = impedance in *ohms*
R = resistance in *ohms*

495

Power in d-c circuits

$$P = I \times E; \qquad P = I^2 \times R; \qquad P = \frac{E^2}{R}$$

Power in a-c circuits

$$P = I \times E \times \text{pf}$$

P = power in *watts*
I = current in *amperes*
E = electromotive force in *volts*
R = resistance in *ohms*
pf = power factor in *percent*

$$\text{Power factor} = \frac{\text{true power (as measured by wattmeter)}}{\text{apparent power (as measured by voltmeter and ammeter)}}$$

Heat produced by electric current

$$\text{Heat} = 0.24 \times I^2 \times R \times T$$

Heat = calories
I = current in *amperes*
R = resistance in *ohms*
T = time in *seconds*
1 wattsecond = 0.24 calorie
1 British thermal unit = 252 calories = 1050 wattseconds

Resistors in series

$$R_{\text{total}} = R_1 + R_2 + R_3 +, \text{ and so forth}$$

Resistors in parallel

$$\frac{1}{R_{\text{total}}} = \frac{1}{R_1} + \frac{1}{R_2} + \frac{1}{R_3} +, \text{ and so forth}$$

R = resistance in *ohms*

Inductors in series (no interaction of fields)

$$L_{\text{total}} = L_1 + L_2 + L_3 +, \text{ and so forth}$$

Inductors in parallel (no interaction of fields)

$$\frac{1}{L_{\text{total}}} = \frac{1}{L_1} + \frac{1}{L_2} + \frac{1}{L_3} +, \text{ and so forth}$$

Inductors in series (fields aiding)

$$L_{\text{total}} = L_1 + L_2 + 2M$$

Inductors in parallel (fields aiding)

$$\frac{1}{L_{\text{total}}} = \frac{1}{L_1 + M} + \frac{1}{L_2 + M}$$

Inductors in series (fields bucking)

$$L_{\text{total}} = L_1 + L_2 - 2M$$

Inductors in parallel (fields bucking)

$$\frac{1}{L_{\text{total}}} = \frac{1}{L_1 - M} + \frac{1}{L_2 - M}$$

L = inductance in *henrys*

M = mutual inductance in *henrys*

Capacitance of a capacitor

$$C = \frac{0.0885 \times K \times A}{T}$$

C = capacitance in pf

K = dielectric constant

A = area of plate in contact with one side of the dielectric, in *square centimeters*

T = thickness of dielectric in *centimeters*

Capacitors in series

$$\frac{1}{C_{\text{total}}} = \frac{1}{C_1} + \frac{1}{C_2} + \frac{1}{C_3} +, \text{ and so forth}$$

Capacitors in parallel

$$C_{\text{total}} = C_1 + C_2 + C_3 +, \text{ and so forth}$$

C = capacitance in *farads*

Reactance and impedance

$$X_{\text{total}} = X_L - X_C$$

X_{total} = total reactance in *ohms*. If it be positive, the reactance is inductive. If it be negative, the reactance is capacitive.

X_L = inductive reactance in *ohms*

X_C = capacitive reactance in *ohms*

$$Z = \sqrt{R^2 + X_L^2}$$

$$Z = \sqrt{R^2 + X_C^2}$$

$$Z = \sqrt{R^2 + (X_L - X_C)^2}$$

Z = impedance in *ohms*

R = resistance in *ohms*

X_L = inductive reactance in *ohms*

X_C = capacitive reactance in *ohms*

Inductive reactance

$$X_L = 2\pi f L$$

X_L = inductive reactance in *ohms*
f = frequency in *cycles per second*
L = inductance in *henrys*
2π = 6.28

$$X_L = \frac{E}{I}; \qquad I = \frac{E}{X_L}; \qquad E = I \times X_L$$

X_L = inductive reactance in *ohms*
I = current in *amperes*
E = electromotive force in *volts*

Capacitive reactance

$$X_C = \frac{1}{2\pi f C}$$

X_C = capacitive reactance in *ohms*
f = frequency in *cycles per second*
C = capacitance in *farads*
2π = 6.28

$$X_C = \frac{E}{I}; \qquad I = \frac{E}{X_C}; \qquad E = I \times X_C$$

X_C = capacitive reactance in *ohms*
I = current in *amperes*
E = electromotive force in *volts*

Resonant frequency

$$f_r = \frac{1}{2\pi\sqrt{L \times C}}$$

f_r = resonant frequency in *cycles per second*
L = inductance in *henrys*
C = capacitance in *farads*
2π = 6.28

R-C time constant

$$t = C \times R$$

t = time in *seconds*
C = capacitance in *farads*
R = resistance in *ohms*

R-L time constant

$$t = \frac{L}{R}$$

t = time in *seconds*
L = inductance in *henrys*
R = resistance in *ohms*

Generator

$$f = \frac{\text{sets of poles} \times \text{speed}}{60}$$

f = frequency in *cycles per second*
speed = revolutions per minute

$$\text{Efficiency (percent)} = \frac{\text{power output}}{\text{power input}} \times 100$$

$$\begin{array}{c}\text{Voltage regulation} \\ \text{(percent)}\end{array} = \frac{\left(\begin{array}{c}\text{Voltage at} \\ \text{no load}\end{array}\right) - \left(\begin{array}{c}\text{Voltage at} \\ \text{full load}\end{array}\right)}{\text{Voltage at full load}} \times 100$$

1 kilowatt = 0.746 horsepower

Motor

$$\text{Speed} = \frac{f \times 120}{\text{poles per phase}}$$

f = frequency in *cycles per second*
speed = revolutions per minute

$$\text{Efficiency (percent)} = \frac{\text{power output}}{\text{power input}} \times 100$$

$$\text{Horsepower} = \frac{\text{Watts}}{746}$$

1 Horsepower = 550 foot-pounds per second

Three-phase configurations

    Delta

        Line voltage = phase voltage
        Line current = 1.73 × phase current

    Wye

        Line voltage = 1.73 × phase voltage
        Line current = phase current

Transformer

$$\frac{E_p}{E_s} = \frac{N_p}{N_s}, \qquad \frac{I_p}{I_s} = \frac{N_s}{N_p}$$

E_p = primary voltage
E_s = secondary voltage
I_p = primary current
I_s = secondary current
N_p = number of turns in primary
N_s = number of turns in secondary

Magnetism

$$B = \frac{\phi}{\text{area}}$$

B = flux density in *gauss*
ϕ = lines of force in *maxwells*
Area (in square centimeters) taken at right angles to the direction of the flux

$$H = \frac{\mathfrak{F}}{\text{length}}$$

H = magnetizing force in *oersteds*
$\mathfrak{F}$ = magnetomotive force in *gilberts*
Length = length of magnetic circuit in centimeters

$$\mu = \frac{B}{H}$$

μ = permeability
B = flux density in *gauss*
H = magnetizing force in *oersteds*

$$\mathfrak{R} = \frac{\mathfrak{F}}{\phi}$$

$\mathfrak{R}$ = reluctance
$\mathfrak{F}$ = magnetomotive force in *gilberts*
ϕ = magnetic flux in *maxwells*

APPENDIX C:
SYMBOLS USED IN
ELECTRICAL DIAGRAMS

C

The graphic symbols used for electrical diagrams come from two general sources. One is from the field of communications, the other from the industrial field. In most instances, the same symbols are used for both types of diagrams. There are, however, some differences, as shown in the following list.

	Communication	Industrial
Amplidyne		
Capacitors		
Capacitor, fixed		
Capacitor, variable		
Cell		
Single cell		
Battery		
Fuse		

	Communication	*Industrial*
Generators		
A-C generator		
Armature and slip rings (if any)		
Field coil		
D-C generator		
Armature, commutator, and brushes		
Field coil		
Ground		
Inductors		
Inductor, air-core		
Inductor, iron-core		
Inductor, powdered iron-core		
Inductor, variable		
Lamps		
Meters		
Ammeter		
Milliammeter		
Microammeter		
Galvanometer		
Voltmeter		
Millivoltmeter		
Wattmeter		
Motors		
A-C motor		
Armature and slip rings (if any)		
Field coil		
D-C motor		
Armature, commutator, and brushes		
Field coil		
Phototube		

	Communication	Industrial
Rectifier, semiconductor diode		
Relay, electromagnetic		
Relay coil		
Contacts, normally-closed		
Contacts, normally-open		
Resistors		
Resistor, fixed		
Resistor, tapped		
Resistor, variable		
Saturable reactor		
Switches		
Switch, single-pole, single throw		
Switch, single-pole, double-throw		
Switch, double-pole, single-throw		
Switch, double-pole, double-throw		
Switch, rotary, single-pole, 5-position		
Switch, push button, normally-open		
Switch, push button, normally-closed		
Transformers		
Transformer, air-core		
Transformer, iron-core		
Transformer, powdered iron-core		
Autotransformer, step-up		
Autotransformer, step-down		
Three-phase configurations		
Delta configuration		
Wye configuration		
Four-wire, wye configuration		
Wires crossing, connected		
Wires crossing, not connected		

APPENDIX D: MATHEMATICAL TABLE OF SQUARES AND SQUARE ROOTS

No.	Square	Square root	No.	Square	Square root	No.	Square	Square root	No.	Square	Square root
1	1	1.000	26	676	5.0990	51	2601	7.1414	76	5776	8.7178
2	4	1.414	27	729	5.1962	52	2704	7.2111	77	5929	8.7750
3	9	1.732	28	784	5.2915	53	2809	7.2801	78	6084	8.8318
4	16	2.000	29	841	5.3852	54	2916	7.3485	79	6241	8.8882
5	25	2.236	30	900	5.4772	55	3025	7.4162	80	6400	8.9443
6	36	2.449	31	961	5.5678	56	3136	7.4833	81	6561	9.0000
7	49	2.646	32	1024	5.6569	57	3249	7.5498	82	6724	9.0554
8	64	2.828	33	1089	5.7446	58	3364	7.6158	83	6889	9.1104
9	81	3.000	34	1156	5.8310	59	3481	7.6811	84	7056	9.1652
10	100	3.162	35	1225	5.9161	60	3600	7.7460	85	7225	9.2195
11	121	3.3166	36	1296	6.0000	61	3721	7.8102	86	7396	9.2736
12	144	3.4641	37	1369	6.0828	62	3844	7.8740	87	7569	9.3274
13	169	3.6056	38	1444	6.1644	63	3969	7.9373	88	7744	9.3808
14	196	3.7417	39	1521	6.2450	64	4096	8.0000	89	7921	9.4340
15	225	3.8730	40	1600	6.3246	65	4225	8.0623	90	8100	9.4868
16	256	4.0000	41	1681	6.4031	66	4356	8.1240	91	8281	9.5394
17	289	4.1231	42	1764	6.4807	67	4489	8.1854	92	8464	9.5917
18	324	4.2426	43	1849	6.5574	68	4624	8.2462	93	8649	9.6437
19	361	4.3589	44	1936	6.6332	69	4761	8.3066	94	8836	9.6954
20	400	4.4721	45	2025	6.7082	70	4900	8.3666	95	9025	9.7468
21	441	4.5826	46	2116	6.7823	71	5041	8.4261	96	9216	9.7980
22	484	4.6904	47	2209	6.8557	72	5184	8.4853	97	9409	9.8489
23	529	4.7958	48	2304	6.9282	73	5329	8.5440	98	9604	9.8995
24	576	4.8990	49	2401	7.0000	74	5476	8.6023	99	9801	9.9499
25	625	5.0000	50	2500	7.0711	75	5625	8.6603	100	10000	10.0000

APPENDIX E:
DECIMAL EQUIVALENTS
OF FRACTIONS

	1/64	0.015625	
1/32		0.031250	
	3/64	0.046875	
1/16		0.062500	
	5/64	0.078125	
3/32		0.093750	
	7/64	0.109375	
1/8		0.125000	
	9/64	0.140625	
5/32		0.156250	
	11/64	0.171875	
3/16		0.187500	
	13/64	0.203125	
7/32		0.218750	
	15/64	0.234375	
1/4		0.250000	
	17/64	0.265625	
9/32		0.281250	
	19/64	0.296875	
5/16		0.312500	
	21/64	0.328125	
11/32		0.343750	
	23/64	0.359375	
3/8		0.375000	
	25/64	0.390625	
13/32		0.406250	
	27/64	0.421875	
7/16		0.437500	
	29/64	0.453125	
15/32		0.468750	
	31/64	0.484375	
1/2		0.500000	

	33/64	0.515625	
17/32		0.531250	
	35/64	0.546875	
9/16		0.562500	
	37/64	0.578125	
19/32		0.593750	
	39/64	0.609375	
5/8		0.625000	
	41/64	0.640625	
21/32		0.656250	
	43/64	0.671875	
11/16		0.687500	
	45/64	0.703125	
23/32		0.718750	
	47/64	0.734375	
3/4		0.750000	
	49/64	0.765625	
25/32		0.781250	
	51/64	0.796875	
13/16		0.812500	
	53/64	0.828125	
27/32		0.843750	
	55/64	0.859375	
7/8		0.875000	
	57/64	0.890625	
29/32		0.906250	
	59/64	0.921875	
15/16		0.937500	
	61/64	0.953125	
31/32		0.968750	
	63/64	0.984375	
1		1.000000	

APPENDIX F:
TABLES OF MEASUREMENT

Tables of Length

English System

12 inches	= 1 foot
3 feet	= 1 yard
1,760 yards ⎫ 5,280 feet ⎬	= 1 mile (statute)
1.15 statute ⎫ miles ⎭	= 1 knot (nautical mile)

Metric System

10 millimeters	= 1 centimeter
100 centimeters ⎫ 1,000 millimeters ⎬	= 1 meter
1,000 meters	= 1 kilometer

1 centimeter	= 0.394 inch
1 inch	= 2.54 centimeters
1 yard	= 0.91 meter
1 mile	= 1.60 kilometers

Tables of Area

English System

144 square inches	= 1 square foot
9 square feet	= 1 square yard
4,840 square yards	= acre
640 acres	= 1 square mile

Metric System

100 square millimeters	= 1 square centimeter
10,000 square centimeters ⎫ 1,000,000 square millimeters ⎬	= 1 square meter
1,000,000 square meters	= 1 square kilometer

1 square centimeter	= 0.155 square inch
1 square inch	= 6.452 square centimeters
1 square yard	= 0.83 square meter

Tables of Volume

English System	*Metric System*

English System	*Metric System*
1,728 cubic inches = 1 cubic foot	1,000 cubic millimeters = 1 cubic centimeter
27 cubic feet　　= 1 cubic yard	1,000 cubic centimeters = 1 cubic decimeter
128 cubic feet　　= 1 cord	1,000 cubic decimeters = 1 cubic meter

1 cubic centimeter = 0.061 cubic inch
1 cubic inch　　　= 16.387 cubic centimeters

Tables of Weight

English System	*Metric System*

7,000 grains ⎫
16 ounces　 ⎬ = 1 pound
　　　　　　⎭
2,000 pounds = 1 ton
2,240 pounds = 1 long ton

1,000 milligrams　　= 1 gram
1,000 grams　　　　⎫
1,000,000 milligrams ⎬ = 1 kilogram
　　　　　　　　　　⎭
1,000 kilograms　　= 1 metric ton

1 gram　= 0.035 ounce
1 ounce = 28.349 grams
1 pound = { 453.6 grams
　　　　　{ 0.453 kilogram

G

APPENDIX G: TABLE OF SINES, COSINES, AND TANGENTS

Angle	Sine	Cosine	Tangent	Angle	Sine	Cosine	Tangent
0°	.0000	1.0000	.0000	45°	.7071	.7071	1.0000
1	.0175	.9998	.0175	46	.7193	.6947	1.0355
2	.0349	.9994	.0349	47	.7314	.6820	1.0724
3	.0523	.9986	.0524	48	.7431	.6691	1.1106
4	.0698	.9976	.0699	49	.7547	.6561	1.1504
5	.0872	.9962	.0875	50	.7660	.6428	1.1918
6	.1045	.9945	.1051	51	.7771	.6293	1.2349
7	.1219	.9925	.1228	52	.7880	.6157	1.2799
8	.1392	.9903	.1405	53	.7986	.6018	1.3270
9	.1564	.9877	.1584	54	.8090	.5878	1.3764
10	.1736	.9848	.1763	55	.8192	.5736	1.4281
11	.1908	.9816	.1944	56	.8290	.5592	1.4826
12	.2079	.9781	.2126	57	.8387	.5446	1.5399
13	.2250	.9744	.2309	58	.8480	.5299	1.6003
14	.2419	.9703	.2493	59	.8572	.5150	1.6643
15	.2588	.9659	.2679	60	.8660	.5000	1.7321
16	.2756	.9613	.2867	61	.8746	.4848	1.8040
17	.2924	.9563	.3057	62	.8829	.4695	1.8807
18	.3090	.9511	.3249	63	.8910	.4540	1.9626
19	.3256	.9455	.3443	64	.8988	.4384	2.0503
20	.3420	.9397	.3640	65	.9063	.4226	2.1445
21	.3584	.9336	.3839	66	.9135	.4067	2.2460
22	.3746	.9272	.4040	67	.9205	.3907	2.3559
23	.3907	.9205	.4245	68	.9272	.3746	2.4751
24	.4067	.9135	.4452	69	.9336	.3584	2.6051
25	.4226	.9063	.4663	70	.9397	.3420	2.7475
26	.4384	.8988	.4877	71	.9455	.3256	2.9042
27	.4540	.8910	.5095	72	.9511	.3090	3.0777
28	.4695	.8829	.5317	73	.9563	.2924	3.2709
29	.4848	.8746	.5543	74	.9613	.2756	3.4874
30	.5000	.8660	.5774	75	.9659	.2588	3.7321
31	.5150	.8572	.6009	76	.9703	.2419	4.0108
32	.5299	.8480	.6249	77	.9744	.2250	4.3315
33	.5446	.8387	.6494	78	.9781	.2079	4.7046
34	.5592	.8290	.6745	79	.9816	.1908	5.1446
35	.5736	.8192	.7002	80	.9848	.1736	5.6713
36	.5878	.8090	.7265	81	.9877	.1564	6.3138
37	.6018	.7986	.7536	82	.9903	.1392	7.1154
38	.6157	.7880	.7813	83	.9925	.1219	8.1443
39	.6293	.7771	.8098	84	.9945	.1045	9.5144
40	.6428	.7660	.8391	85	.9962	.0872	11.43
41	.6561	.7547	.8693	86	.9976	.0698	14.30
42	.6691	.7431	.9004	87	.9986	.0523	19.08
43	.6820	.7314	.9325	88	.9994	.0349	28.64
44	.6947	.7193	.9657	89	.9998	.0175	57.29

INDEX